FINITE MATHEMATICS

Second Custom Edition for Northeastern University

LIAL · GREENWELL · RITCHEY

Taken from:

Finite Mathematics, Ninth Edition
by Margaret L. Lial, Raymond N. Greenwell, and Nathan P. Ritchey

Custom Publishing

New York Boston San Francisco
London Toronto Sydney Tokyo Singapore Madrid
Mexico City Munich Paris Cape Town Hong Kong Montreal

Cover photograph courtesy of Brand X Pictures.

Taken from:

Finite Mathematics, Ninth Edition
by Margaret L. Lial, Raymond N. Greenwell, and Nathan P. Ritchey

Published by Addison-Wesley
Boston, Massachusetts 02116

This special edition published in cooperation with Pearson Custom Publishing.

Printed in the United States of America

10 9 8 7 6 5 4 3 2

2008360910

KA/JW

Pearson
Custom Publishing
is a division of

PEARSON

www.pearsonhighered.com

ISBN 10: 0-555-04556-0
ISBN 13: 978-0-555-04556-5

Contents

6 Logic

The rules of a game often include complex conditional statements, such as "if you roll doubles, you can roll again, but if you roll doubles twice in a row, you lose a turn." As exercises in this chapter illustrate, logical analysis of complex statements helps us clarify not only the rules of games but any precise use of language, from legal codes to medical diagnoses.

In 1943, Thomas Watson, head of IBM, made the now-infamous prediction, "I think there is a world market for maybe five computers."* In 1977, Ken Olson, founder of Digital Equipment Corp., prophesied that "There is no reason anyone would want a computer in their home."† Perhaps such predictions were so wrong because of the difficulty, in the early days of computing, of foreseeing later uses of the computer, such as sending e-mail or looking up a movie review. It's rather amazing that such modern conveniences are made possible by a machine made of components based on mathematical logic. Even when you're using your computer for some seemingly nonmathematical activity, components known as gates, based on logical operators that you'll learn about in this chapter, are making it all possible.

6.1 Statements

? THINK ABOUT IT **How could we express the negation of statements found in an income tax guide or a medical reference?**

You will be asked to answer this question in one of the exercises at the end of this section.

This section introduces the study of *symbolic logic*, which uses formal mathematics with symbols to represent statements and arguments in everyday language. Logic can help us determine the truth value of complex statements, as well as determine whether a conclusion necessarily follows from a set of assumptions.

Many kinds of sentences occur in ordinary language, including factual statements, opinions, commands, and questions. Symbolic logic discusses statements and opinions that may be true or false, but not commands or questions, as we shall see.

Statements A **statement** is defined as a declarative sentence that is either true or false, but not both simultaneously. For example, both of the following are statements:

Mount McKinley is the tallest mountain in North America.

$$15 - 9 = 10.$$

The first sentence happens to be true, while the second is false. None of the following sentences are considered statements in logic, however, because they cannot be identified as being either true or false:

Follow the yellow brick road.

Are we having fun yet?

This sentence is false.

The first sentence is a command or suggestion, while the second is a question. The third sentence is a paradox: If the sentence is true, then the sentence itself tells us that it must be false. But if what it says is false, then the sentence must be true. We avoid such paradoxes by disallowing statements that refer to themselves.

*http://research.microsoft.com/acm97/gb/sld010.htm.

†http://research.microsoft.com/acm97/gb/tsld047.htm.

When one or more simple statements are combined with **logical connectives** such as *and*, *or*, *not*, and *if . . . then*, the result is called a **compound statement**, while the simple statements that make up the compound statement are called **component statements**.

EXAMPLE 1

Compound Statements

Decide whether each statement is compound.

(a) George Washington was the first U.S. president, and John Adams was his vice president.

▶**Solution** This statement is compound using the connective *and.* The component statements are "George Washington was the first U.S. president" and "John Adams was his vice president."

(b) If what you've told me is true, then we are in great peril.

▶**Solution** This statement is also compound. The component statements "what you've told me is true" and "we are in great peril" are linked with the connective *if . . . then.*

(c) We drove across New Mexico toward the town with the curious name Truth or Consequences.

▶**Solution** This statement is not compound. Even though *or* is a connective, here it is part of the name of the city. It is not connecting two statements.

(d) The money is not there.

▶**Solution** Most logicians consider this statement to be compound, even though it has just one component statement, and we will do so in this book. The connective *not* is applied to the component statement "The money is there."

We will study the *and*, *or*, and *not* connectives in detail in this section, and return to the *if . . . then* connective in the third section of this chapter.

Negation The **negation** of the statement "I play the guitar" is "I do not play the guitar." There are equivalent ways to say the same thing, such as "It is not true that I play the guitar." The negation of a true statement is false, and the negation of a false statement is true.

EXAMPLE 2

Negation

Give the negation of each statement.

(a) California is the most populous state in the country.

▶**Solution** Form the negation using the word *not*: "California is not the most populous state in the country."

(b) It is not raining today.

▶**Solution** If a statement already has the word *not*, we can remove it to form the negation: "It is raining today."

Notice how the negation of the true statement in Example 2(a) is a false statement. If at some time in the future the statement in Example 2(a) is no longer true, then the negation would become true.

EXAMPLE 3 **Negation**

Write the negation of each inequality.

(a) $x > 11$

▶**Solution** The negation of "x is greater than 11" is "x is *not* greater than 11," or $x \leq 11$.

(b) $4x + 9y \leq 36$

▶**Solution** The negation is $4x + 9y > 36$.

▶FOR REVIEW

Recall the following inequalities from Chapter R.

Symbolism	Meaning
$a < b$	a is less than b.
$a > b$	a is greater than b.
$a \leq b$	a is less than or equal to b.
$a \geq b$	a is greater than or equal to b.

Some calculators have logic functions that allow the user to test the truth or falsity of statements involving $=$, $\neq$, $>$, $\geq$, $<$, and $\leq$. For example, functions from the TEST menu of a TI-83/84 Plus calculator, illustrated in Figure 1, return a 1 when a statement is true and 0 when a statement is false. Figure 2 shows the input of 4 > 9 and the corresponding output of 0, indicating that the statement is false. Verify that the statement $5 > 3$ has a corresponding output of 1 on a TI-83/84 Plus calculator.

FIGURE 1

FIGURE 2

Symbols

To simplify work with logic, symbols are used. Statements are represented with letters, such as p, q, or r, while several symbols for connectives are shown in the following table. The table also gives the type of compound statement having the given connective.

Connective	Symbol	Type of Statement
and	$\wedge$	Conjunction
or	$\vee$	Disjunction
not	$\sim$	Negation

The symbol $\sim$ represents the connective *not.* If p represents the statement "Bill Clinton was president in 1995" then $\sim p$ represents "Bill Clinton was not president in 1995." The statement $\sim p$ could also be translated as "It is not true that Bill Clinton was president in 1995." There is usually more than one way to express a negation, and so your answer may not always agree exactly with ours. We recommend avoiding convoluted wording.

In applications, choosing meaningful letters will help you remember what the letter represents. While p may be perfectly good for representing a generic statement, a statement such as "Django is a good dog" might be better represented by the letter d.

EXAMPLE 4 **Symbolic Statements**

Let h represent "My backpack is heavy," and r represent "It's going to rain." Write each symbolic statement in words.

(a) $h \wedge r$

▶**Solution** From the table, $\wedge$ represents *and*, so the statement represents

"My backpack is heavy and it's going to rain."

(b) $\sim h \lor r$

▶**Solution** The *not* applies only to the first symbol, not the entire expression:

"My backpack is not heavy or it's going to rain."

(c) $\sim (h \lor r)$

▶**Solution** Because of the parentheses, the *not* applies to the entire expression:

"It is not the case that either my backpack is heavy or it's going to rain."

(d) $\sim (h \land r)$

▶**Solution**

"It is not the case that both my backpack is heavy and it's going to rain."

The statement in Example 4(c) is usually translated, "Neither h nor r," as in "Neither is my backpack heavy nor is it going to rain."

The negation of the negation of a statement is simply the statement itself. For example, the negation of the statement in Example 4(d) is $h \land r$, or "My backpack is heavy and it's going to rain." Symbolically, $\sim(\sim p)$ is equivalent to p.

We can represent the fact that the negation of a true statement is false and the negation of a false statement is true using a table known as a **truth table**, which shows all possible combinations of truth values for the component statements, as well as the corresponding truth value for the compound statement under consideration. Here is the truth table for negation.

TRUTH TABLE FOR THE NEGATION not p

p	$\sim p$
T	F
F	T

This truth table can be thought of as the definition of negation. Later in this chapter we will see how more complicated truth tables allow us to analyze complicated statements.

Conjunction The word logicians use for "*p and q*," denoted $p \land q$, is **conjunction**. In everyday language, *and* conveys the idea that both component statements are true. For example, the statement

My birthday is in April and yours is in May

would be true only if my birthday is indeed in April and yours is in May. If either part were false, the statement would be considered false. We represent this definition of $p \land q$ symbolically using the following truth table for conjunction. Notice that there are four possible combinations of truth values for p and q.

TRUTH TABLE FOR THE CONJUNCTION p and q

p	q	$p \wedge q$
T	T	T
T	F	F
F	T	F
F	F	F

Although the order of the rows in a truth table is arbitrary, you should follow the order that we use, which is not only standard but also organized and easy to remember. A different method can make it hard to compare answers, and with a disorganized method one could easily miss a case or even have duplicate cases. Notice that in the columns labeled with a single variable (p or q), the rightmost column (the q column) alternates T, F, T, F. The next column to the left (the p column) has two T's followed by two F's. When we introduce truth tables with three variables, the leftmost column will have four T's followed by four F's.

EXAMPLE 5 **Truth Value**

Let p represent "$5 > 3$" and let q represent "$6 < 0$." Find the truth value of $p \wedge q$.

▶**Solution**

METHOD 1 Truth Tables Here p is true and q is false. Looking in the second row of the conjunction truth table shows that $p \wedge q$ is false.

METHOD 2 Graphing Calculator This example can also be completed with the help of a TI-83/84 Plus calculator. Using the `LOGIC` menu, illustrated in Figure 3, we input `5 > 3 and 6 < 0` into the calculator. The output is zero, shown in Figure 4, which means the statement is false.

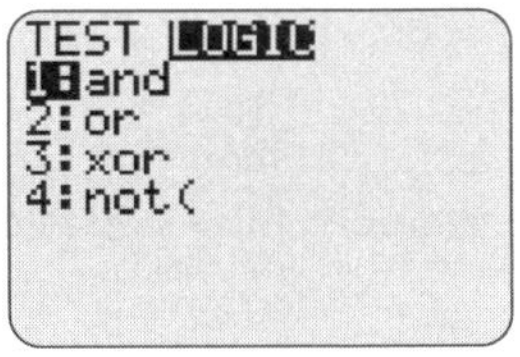

FIGURE 3

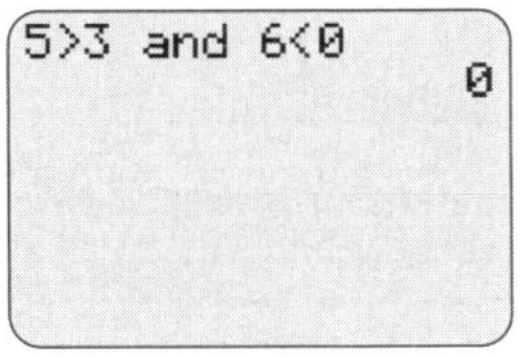

FIGURE 4

There's another word that has the same logical meaning as the word *and*, namely *but*. For example, the statement

I was not ready yesterday, but I am ready today

conveys the same logical meaning as "I was not ready yesterday, and I am ready today," with the additional idea of contrast between my status on the two days. Contrast may be relevant in normal conversation, but it makes no difference in logic.

Disjunction In ordinary language, the word *or* can be ambiguous. For example, the statement

Those with a passport or driver's license will be admitted

means that anyone will be admitted who has a passport or a driver's license or both. On the other hand, the statement

You can have a piece of cake or a piece of fruit

probably means you cannot have both.

In logic, the word *or* has the first meaning, known as *inclusive disjunction* or just **disjunction**. It is written with the symbol $\vee$, so that $p \vee q$ means "p or q or both." It is only false when both component statements are false. The truth table for disjunction is given below.

TRUTH TABLE FOR THE DISJUNCTION p or q

p	q	$p \vee q$
T	T	T
T	F	T
F	T	T
F	F	F

NOTE In English, the *or* is often interpreted as exclusive disjunction, not the way it is used in logic. For more on the exclusive disjunction, see Exercises 35–38 in the next section.

Compound Statements

The next two examples show how we can find the truth value of compound statements.

EXAMPLE 6

Truth Value of a Compound Statement

Suppose p is false, q is true, and r is false. What is the truth value of the compound statement $\sim p \wedge (q \vee \sim r)$?

▶Solution Here parentheses are used to group q and $\sim r$ together. Work first inside the parentheses. Since r is false, $\sim r$ will be true. Since $\sim r$ is true and q is true and an *or* statement is true when either component is true, $q \vee \sim r$ must be true. An *and* statement is only true when both components are true. Since $\sim p$ is true and $q \vee \sim r$ is true, the statement $\sim p \wedge (q \vee \sim r)$ is true.

The preceding paragraph may be interpreted using a short-cut symbolic method, letting T represent a true statement and F represent a false statement:

$$\sim p \wedge (q \vee \sim r)$$
$$\sim \mathrm{F} \wedge (\mathrm{T} \vee \sim \mathrm{F})$$
$$\mathrm{T} \wedge (\mathrm{T} \vee \mathrm{T}) \qquad \sim\mathrm{F} \text{ gives } \mathrm{T}.$$
$$\mathrm{T} \wedge \mathrm{T} \qquad \mathrm{T} \vee \mathrm{T} \text{ gives } \mathrm{T}.$$
$$\mathrm{T}. \qquad \mathrm{T} \wedge \mathrm{T} \text{ gives } \mathrm{T}.$$

The T in the final row indicates that the compound statement is true.

EXAMPLE 7

Mathematical Statements

Let p represent the statement $3 > 2$, q represent $5 < 4$, and r represent $3 < 8$. Decide whether the following statements are *true* or *false.*

(a) $\sim p \wedge \sim q$

▶Solution Since p is true, $\sim p$ is false. By the *and* truth table, if one part of an "and" statement is false, the entire statement is false. This makes $\sim p \wedge \sim q$ false.

(b) $\sim(p \wedge q)$

▶**Solution** First, work within the parentheses. Since p is true and q is false, $p \wedge q$ is false by the *and* truth table. Next, apply the negation. The negation of a false statement is true, making $\sim(p \wedge q)$ a true statement.

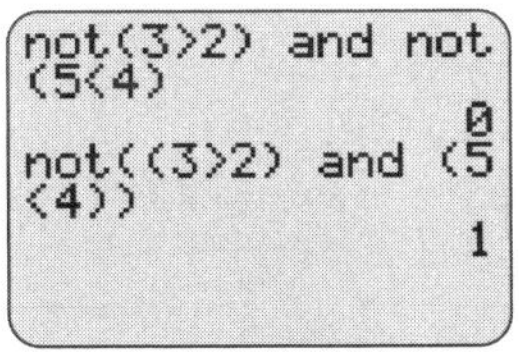

FIGURE 5

Figure 5 shows the result of using a TI-83/84 Plus to solve parts (a) and (b) of this example.

(c) $(\sim p \wedge r) \vee (\sim q \wedge \sim p)$

▶**Solution** Here p is true, q is false, and r is true. This makes $\sim p$ false and $\sim q$ true. By the *and* truth table, the statement $\sim p \wedge r$ is false, and the statement $\sim q \wedge \sim p$ is also false. Finally,

$$\begin{array}{ccc} (\sim p \wedge r) & \vee & (\sim q \wedge \sim p) \\ \downarrow & & \downarrow \\ \text{F} & \vee & \text{F}, \end{array}$$

which is false by the *or* truth table. (For an alternate solution, see Example 3(b) in the next section.)

NOTE The expression $\sim p \wedge \sim q$ in part (a) is often expressed in English as "Neither p nor q." We saw at the end of Example 4 that the expression $\sim(p \vee q)$ can also be expressed as "Neither p nor q." As we shall see in the next section, the expressions $\sim p \wedge \sim q$ and $\sim(p \vee q)$ are equivalent.

▶ 6.1 Exercises

Decide whether each of the following is a statement. If it is a statement, decide whether or not it is compound.

1. Montevideo is the capital of Uruguay.
2. John Marshall was not the first chief justice of the United States.
3. Don't cry for me, Argentina.
4. Do unto others as you would have them do unto you.
5. $2 + 2 = 5$ and $3 + 3 = 7$
6. $x < 7$ or $x > 14$
7. Where's the beef?
8. Is that all there is?
9. I am not a crook.
10. China has a population of more than 1 billion, and so does India.
11. She enjoyed the comedy team of Penn and Teller.
12. The New Hampshire motto is "Live free or die."
13. If ever I would leave you, it wouldn't be in summer.
14. If it's past 8:00, then we are late.

Write a negation for each statement.

15. My favorite flavor is chocolate.

16. This is not the time to complain.

Give a negation of each inequality.

17. $y > 12$

18. $x < -6$

19. $q \geq 5$

20. $r \leq 19$

21. Try to negate the sentence "The exact number of words in this sentence is 10" and see what happens. Explain the problem that arises.

22. Explain why the negation of "$r > 4$" is not "$r < 4$."

Let b represent the statement "I'm getting better" *and d represent the statement* "My parrot is dead." *Translate each symbolic statement into words.*

23. $\sim b$

24. $\sim d$

25. $\sim b \vee d$

26. $b \wedge \sim d$

27. $\sim(b \wedge \sim d)$

28. $\sim(b \vee d)$

Use the concepts introduced in this section to answer Exercises 29–34.

29. If q is false, what must be the truth value of $(p \wedge \sim q) \wedge q$?

30. If q is true, what must be the truth value of $q \vee (q \wedge \sim p)$?

31. If $p \wedge q$ is true, then q must be _____.

32. If $p \vee q$ is false, and p is false, then q must be _____.

33. If $\sim(p \vee q)$ is true, what must be the truth values of each of the component statements?

34. If $\sim(p \wedge q)$ is false, what must be the truth values of the component statements?

Let p represent a false statement and let q represent a true statement. Find the truth value of each compound statement.

35. $\sim p$

36. $\sim q$

37. $p \vee q$

38. $p \wedge q$

39. $p \vee \sim q$

40. $\sim p \wedge q$

41. $\sim p \vee \sim q$

42. $p \wedge \sim q$

43. $\sim(p \wedge \sim q)$

44. $\sim(\sim p \vee \sim q)$

45. $\sim[\sim p \wedge (\sim q \vee p)]$

46. $\sim[(\sim p \wedge \sim q) \vee \sim q]$

47. Is the statement $3 \geq 1$ a conjunction or a disjunction? Why?

48. Why is the statement $6 \geq 2$ true? Why is $6 \geq 6$ true?

Let p represent a true statement, and q and r represent false statements. Find the truth value of each compound statement.

49. $(p \wedge r) \vee \sim q$

50. $(q \vee \sim r) \wedge p$

51. $p \wedge (q \vee r)$

52. $(\sim p \wedge q) \vee \sim r$

53. $\sim(p \wedge q) \wedge (r \vee \sim q)$

54. $(\sim p \wedge \sim q) \vee (\sim r \wedge q)$

55. $\sim[(\sim p \wedge q) \vee r]$

56. $\sim[r \vee (\sim q \wedge \sim p)]$

Let p represent the statement $2 > 7$, *let q represent the statement* $8 \leq 6$, *and let r represent the statement* $19 \leq 19$. *Find the truth value of each compound statement.*

57. $p \wedge r$

58. $p \vee \sim q$

59. $\sim q \vee \sim r$

60. $\sim p \wedge \sim r$

61. $(p \wedge q) \vee r$

62. $\sim p \vee (\sim r \vee \sim q)$

63. $(\sim r \wedge q) \vee \sim p$

64. $\sim(p \vee \sim q) \vee \sim r$

Applications

BUSINESS AND ECONOMICS

Credit Cards *The following excerpts appear in an agreement accompanying an American Express credit card.*

a. *All payments must be sent to the payment address shown on your billing statement and must include the remittance coupon from your billing statement.*

b. *Please read and keep this Agreement.*

c. *The "Prime Rate" is determined once with respect to each billing period.*

d. *We may charge a fee of $35 in each billing period if the New Balance on your statement exceeds your credit line.*

65. Which of these excerpts are statements?

66. Which of these excerpts are compound statements?

67. Write the negation of excerpt d.

Income Tax *The following excerpts appear in a guide for preparing income tax reports.**

a. *Make sure you have received a Form W-2 from each employer for whom you worked in 2005.*

b. *Separate property brought into a community property state generally retains its character as separately owned property.*

c. *Tax rates are lower for a head of household than for a person filing as single and the standard deduction is higher.*

d. *You do not reduce the exemptions because of the shorter taxable year.*

68. Which of these excerpts are statements?

69. Which of these excerpts are compound statements?

70. Write the negation of excerpt d.

71. Define p and q to symbolically represent excerpt c.

LIFE SCIENCES

Medicine *The following excerpts appear in a home medical reference book.†*

a. *Can you climb one or two flights of stairs without shortness of breath or heaviness or fatigue in your legs?*

b. *Regularly doing exercises that concentrate on strengthening particular muscle groups and improving overall flexibility can help prevent back pain and keep you mobile.*

c. *If you answered yes to all of the questions above, you are reasonably fit.*

d. *These chemical compounds act as natural antidepressants and they can help you feel more relaxed.*

e. *You may find that exercise helps you cope with stress.*

72. Which of these excerpts are statements?

73. Which of these excerpts are compound statements?

74. Write the negation of excerpt e.

SOCIAL SCIENCES

Law *The following excerpts appear in a guide to common laws.‡*

a. *If a mortgage lender denies you a loan or a seller refuses to sell to you and you believe that you're being discriminated against, call the Department of Housing and Urban Development to file a formal complaint.*

b. *Don't be pressured into signing a contract.*

c. *The court won't do it for you, and hiring an attorney is usually not cost effective, given the small amount of money involved.*

d. *You can't marry unless you're at least 18 years old or unless you have the permission of your parents or guardian.*

e. *Most legal problems are matters of civil law.*

75. Which of these excerpts are statements?

76. Which of these excerpts are compound statements?

77. Write the negation of excerpt e.

78. *Philosophy* Read each of the following quotes from ancient philosophers.§ Provide an argument why these quotes may or may not be called statements.

a. "A friend is a friend of someone."—Socrates

b. "Every art, and every science reduced to a teachable form, and in like manner every action and moral choice, aims, it is thought, at some good: for which reason a common and

*J. K. Lasser Institute, *Your Income Tax 2006*, New York: John Wiley & Sons, pp. 6, 15, 22, 25.
†Goldman, D. R., ed., *American College of Physicians Complete Home Medical Guide*, 2nd ed., New York: DK Publishing, 2003, p. 57.
‡Ventura, John, *Law for Dummies*, Foster City, CA: IDG Books Worldwide, 1996.
§Frost, S. E., ed., *Masterworks of Philosophy*, New York: McGraw-Hill, 1946.

by no means a bad description of what the Chief Good is, 'that which all things aim at.'"—Aristotle

c. "Furthermore, Friendship helps the young to keep from error: the old, in respect of attention and such deficiencies in action as their weakness makes them liable to; and those who are in their prime, in respect of noble deeds, because they are thus more able to devise plans and carry them out."—Aristotle

79. *Bible* Read each of the following quotes from the biblical book Proverbs.* Provide an argument why these quotes may or may not be called statements.

a. "A gentle answer turns away wrath."—Proverbs 15:1

b. "The hot-tempered stir up dissension, but those who are patient calm a quarrel."—Proverbs 15:18

c. "When justice is done, it brings joy to the righteous but terror to evildoers."—Proverbs 21:15

d. "Do not exploit the poor because they are poor and do not crush the needy in court."—Proverbs 22:22

e. "Apply your heart to instruction and your ears to words of knowledge."—Proverbs 23:12

GENERAL INTEREST

Football For Exercises 80–85, let n represent the statement "New England won the Super Bowl" *and b represent the statement* "Tom Brady is the best quarterback." *Convert each compound statement into symbols.*

80. New England won the Super Bowl but Tom Brady is not the best quarterback.

81. New England did not win the Super Bowl or Tom Brady is not the best quarterback.

82. New England did not win the Super Bowl or Tom Brady is the best quarterback.

83. New England did not win the Super Bowl but Tom Brady is the best quarterback.

84. Neither did New England win the Super Bowl nor is Tom Brady the best quarterback.

85. Either New England won the Super Bowl or Tom Brady is the best quarterback, and it is not the case that both New England won the Super Bowl and Tom Brady is the best quarterback.

86. Suppose the statements that New England won the Super Bowl and that Tom Brady is the best quarterback are both true. Which of Exercises 80–85 are true statements?

87. Suppose the statements that New England won the Super Bowl and that Tom Brady is the best quarterback are both false. Which of Exercises 80–85 are true statements?

6.2 Truth Tables and Equivalent Statements

? THINK ABOUT IT **When using a search engine on the Internet, you are asked to supply key words. How does the search engine connect these key words?**

You will be asked to explore this question in one of the exercises at the end of this section.

In the previous section, we created truth tables for some simple logical expressions. We will now create truth tables for more complex statements, such as those you might read or hear, and determine for what values of the individual component statements the complex statement is true. As described in the previous section, we will use the following standard format for listing the possible truth values in compound statements involving two statements.

p	q	Compound Statement
T	T	
T	F	
F	T	
F	F	

**TNIV Bible*, Zondervan, 2005.

EXAMPLE 1 Truth Tables

(a) Construct a truth table for $(\sim p \wedge q) \vee \sim q$.

Solution Begin by listing all possible combinations of truth values for p and q, as above. Then find the truth values of $\sim p \wedge q$. Start by listing the truth values of $\sim p$, which are the opposite of those of p.

p	q	$\sim p$
T	T	F
T	F	F
F	T	T
F	F	T

Use only the "$\sim p$" column and the "q" column, along with the *and* truth table, to find the truth values of $\sim p \wedge q$. List them in a separate column.

p	q	$\sim p$	$\sim p \wedge q$
T	T	F	F
T	F	F	F
F	T	T	T
F	F	T	F

Next include a column for $\sim q$.

p	q	$\sim p$	$\sim p \wedge q$	$\sim q$
T	T	F	F	F
T	F	F	F	T
F	T	T	T	F
F	F	T	F	T

Finally, make a column for the entire compound statement. To find the truth values, use *or* to combine $\sim p \wedge q$ with $\sim q$.

p	q	$\sim p$	$\sim p \wedge q$	$\sim q$	$(\sim p \wedge q) \vee \sim q$
T	T	F	F	F	F
T	F	F	F	T	T
F	T	T	T	F	T
F	F	T	F	T	T

(b) Suppose both p and q are true. Find the truth value of $(\sim p \wedge q) \vee \sim q$.

Solution Look in the first row of the final truth table above, where both p and q have truth value T. Read across the row to find that the compound statement is false.

EXAMPLE 2 **Truth Table**

Construct a truth table for the following statement:

I'm bringing the food, or Heather Peck's bringing the food and I'm not.

▶**Solution** If we let i represent "I'm bringing the food" and h represent "Heather Peck is bringing the food," the statement can be represented symbolically as $i \vee (h \wedge \sim i)$. Proceed as shown in the following truth table.

i	h	$\sim i$	$h \wedge \sim i$	$i \vee (h \wedge \sim i)$
T	T	F	F	T
T	F	F	F	T
F	T	T	T	T
F	F	T	F	F

Notice from the truth table above that the only circumstances under which the original statement is false is when both statements "I'm bringing the food" and "Heather Peck is bringing the food" are false.

If a compound statement involves three component statements p, q, and r, we will use the following format in setting up the truth table.

p	q	r	**Compound Statement**
T	T	T	
T	T	F	
T	F	T	
T	F	F	
F	T	T	
F	T	F	
F	F	T	
F	F	F	

As we mentioned in the previous section, the rightmost column (the r column) alternates T, F, T, F. The next column to the left (the q column) has two T's followed by two F's, and then repeats this pattern. The leftmost column has four T's followed by four F's.

EXAMPLE 3 **Truth Tables**

(a) Construct a truth table for $(\sim p \wedge r) \vee (\sim q \wedge \sim p)$.

▶**Solution** This statement has three component statements, p, q, and r. The truth table thus requires eight rows to list all possible combinations of truth values of p, q, and r. The final truth table, however, can be found in much the same way as the ones above.

p	q	r	$\sim p$	$\sim p \wedge r$	$\sim q$	$\sim q \wedge \sim p$	$(\sim p \wedge r) \vee (\sim q \wedge \sim p)$
T	T	T	F	F	F	F	F
T	T	F	F	F	F	F	F
T	F	T	F	F	T	F	F
T	F	F	F	F	T	F	F
F	T	T	T	T	F	F	T
F	T	F	T	F	F	F	F
F	F	T	T	T	T	T	T
F	F	F	T	F	T	T	T

(b) Suppose p is true, q is false, and r is true. Find the truth value of $(\sim p \wedge r) \vee (\sim q \wedge \sim p)$.

▶**Solution** By the third row of the truth table in part (a), the compound statement is false. (This is an alternate method for working part (c) of Example 7 of the previous section.)

NOTE One strategy for problem solving is noticing a pattern and using *inductive reasoning*, or reasoning that uses particular facts to find a general rule. This strategy is used in the next example.

EXAMPLE 4 **Counting**

If n is a counting number, and a logical statement is composed of n component statements, how many rows will appear in the truth table for the compound statement?

▶**Solution** To answer this question, let us examine some of the earlier truth tables in this section. The truth table for the negation has one statement and two rows. The truth tables for the conjunction and the disjunction have two component statements, and each has four rows. The truth table in Example 3(a) has three component statements and eight rows. Summarizing these in a table shows a pattern seen earlier.

Number of Statements	Number of Rows
1	$2 = 2^1$
2	$4 = 2^2$
3	$8 = 2^3$

Inductive reasoning leads us to the conjecture that, if a logical statement is composed of n component statements, it will have 2^n rows. This can be proved using ideas in Chapter 8.

Intuitively, it's not hard to see why adding a statement doubles the number of rows. For example, if we wanted to construct a truth table with the four statements p, q, r, and s, we could start with the truth table for p, q, and r, which has eight rows. We must let s have the value of T for each of these rows, and we must also let s have the value of F for each of these rows, giving a total of 16 rows.

A logical statement having n component statements will have 2^n rows in its truth table.

Alternative Method for Constructing Truth Tables

After making a reasonable number of truth tables, some people prefer the shortcut method shown in Example 5, which repeats Examples 1 and 3.

EXAMPLE 5 **Truth Tables**

Construct the truth table for each statement.

(a) $(\sim p \wedge q) \vee \sim q$

▶**Solution** Start by inserting truth values for $\sim p$ and for q.

p	q	$(\sim p$	$\wedge$	$q)$	$\vee$	$\sim q$
T	T	F		T		
T	F	F		F		
F	T	T		T		
F	F	T		F		

Next, use the *and* truth table to obtain the truth values of $\sim p \wedge q$.

p	q	$(\sim p$	$\wedge$	$q)$	$\vee$	$\sim q$
T	T	F	F	T		
T	F	F	F	F		
F	T	T	T	T		
F	F	T	F	F		

Now disregard the two preliminary columns of truth values for $\sim p$ and for q, and insert truth values for $\sim q$.

p	q	$(\sim p$	$\wedge$	$q)$	$\vee$	$\sim q$
T	T		F			F
T	F		F			T
F	T		T			F
F	F		F			T

Finally, use the *or* truth table.

p	q	$(\sim p$	$\wedge$	$q)$	$\vee$	$\sim q$
T	T		F		F	F
T	F		F		T	T
F	T		T		T	F
F	F		F		T	T

These steps can be summarized as follows.

p	q	$(\sim p$	$\wedge$	$q)$	$\vee$	$\sim q$
T	T	F	F	T	F	F
T	F	F	F	F	T	T
F	T	T	T	T	T	F
F	F	T	F	F	T	T
		①	②	①	④	③

The circled numbers indicate the order in which the various columns of the truth table were found.

(b) $(\sim p \wedge r) \vee (\sim q \wedge \sim p)$

▶**Solution** Work as follows.

p	q	r	$(\sim p$	$\wedge$	$r)$	$\vee$	$(\sim q$	$\wedge$	$\sim p)$
T	T	T	F	F	T	F	F	F	F
T	T	F	F	F	F	F	F	F	F
T	F	T	F	F	T	F	T	F	F
T	F	F	F	F	F	F	T	F	F
F	T	T	T	T	T	T	F	F	T
F	T	F	T	F	F	F	F	F	T
F	F	T	T	T	T	T	T	T	T
F	F	F	T	F	F	T	T	T	T
			①	②	①	⑤	③	④	③

Equivalent Statements

One application of truth tables is illustrated by showing that two statements are equivalent; by definition, two statements are **equivalent** if they have the same truth value in *every* possible situation. The columns of each truth table that were the last to be completed will be exactly the same for equivalent statements.

EXAMPLE 6 **Equivalent Statements**

Are the statements

$$\sim p \wedge \sim q \qquad \text{and} \qquad \sim(p \vee q)$$

equivalent?

Solution To find out, make a truth table for each statement, with the following results.

p	q	$\sim p \wedge \sim q$
T	T	F
T	F	F
F	T	F
F	F	T

p	q	$\sim(p \vee q)$
T	T	F
T	F	F
F	T	F
F	F	T

Since the truth values are the same in all cases, as shown in the columns in color, the statements $\sim p \wedge \sim q$ and $\sim(p \vee q)$ are equivalent. Equivalence is written with a three-bar symbol, $\equiv$. Using this symbol, $\sim p \wedge \sim q \equiv \sim(p \vee q)$.

In the same way, the statements $\sim p \vee \sim q$ and $\sim(p \wedge q)$ are equivalent. We call these equivalences **De Morgan's Laws**.*

DE MORGAN'S LAWS

For any statements p and q,

$$\sim(p \vee q) \equiv \sim p \wedge \sim q$$

$$\sim(p \wedge q) \equiv \sim p \vee \sim q.$$

DeMorgan's Laws can be used to find the negations of certain compound statements.

EXAMPLE 7

Negation

Find the negation of each statement, applying De Morgan's Laws to simplify.

(a) It was a dark and stormy night.

Solution If we first rephrase the statement as "The night was dark and the night was stormy," we can let d represent "the night was dark" and s represent "the night was stormy." The original compound statement can then be written $d \wedge s$. The negation of $d \wedge s$ is $\sim(d \wedge s)$. Using the second of De Morgan's Laws, $\sim(d \wedge s) \equiv \sim d \vee \sim s$. In words this reads:

The night was not dark or the night was not stormy.

In retrospect, this should be obvious. If it's not true that the night was dark and stormy, then either the night wasn't dark or it wasn't stormy.

*Augustus De Morgan (1806–1871) was born in India, where his father was serving as an English army officer. After studying mathematics at Cambridge, he prepared for a career in law because his performance on the challenging Tripos Exam in mathematics was not very good. Nevertheless, De Morgan was offered a position as chair of the department of mathematics at London University. He wrote and taught mathematics with great clarity and dedication, and he was also an excellent flutist.

(b) Either John will play the guitar or George will not sing.

▶**Solution** In this book, we interpret "either . . . or" as disjunction; the word "either" just highlights that a disjunction is about to occur. Using the first of De Morgan's Laws, $\sim(j \vee \sim g) \equiv \sim j \wedge \sim(\sim g)$. Using the observation from the last section that the negation of a negation of a statement is simply the statement, we can simplify the last statement to $\sim j \wedge g$. In words this reads:

John will not play the guitar and George will sing.

(c) $(\sim p \wedge q) \vee \sim r$

▶**Solution** After negating, we apply the first of De Morgan's Laws, changing $\vee$ to $\wedge$:

$$\sim[(\sim p \wedge q) \vee \sim r] \equiv \sim(\sim p \wedge q) \wedge \sim(\sim r).$$

The last term simplifies to r. Applying the second of De Morgan's Laws to the first term yields

$$\sim(\sim p \wedge q) \wedge r \equiv (p \vee \sim q) \wedge r,$$

where we have replaced $\sim(\sim p)$ with p. (From now on, we will make this simplification without mentioning it.)

6.2 Exercises

Give the number of rows in the truth table for each compound statement.

1. $p \vee \sim r$

2. $p \wedge (r \wedge \sim s)$

3. $(\sim p \wedge q) \vee (\sim r \vee \sim s) \wedge r$

4. $[(p \vee q) \wedge (r \wedge s)] \wedge (t \vee \sim p)$

5. $[(\sim p \wedge \sim q) \wedge (\sim r \wedge s \wedge \sim t)] \wedge (\sim u \vee \sim v)$

6. $[(\sim p \wedge \sim q) \vee (\sim r \vee \sim s)] \vee [(\sim m \wedge \sim n) \wedge (u \wedge \sim v)]$

7. If the truth table for a certain compound statement has 64 rows, how many distinct component statements does it have?

8. Is it possible for the truth table of a compound statement to have exactly 48 rows? Why or why not?

Construct a truth table for each compound statement.

9. $\sim p \wedge q$

10. $\sim p \vee \sim q$

11. $\sim(p \wedge q)$

12. $p \vee \sim q$

13. $(q \vee \sim p) \vee \sim q$

14. $(p \wedge \sim q) \wedge p$

15. $\sim q \wedge (\sim p \vee q)$

16. $\sim p \vee (\sim q \wedge \sim p)$

17. $(p \vee \sim q) \wedge (p \wedge q)$

18. $(\sim p \wedge \sim q) \vee (\sim p \vee q)$

19. $(\sim p \wedge q) \wedge r$

20. $r \vee (p \wedge \sim q)$

21. $(\sim p \wedge \sim q) \vee (\sim r \vee \sim p)$

22. $(\sim r \vee \sim p) \wedge (\sim p \vee \sim q)$

23. $\sim(\sim p \wedge \sim q) \vee (\sim r \vee \sim s)$

24. $(\sim r \vee s) \wedge (\sim p \wedge q)$

Write the negation of each statement, applying De Morgan's Laws to simplify.

25. It's summertime, and the living is easy.

26. Rachel Reeve was elected president and Joanne Ha was elected treasurer.

27. Either the door was unlocked or the thief broke a window.

28. Sue brings the wrong book or she forgets the notes.

29. I'm ready to go, but Emily Portwood isn't.

30. You may say I'm a dreamer, but I'm not the only one.

31. $12 > 4$ or $8 = 9$

32. $2 + 3 = 5$ and $12 + 13 = 15$

33. Larry or Moe is out sick today.

34. Jack and Jill went up the hill.

35. Complete the truth table for *exclusive disjunction.* The symbol $\underline{\vee}$ represents "one or the other is true, but not both."

p	q	$p \underline{\vee} q$
T	T	
T	F	
F	T	
F	F	

Exclusive disjunction

Decide whether the following compound statements are true *or* false. *Remember from Exercise 35 that* $\underline{\vee}$ *is the exclusive disjunction; that is, assume "either p or q is true, but not both."*

36. $(3 + 1 = 4) \underline{\vee} (2 + 5 = 7)$

37. $(3 + 1 = 4) \underline{\vee} (2 + 5 = 9)$

38. $(3 + 1 = 7) \underline{\vee} (2 + 5 = 7)$

39. Let p represent $2\sqrt{6} - 4\sqrt{5} > -1$, q represent $\dfrac{14 - 7\sqrt{8}}{2.5 - \sqrt{5}} > -22$, and s represent $\dfrac{7 - \dfrac{5}{\sqrt{3}}}{\sqrt{8} - 2} < \dfrac{\sqrt{3}}{\sqrt{2}}$. Use the LOGIC menu on a graphing calculator to find the truth value of each statement.

a. $p \wedge q$ **b.** $\sim p \wedge q$ **c.** $\sim(p \vee q)$ **d.** $(s \wedge \sim p) \vee (\sim s \wedge q)$

Applications

BUSINESS AND ECONOMICS

40. *Income Tax* The following statement appears in a guide for preparing income tax reports.* Use one of De Morgan's Laws to write the negation of this statement.

> The plan must clearly state that its purpose is to provide qualifying tax-free benefits, and a specific payment schedule must be provided for different types of injuries.

41. *Warranty* The statement below appears in a tire manufacturer's warranty.† Describe the use of *and/or* in this situation.

> This warranty is in addition to and/or may be limited by any other applicable written warranty concerning special tires or situations you may have received.

42. *Warranty* The following statement appears in a tire manufacturer's warranty.‡ Negate this statement. Does the new statement make sense?

> This warranty gives you specific legal rights and you may also have other rights, which vary from state to state.

43. *Guarantees* The guarantee on a brand of vacuum cleaner reads: "You will be completely satisfied or we will refund your money without asking any questions." Let s represent

*J. K. Lasser Institute, *Your Income Tax 2006*, New York: John Wiley & Sons, p. 58.

†"Passenger and Light Truck Tire Limited Warranty with Tire Maintenance and Safety Manual," Nashville, TN, Bridgestone/Firestone, Inc., 1997.

‡Ibid.

"You will be completely satisfied," r represent "We will refund your money," and q represent "We will ask you questions." Write the guarantee symbolically and then construct a truth table for the statement, putting the variables in the order s, r, q. Under what conditions would the guarantee be false?

LIFE SCIENCES

44. *Medicine* The following statements appear in a home medical reference book.* Define p and q so that the statements can be written symbolically. Then negate each statement.

a. Tissue samples may be taken from almost anywhere in the body, and the procedure used depends on the site.

b. The procedure can be carried out quickly in the doctor's office and is not painful.

c. Fluid samples may be examined for infection, or the cells in the fluid may be separated and examined to detect other abnormalities.

45. *Fisheries Management* The following statement is found in the mission of the Pennsylvania Fish and Boat Commission.[†] Define p and q so that the statement can be written symbolically. Then negate the statement.

> The Pennsylvania Fish and Boat Commission is sensitive to the needs of the physically challenged and works to make our facilities accessible.

SOCIAL SCIENCES

46. *Law* Attorneys sometimes use the phrase "and/or." This phrase corresponds to which usage of the word *or*: inclusive or exclusive?

47. *Law* The following statement appears in a guide to common laws.[‡] Define p and q so that the statement can be written symbolically. Then negate the statement.

> The court won't do it for you, and hiring an attorney is usually not cost effective.

48. *Presidential Quote* Use both of De Morgan's Laws to rewrite the negation of the following quote made by President John F. Kennedy at Vanderbilt University on March 18, 1963:

> Liberty without learning is always in peril, and learning without liberty is always in vain.[§]

49. *Politician* Senator Pompous B. Blowhard made the following campaign promise: "I will cut taxes and eliminate the deficit, or I will not run for reelection." Let c represent "I will cut taxes," let e represent "I will eliminate the deficit," and let r represent "I will run for reelection." Write the promise symbolically and then construct a truth table for the statement, putting the variables in the order c, e, r. Under what conditions would the promise be false?

GENERAL INTEREST

50. *Yahtzee®* The following statement appears in the instructions for the Milton Bradley game Yahtzee.®[||] Negate the statement.

> You could reroll the die again for your Large Straight or set aside the 2 Twos and roll for your Twos or for 3 of a Kind.

51. *Logic Puzzles* Raymond Smullyan is one of today's foremost writers of logic puzzles. Smullyan proposed a question, based on the classic Frank Stockton short story, in which a prisoner must make a choice between two doors: Behind one is a beautiful lady, and behind the other is a hungry tiger.[#] What if each door has a sign, and the prisoner knows that only one sign is true? The sign on Door 1 reads: In this room there is a lady and in the other room there is a tiger. The sign on Door 2 reads: In one of these rooms there is a lady and in one of these rooms there is a tiger. With this information, determine what is behind each door.

52. Describe how a search engine on the Internet uses key words and logical connectives to locate information.

*Goldman, David R., ed., *American College of Physicians Complete Home Medical Guide,* 2nd ed., New York: DK Publishing, 2003, p. 223.

†"2000 Pennsylvania Summary of Fishing Regulations and Laws," Peter A. Colangelo, Executive Director, Commonwealth of Pennsylvania Fish and Boat Commission, Harrisburg, PA.

‡Ventura, John, *Law for Dummies,* Foster City, CA: IDG Books Worldwide, 1996.

§http://www.chiasmus.com/mastersofchiasmus/kennedy.shtml.

||Milton Bradley Company, East Longmeadow, MA, 1996.

#Smullyan, Raymond, *The Lady or the Tiger? And Other Logic Puzzles, Including a Mathematical Novel That Features Godel's Great Discovery*, New York: Knopf, 1982.

6.3 The Conditional and Circuits

? THINK ABOUT IT **How can logic be used in the design of electrical circuits?**

This question will be answered in this section.

Conditionals

A **conditional** statement is a compound statement that uses the connective *if . . . then*, or anything equivalent. For example, here are a few conditional statements:

If it rains, *then* I carry my umbrella.

If the president comes, *then* security will be tight.

If the check doesn't arrive today, I will call to find out why.

Notice in the last statement that the word *then* was implied but not explicitly stated; the sentence is equivalent to the statement "If the check doesn't arrive today, *then* I will call to find out why." In each of these statements, the component after the word *if* gives a condition under which the last component is true. The last component is the statement coming after the word *then* (or, in the third statement, the implied *then*). There may be other conditions under which the last component is true. In the second statement above, for example, it might be true that even if the president doesn't come, security will be tight because the vice president is coming.

The conditional is written with an arrow, so that "if p, then q" is symbolized as

$$p \rightarrow q.$$

We read $p \rightarrow q$ as "p implies q" or "if p, then q." In the conditional $p \rightarrow q$, the statement p is the **antecedent**, while q is the **consequent**.

There are many equivalent forms of the conditional. For example, the statement

Winners never quit

can be rephrased as

If you are a winner, then you never quit.

As we will see in the last section of this chapter, the statement "Winners never quit" can also be translated as "All winners are nonquitters," but we will avoid that approach for now. There are other forms of the conditional that we will study in the next section.

Just as we defined conjunction and disjunction using a truth table, we will now do the same for the conditional. To see how such a table should be set up, we will analyze the following statement that might be made by a cereal company:

If you eat Wheat Crunchies, then you'll be full of energy.

Let e represent "You eat Wheat Crunchies" and f represent "You'll be full of energy." As before, there are four possible combinations of truth values for the two component statements, as illustrated by the following table.

Possibility	Eat Wheat Crunchies?	Full of Energy?	
1	Yes	Yes	e is T, f is T
2	Yes	No	e is T, f is F
3	No	Yes	e is F, f is T
4	No	No	e is F, f is F

Let's consider each of these possibilities.

1. If you eat Wheat Crunchies and indeed find that you are full of energy, then you must conclude that the company's claim is true, so place T in the first row of the truth table. This does not necessarily mean that eating Wheat Crunchies caused you to be full of energy; perhaps you are full of energy for some other reason unrelated to what you ate for breakfast.
2. If you eat Wheat Crunchies and are not full of energy, then the company's claim is false, so place F in the second row of the truth table.
3. If you don't eat Wheat Crunchies and yet find that you are full of energy, this doesn't invalidate the company's claim. They only promised results if you ate Wheat Crunchies; they made no claims about what would happen if you don't eat Wheat Crunchies. We will therefore place T in the third row of the truth table.
4. If you don't eat Wheat Crunchies and are not full of energy, you can't very well blame the company. Because they can still claim that their promise is true, we will place T in the fourth row.

The discussion above leads to the following truth table.

TRUTH TABLE FOR THE CONDITIONAL if p, then q

p	q	$p \rightarrow q$
T	T	T
T	F	F
F	T	T
F	F	T

The truth table for the conditional leads to some counterintuitive conclusions, because *if . . . then* sometimes has other connotations in English, as the following examples illustrate.

EXAMPLE 1

Truth Value

Suppose you get a 61 on the test and you pass the course. Find the truth value of the following statement:

If you get a 70 or higher on the test, then you pass the course.

▶Solution The first component of this conditional is false, while the second is true. According to the third line of the truth table for the conditional, this statement is true.

This result may surprise you. Even though the statement only says what happens if you get a 70 or higher on the test, perhaps you also infer from the statement that if you don't get 70 or higher, you won't pass the course. This is a common interpretation of *if . . . then*, but it is not consistent with our truth table above. This interpretation is called the *biconditional* and will be discussed in the next section.

EXAMPLE 2

Truth Value

Find the truth value of each of the following statements.

(a) If the earth is shaped like a cube, then elephants are smaller than mice.

▶Solution Both components of this conditional are false. According to the fourth line of the truth table for the conditional, this statement is true.

(b) If the earth is shaped like a cube, then George Washington was the first president of the United States.

▶Solution The first component of this conditional is false, while the second is true. According to the third line of the truth table for the conditional, this statement is true, even though it may seem like an odd statement.

The following observations come from the truth table for $p \rightarrow q$.

SPECIAL CHARACTERISTICS OF CONDITIONAL STATEMENTS

1. $p \rightarrow q$ is false only when the antecedent is *true* and the consequent is *false.*
2. If the antecedent is *false,* then $p \rightarrow q$ is automatically *true.*
3. If the consequent is *true,* then $p \rightarrow q$ is automatically *true.*

EXAMPLE 3

Conditional Statement

Write *true* or *false* for each statement. Here T represents a true statement, and F represents a false statement.

(a) $T \rightarrow (6 = 3)$

▶Solution Since the antecedent is true, while the consequent, $6 = 3$, is false, the given statement is false by the first point mentioned above.

(b) $(5 < 2) \rightarrow F$

▶Solution The antecedent is false, so the given statement is true by the second observation.

(c) $(3 \neq 2 + 1) \rightarrow T$

▶**Solution** The consequent is true, making the statement true by the third characteristic of conditional statements.

EXAMPLE 4 **Truth Value**

Given that p, q, and r are all false, find the truth value of the statement $(p \to \sim q) \to (\sim r \to q)$.

▶**Solution** Using the shortcut method explained in Example 6 of Section 6.1, we can replace p, q, and r with F (since each is false) and proceed as before, using the negation and conditional truth tables as necessary.

$$(p \to \sim q) \to (\sim r \to q)$$
$$(\mathrm{F} \to \sim \mathrm{F}) \to (\sim \mathrm{F} \to \mathrm{F})$$
$$(\mathrm{F} \to \mathrm{T}) \to (\mathrm{T} \to \mathrm{F}) \quad \text{Use the negation truth table.}$$
$$\mathrm{T} \to \mathrm{F} \quad \text{Use the conditional truth table.}$$
$$\mathrm{F}$$

The statement $(p \to \sim q) \to (\sim r \to q)$ is false when p, q, and r are all false.

Truth tables for compound statements involving conditionals are found using the techniques described in the previous section. The next example shows how this is done.

EXAMPLE 5 **Truth Table**

Construct a truth table for each statement.

(a) $(\sim p \to \sim q) \wedge \sim(p \vee \sim q)$

▶**Solution** First insert the truth values of $\sim p$ and of $\sim q$. Then find the truth values of $\sim p \to \sim q$.

p	q	$\sim p$	$\sim q$	$\sim p \to \sim q$
T	T	F	F	T
T	F	F	T	T
F	T	T	F	F
F	F	T	T	T

Next use p and $\sim q$ to find the truth values of $p \vee \sim q$ and then $\sim(p \vee \sim q)$.

p	q	$\sim p$	$\sim q$	$\sim p \to \sim q$	$p \vee \sim q$	$\sim(p \vee \sim q)$
T	T	F	F	T	T	F
T	F	F	T	T	T	F
F	T	T	F	F	F	T
F	F	T	T	T	T	F

Finally, add the column for the statement $(\sim p \rightarrow \sim q) \wedge \sim(p \vee \sim q)$.

p	q	$\sim p$	$\sim q$	$\sim p \rightarrow \sim q$	$p \vee \sim q$	$\sim(p \vee \sim q)$	$(\sim p \rightarrow \sim q) \wedge \sim(p \vee \sim q)$
T	T	F	F	T	T	F	F
T	F	F	T	T	T	F	F
F	T	T	F	F	F	T	F
F	F	T	T	T	T	F	F

(b) $(p \rightarrow q) \rightarrow (\sim p \vee q)$

▶**Solution** Go through steps similar to the ones above.

p	q	$p \rightarrow q$	$\sim p$	$\sim p \vee q$	$(p \rightarrow q) \rightarrow (\sim p \vee q)$
T	T	T	F	T	T
T	F	F	F	F	T
F	T	T	T	T	T
F	F	T	T	T	T

As the truth table in Example 5(a) shows, the statement $(\sim p \rightarrow \sim q) \wedge \sim(p \vee \sim q)$ is always false, regardless of the truth values of the components. Such a statement is called a **contradiction**. Most contradictions consist of two statements that cannot simultaneously be true that are united with a conjunction. The simplest such statement is $p \wedge \sim p$.

Notice from the truth table in Example 5(b) that the statement $(p \rightarrow q) \rightarrow (\sim p \vee q)$ is always true, regardless of the truth values of the components. Such a statement is called a **tautology**. Other examples of tautologies (as can be checked by forming truth tables) include $p \vee \sim p$, $p \rightarrow p$, $(\sim p \vee \sim q) \rightarrow \sim(q \wedge p)$, and so on. An important point here is that there are compound statements (such as $p \vee \sim p$) that are true (or false) independent of the truth values of the component parts; the truth depends on the logical structure alone. By the way, the truth tables in Example 5 also could have been found by the alternative method shown in the previous section.

Notice from the third and fifth columns of the truth table in Example 5(b) that $p \rightarrow q$ and $\sim p \vee q$ are equivalent.

WRITING A CONDITIONAL AS AN *or* STATEMENT

$p \rightarrow q$ is equivalent to $\sim p \vee q$.

EXAMPLE 6 **Equivalent Statements**

Write the following statement without using *if . . . then.*

If this is your first time, we're glad you're here.

►**Solution** Letting f represent "This is your first time" and g represent "We're glad you're here," the conditional may be restated as $\sim f \vee g$, or in words:

This is not your first time or we're glad you're here.

Since

$$p \rightarrow q \equiv \sim p \vee q,$$

we can take the negation of both sides of this equivalence to get

$$\sim(p \rightarrow q) \equiv \sim(\sim p \vee q).$$

By applying De Morgan's Law to the right side, we have

$$\begin{aligned} \sim(p \rightarrow q) &\equiv \sim(\sim p) \wedge \sim q \\ &\equiv p \wedge \sim q. \end{aligned}$$

This gives us an equivalent form for the negation of the conditional.

NEGATION OF $p \rightarrow q$

The negation of $p \rightarrow q$ is $p \wedge \sim q$.

EXAMPLE 7

Negation

Write the negation of each statement.

(a) If you go to the left, I'll go to the right.

►**Solution** Let y represent "You go to the left" and i represent "I'll go to the right." Then the original statement is represented $y \rightarrow i$. As we showed earlier, the negation of this is $y \wedge \sim i$, which can be translated into words as

You go to the left and I won't go to the right.

(b) It must be alive if it is breathing.

►**Solution** First, we must restate the given statement in *if . . . then* form:

If it is breathing, then it must be alive.

Based on our earlier discussion, the negation is

It is breathing and it is not alive.

A common error occurs when students try to write the negation of a conditional statement as another conditional statement. As seen in Example 7, the negation of a conditional statement is written as a conjunction.

Circuits

One of the first nonmathematical applications of symbolic logic was seen in the master's thesis of Claude Shannon in 1937. Shannon showed how the logic developed almost a century earlier by British mathematician George Boole could be used as an aid in designing electrical **circuits**. His work was immediately taken up by the designers of computers. These computers, then in the developmental stage, could be simplified and built for less money using the ideas of Shannon.

FIGURE 6

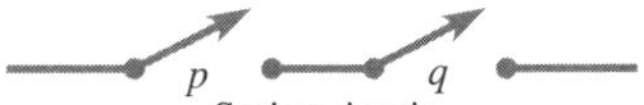

Series circuit

FIGURE 7

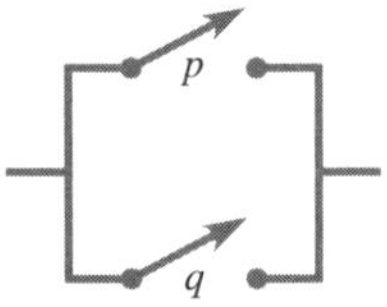

Parallel circuit

FIGURE 8

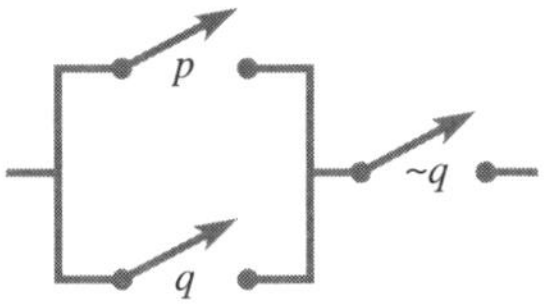

FIGURE 9

To see how Shannon's ideas work, look at the electrical switch shown in Figure 6. We assume that current will flow through this switch when it is closed and not when it is open.

Figure 7 shows two switches connected in **series**; in such a circuit, current will flow only when both switches are closed. Note how closely a series circuit corresponds to the conjunction $p \wedge q$. We know that $p \wedge q$ is true only when both p and q are true.

A circuit corresponding to the disjunction $p \vee q$ can be found by drawing a **parallel** circuit, as in Figure 8. Here, current flows if either p *or* q is closed or if both p *and* q are closed.

The circuit in Figure 9 corresponds to the statement $(p \vee q) \wedge \sim q$, which is a compound statement involving both a conjunction and a disjunction.

The way that logic is used to simplify an electrical circuit depends on the idea of equivalent statements, from Section 6.2. Recall that two statements are equivalent if they have exactly the same truth table final column. The symbol $\equiv$ is used to indicate that the two statements are equivalent. Some of the equivalent statements that we shall need are shown in the following box.

EQUIVALENT STATEMENTS

1. a. $p \vee q \equiv q \vee p$ — Commutative Laws
b. $p \wedge q \equiv q \wedge p$

2. a. $p \vee (q \vee r) \equiv (p \vee q) \vee r$ — Associative Laws
b. $p \wedge (q \wedge r) \equiv (p \wedge q) \wedge r$

3. a. $p \vee (q \wedge r) \equiv (p \vee q) \wedge (p \vee r)$ — Distributive Laws
b. $p \wedge (q \vee r) \equiv (p \wedge q) \vee (p \wedge r)$

4. a. $\sim(p \wedge q) \equiv \sim p \vee \sim q$ — De Morgan's Laws
b. $\sim(p \vee q) \equiv \sim p \wedge \sim q$

5. a. $p \vee p \equiv p$ — Idempotent Laws
b. $p \wedge p \equiv p$

6. a. $(p \wedge q) \vee p \equiv p$ — Absorption Laws
b. $(p \vee q) \wedge p \equiv p$

7. $\sim(\sim p) \equiv p$ — Double Negative

8. $p \rightarrow q \equiv \sim p \vee q$ — Conditional as an "or"

9. $p \rightarrow q \equiv \sim q \rightarrow \sim p$ — Contrapositive

If T represents any true statement and F represents any false statement, then

10. a. $p \vee \text{T} \equiv \text{T}$ — Identity Laws
b. $p \wedge \text{T} \equiv p$
c. $p \vee \text{F} \equiv p$
d. $p \wedge \text{F} \equiv \text{F}$

11. a. $p \vee \sim p \equiv \text{T}$ — Negation Laws
b. $p \wedge \sim p \equiv \text{F}$

This list may seem formidable, but if we break it down, it turns out to be not so bad.

First, notice that the first six equivalences come in pairs, in which $\vee$ is replaced with $\wedge$ and vice versa. This illustrates the **Principle of Duality**, which states that if a logical equivalence contains no logical operators other than $\vee$, $\wedge$, and $\sim$ (that is, it does not contain $\rightarrow$), then we may replace $\vee$ with $\wedge$ and vice versa, and replace T with F and vice versa, and the new equivalence is still true. This means that we need to keep track of just one from each pair of equivalences, although we have put both in the table for completeness.

Next, notice that equivalences 1 and 2 are just the familiar Commutative and Associative laws for addition or multiplication, which means you can rearrange the order and rearrange the parentheses in an expression involving only $\wedge$ or only $\vee$. You are asked to prove these using a truth table in Exercises 64–67.

Equivalence 3 is just the familiar Distributive Law of multiplication over addition. In normal multiplication, however, you *cannot* distribute $2 + (3 \times 5)$ to get $(2 + 3) \times (2 + 5)$. In other words, you cannot distribute addition over multiplication. But you can distribute $\wedge$ over $\vee$ and vice versa. You are asked to prove these using a truth table in Exercises 68 and 69.

De Morgan's Laws, given as equivalence 4, were discussed in the previous section.

Equivalence 5, the Idempotent Laws, may seem trivially true, so you may wonder why we bother stating them. The reason is that they are useful in simplifying circuits, as we shall see in Example 8. Equivalence 7, the Double Negative, is similar in this regard.

Equivalence 6, the Absorption Laws, may be less obvious, but they, too, are useful in simplifying circuits, as we shall see in Example 7. Their proofs are in Exercises 70 and 71.

Equivalence 8, the Conditional as an "or" statement, was discussed earlier in this section. Equivalence 9, the Contrapositive, is very important and will be discussed in the next section.

Equivalences 10 and 11, the Identity and Negation laws, should be fairly obvious. We only list them here because, just like equivalences 5 and 7, they can help us simplify circuits. Notice that the Principle of Duality applies to these equivalences.

Circuits can be used as models of compound statements, with a closed switch corresponding to T, while an open switch corresponds to F. The method for simplifying circuits is explained in the following example.

EXAMPLE 8 **Circuit**

Simplify the circuit in Figure 10.

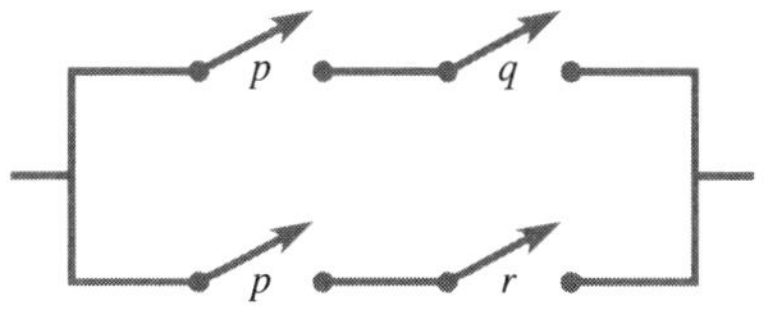

FIGURE 10

▶**Solution** At the top of Figure 10, p and q are connected in series, and at the bottom, p and r are connected in series. These are interpreted as the compound statements $p \wedge q$ and $p \wedge r$, respectively. These two conjunctions are connected in parallel, as indicated by the figure treated as a whole. Therefore, we write the disjunction of the two conjunctions:

$$(p \wedge q) \vee (p \wedge r).$$

(Think of the two switches labeled "p" as being controlled by the same handle.) By the Distributive Law (equivalence statement 2),

$$(p \wedge q) \vee (p \wedge r) \equiv p \wedge (q \vee r),$$

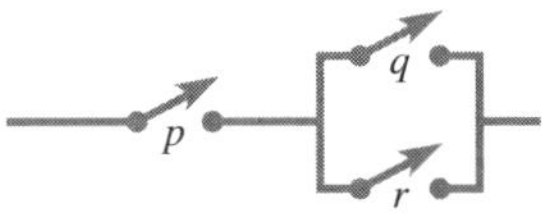

FIGURE 11

which has the circuit of Figure 11. This new circuit is logically equivalent to the one in Figure 10, and yet contains only three switches instead of four—which might well lead to a large savings in manufacturing costs.

EXAMPLE 9 **Circuit**

Simplify the circuit in Figure 12.

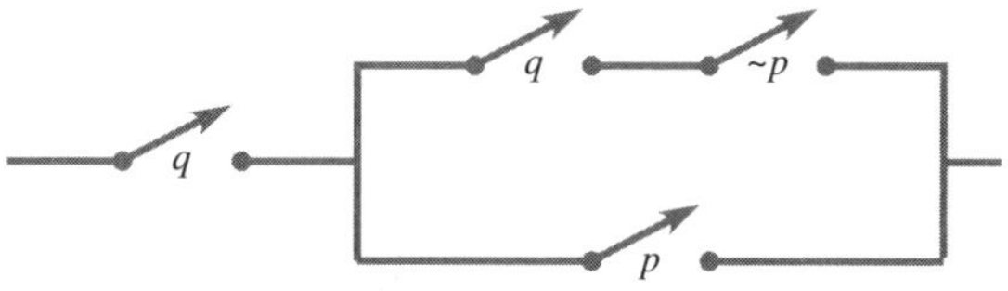

FIGURE 12

▶Solution The diagram shows q in series with a circuit containing two parallel components, one containing q and $\sim p$, and the other containing p. Thus the circuit can be represented by the logical statement

$$q \wedge [(q \wedge \sim p) \vee p].$$

Notice that for the expression in brackets, we can use the Commutative Law to put the p first, and then we can use the Distributive Law, which leads to a series of simplifications.

$$\begin{aligned}
q \wedge [(q \wedge \sim p) \vee p] &\equiv q \wedge [p \vee (q \wedge \sim p)] && \textbf{Commutative Law}\\
&\equiv q \wedge [(p \vee q) \wedge (p \vee \sim p)] && \textbf{Distributive Law}\\
&\equiv q \wedge [(p \vee q) \wedge \mathrm{T}] && \textbf{Idempotent Law}\\
&\equiv q \wedge (p \vee q) && \textbf{Identity Law}\\
&\equiv (p \vee q) \wedge q && \textbf{Commutative Law}\\
&\equiv (q \vee p) \wedge q && \textbf{Commutative Law}\\
&\equiv q && \textbf{Absorption Law}
\end{aligned}$$

The circuit simplified to a single switch! We hope this convinces you of the power of logic. Notice that we used the Commutative Law three times simply to get terms into a form in which another equivalence applies. This could be avoided if we just added more equivalences to our previous list, such as the Distributive Laws in the reverse order. This would, however, make our list even longer, so we will not do that, even though it means that we need an occasional extra step in our proofs.

EXAMPLE 10 **Circuit**

Draw a circuit for $p \rightarrow (q \wedge \sim r)$.

Solution By the equivalent statement Conditional as an "or," $p \rightarrow q$ is equivalent to $\sim p \vee q$. This equivalence gives $p \rightarrow (q \wedge \sim r) \equiv \sim p \vee (q \wedge \sim r)$, which has the circuit diagram in Figure 13.

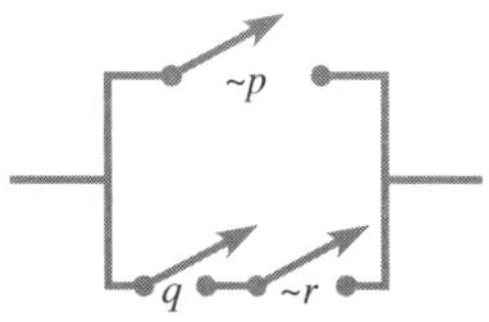

FIGURE 13

6.3 Exercises

In Exercises 1–6, decide whether each statement is true *or* false.

1. If the antecedent of a conditional statement is false, the conditional statement is true.
2. If the consequent of a conditional statement is true, the conditional statement is true.
3. If q is true, then $(p \wedge q) \rightarrow q$ is true.
4. If p is true, then $\sim p \rightarrow (q \vee r)$ is true.
5. Given that $\sim p$ is true and q is false, the conditional $p \rightarrow q$ is true.
6. Given that $\sim p$ is false and q is false, the conditional $p \rightarrow q$ is true.
7. In a few sentences, explain how we determine the truth value of a conditional statement.
8. Explain why the statement "If $3 = 5$, then $4 = 6$" is true.

Tell whether each conditional is true *or* false. *Here* T *represents a true statement and* F *represents a false statement.*

9. $F \rightarrow (4 \neq 7)$
10. $(6 \geq 6) \rightarrow F$
11. $(4 = 11 - 7) \rightarrow (8 > 0)$
12. $(4^2 \neq 16) \rightarrow (4 - 4 = 8)$

Let d represent "She dances tonight," *let s represent* "He sings loudly," *and let e represent* "I'm leaving early." *Express each compound statement in words.*

13. $d \rightarrow (e \wedge s)$
14. $(d \wedge s) \rightarrow e$
15. $\sim s \rightarrow (d \vee \sim e)$
16. $(\sim d \vee \sim e) \rightarrow \sim s$

Let d represent "My dog ate my homework," *let f represent* "I receive a failing grade," *and let g represent* "I'll run for governor." *Express each compound statement in words.*

17. My dog ate my homework, or if I receive a failing grade, then I'll run for governor.
18. I'll run for governor, and if I receive a failing grade, then my dog did not eat my homework.
19. I'll run for governor if I don't receive a failing grade.
20. I won't receive a failing grade if my dog didn't eat my homework.

Find the truth value of each statement. Assume that p and r are false, and q is true.

21. $\sim r \rightarrow q$
22. $\sim p \rightarrow \sim r$
23. $\sim r \rightarrow p$

24. $\sim q \rightarrow r$

25. $\sim p \rightarrow (q \wedge r)$

26. $(\sim r \vee p) \rightarrow p$

27. $\sim q \rightarrow (p \wedge r)$

28. $(\sim p \wedge \sim q) \rightarrow (p \wedge \sim r)$

29. $(p \rightarrow \sim q) \rightarrow (\sim p \wedge \sim r)$

30. $(p \rightarrow \sim q) \wedge (p \rightarrow r)$

31. Explain why, if we know that p is true, we also know that

$$[r \vee (p \vee s)] \rightarrow (p \vee q)$$

is true, even if we are not given the truth values of q, r, and s.

32. Construct a true statement involving a conditional, a conjunction, a disjunction, and a negation (not necessarily in that order), that consists of component statements p, q, and r, with all of these component statements false.

33. Using the table of equivalent statements rather than a truth table, explain why the statement $(\sim p \rightarrow \sim q) \wedge \sim(p \vee \sim q)$ in Example 5(a) must be a contradiction.

Construct a truth table for each statement. Identify any tautologies or contradictions.

34. $\sim q \rightarrow p$

35. $p \rightarrow \sim q$

36. $(p \vee \sim p) \rightarrow (p \wedge \sim p)$

37. $(\sim q \rightarrow \sim p) \rightarrow \sim q$

38. $(p \vee q) \rightarrow (q \vee p)$

39. $(p \wedge \sim q) \wedge (p \rightarrow q)$

40. $(\sim p \rightarrow \sim q) \rightarrow (p \wedge q)$

41. $r \rightarrow (p \wedge \sim q)$

42. $[(r \vee p) \wedge \sim q] \rightarrow p$

43. $(\sim r \rightarrow s) \vee (p \rightarrow \sim q)$

44. $(\sim p \wedge \sim q) \rightarrow (\sim r \rightarrow \sim s)$

45. What is the minimum number of times that F must appear in the final column of a truth table for us to be assured that the statement is not a tautology?

Write each statement as an equivalent statement that does not use the if . . . then *connective. Remember that $p \rightarrow q$ is equivalent to $\sim p \vee q$.*

46. If your eyes are bad, your whole body will be full of darkness.

47. If they could see me now, they'd never believe it.

48. I'd buy that car if I had the money.

49. I would watch out if I were you.

Write the negation of each statement. Remember that the negation of $p \rightarrow q$ is $p \wedge \sim q$.

50. If you build it, he will come.

51. If I can make it there, I'll make it anywhere.

52. If you don't love me, I won't be happy.

53. If he's my brother, then he's not heavy.

Use truth tables to decide which of the pairs of statements are equivalent.

54. $p \rightarrow q; \sim p \vee q$

55. $\sim(p \rightarrow q); p \wedge \sim q$

56. $p \rightarrow q; q \rightarrow p$

57. $q \rightarrow p; \sim p \rightarrow \sim q$

58. $p \rightarrow \sim q; \sim p \vee \sim q$

59. $p \rightarrow q; \sim q \rightarrow \sim p$

60. $p \wedge \sim q; \sim q \rightarrow \sim p$

61. $\sim p \wedge q; \sim p \rightarrow q$

In some approaches to logic, the only connectives are $\sim$ and $\rightarrow$, and the other logical connectives are defined in terms of these. Verify this by using a truth table to demonstrate the following equivalences.*

62. $p \wedge q \equiv \sim(p \rightarrow \sim q)$

63. $p \vee q \equiv \sim p \rightarrow q$

*For example, see Stefan Bilaniuk, *A Problem Course in Mathematical Logic*, http://euclid.trentu.ca/math/sb/pcml/welcome.html.

In Exercises 64–71, construct a truth table to prove each law.

64. $p \vee q \equiv q \vee p$, the Commutative Law for $\vee$

65. $p \wedge q \equiv q \wedge p$, the Commutative Law for $\wedge$

66. $p \vee (q \vee r) \equiv (p \vee q) \vee r$, the Associative Law for $\vee$

67. $p \wedge (q \wedge r) \equiv (p \wedge q) \wedge r$, the Associative Law for $\wedge$

68. $p \vee (q \wedge r) \equiv (p \vee q) \wedge (p \vee r)$, the Distributive Law for $\vee$ over $\wedge$

69. $p \wedge (q \vee r) \equiv (p \wedge q) \vee (p \wedge r)$, the Distributive Law for $\wedge$ over $\vee$

70. $(p \wedge q) \vee p \equiv p$, the first Absorption Law

71. $(p \vee q) \wedge p \equiv p$, the second Absorption Law

Write a logical statement representing each circuit. Simplify each circuit when possible.

72.

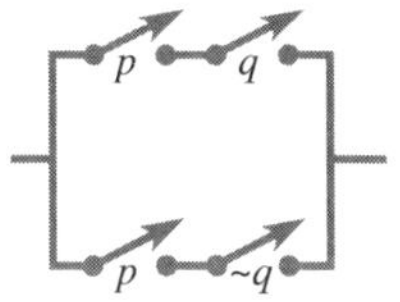

73.

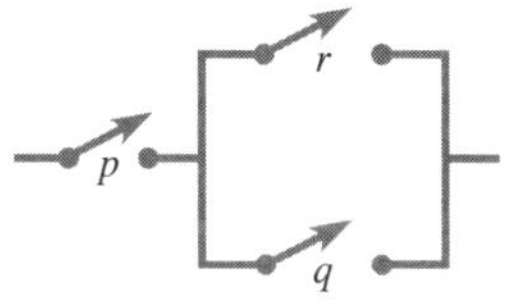

74.

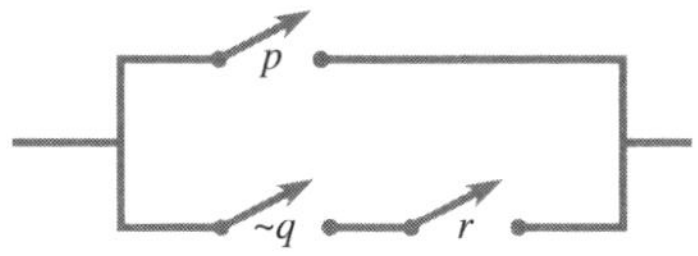

75.

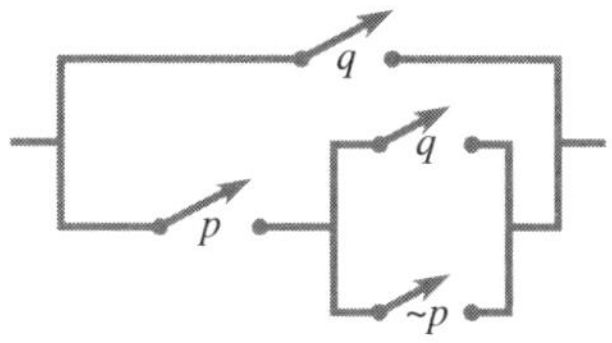

76.

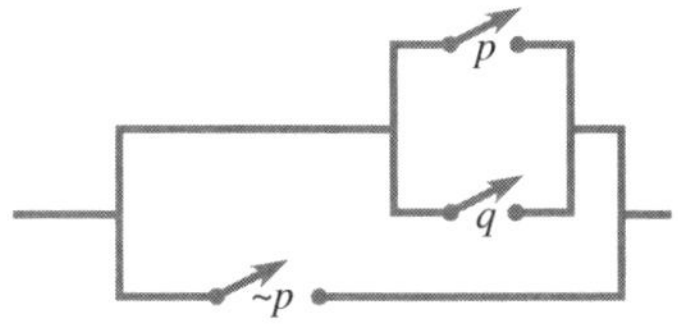

77.

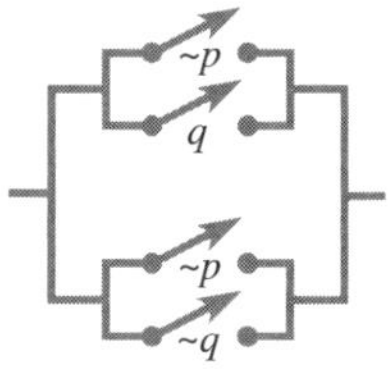

78.

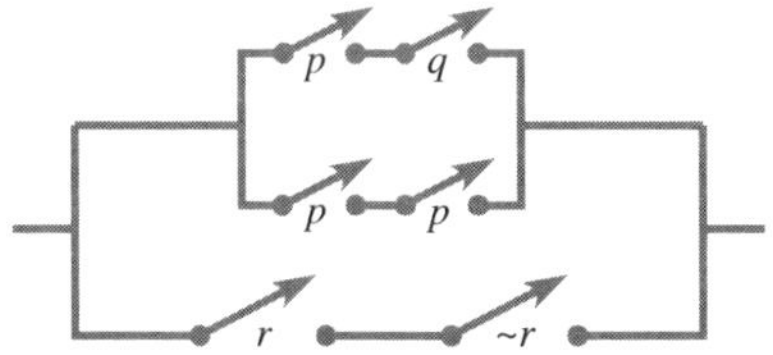

79.

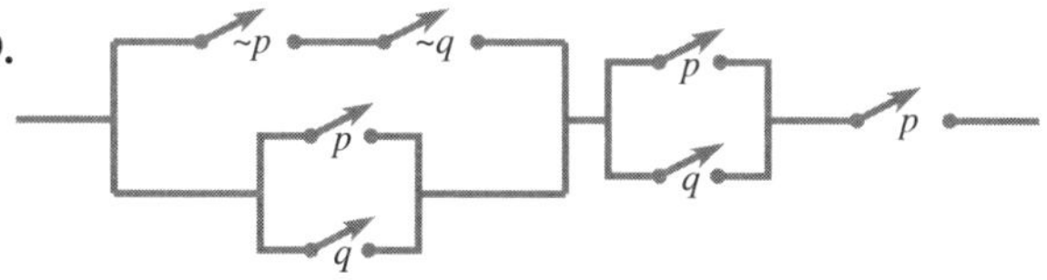

Draw circuits representing the following statements as they are given. Simplify if possible.

80. $p \wedge (q \vee {\sim}p)$

81. $({\sim}p \wedge {\sim}q) \wedge {\sim}r$

82. $(p \vee q) \wedge ({\sim}p \wedge {\sim}q)$

83. $({\sim}q \wedge {\sim}p) \vee ({\sim}p \vee q)$

84. $[(p \vee q) \wedge r] \wedge {\sim}p$

85. $[({\sim}p \wedge {\sim}r) \vee {\sim}q] \wedge ({\sim}p \wedge r)$

86. ${\sim}q \rightarrow ({\sim}p \rightarrow q)$

87. ${\sim}p \rightarrow ({\sim}p \vee {\sim}q)$

88. $[(p \wedge q) \vee p] \wedge [(p \vee q) \wedge q]$

89. $[(p \wedge q) \vee (p \wedge q)] \vee (p \wedge r)$

90. Explain why the circuit

will always have exactly one open switch. What does this circuit simplify to?

91. Refer to Figures 10 and 11 in Example 8. Suppose the cost of the use of one switch for an hour is 3¢. By using the circuit in Figure 11 rather than the circuit in Figure 10, what is the savings for a year of 365 days, assuming that the circuit is in continuous use?

Applications

BUSINESS AND ECONOMICS

92. *Income Tax* The following statements appear in a guide for preparing income tax reports.* Rewrite each statement with an equivalent statement using *or.* Then write the negation of each statement.

a. If two or more employers provide you with group-term insurance coverage, you get only one $50,000 exclusion.

b. If the value of benefits was less than $105 in any month, the unused exclusion cannot be carried over to later months.

c. If the plan provides for health benefits, a special rule applies to determine whether the plan is discriminatory.

LIFE SCIENCES

93. *Medicine* The following statement appears in a home medical reference book.† Rewrite this statement replacing the *if... then* with an *or* statement. Then negate the statement.

If you are wheezing persistently, you should see your doctor.

SOCIAL SCIENCES

94. *Law* The following statements appear in a guide to common laws.‡ Write an equivalent statement using *or.* Then negate the statement.

a. If you are married, you can't get married again.

b. If your job is going to cost more than $500, your contractor is legally required to put it in writing.

c. If your application for citizenship is denied, you can appeal in federal court.

6.4 More on the Conditional

? THINK ABOUT IT **Is it possible to rewrite statements in a tax guide in one or more ways?**

This question will be answered in one of the exercises in this section.

As we saw in the previous section, the conditional can be written in several ways. For example, the word *then* might not be explicitly stated, so the statement "If I'm here, then you're safe" might become "If I'm here, you're safe." Also, the

*J. K. Lasser Institute, *Your Income Tax 2006*, New York: John Wiley & Sons, pp. 58, 63, 68–69.
†Goldman, David R., ed., *American College of Physicians Complete Home Medical Guide*, 2nd ed., New York: DK Publishing, 2003, p. 146.
‡Ventura, John, *Law for Dummies*, Foster City, CA: IDG Books Worldwide, 1996.

consequent (the *then* component) might come before the antecedent (the *if* component), as in "You're safe if I'm here." In this section we will examine other translations of the conditional.

Alternative Forms of the Conditional Another way to express the conditional is with the word *sufficient*. For example, the statement

If it rains in the valley, then snow is falling on the mountain

can be written

Rain in the valley is sufficient for snow to fall on the mountain.

This statement claims that one condition guaranteeing snow to fall on top of the mountain is rain in the valley. The statement doesn't claim that this is the only condition under which snow falls on the mountain; perhaps snow falls on the mountain on days when it's perfectly clear in the valley. On the other hand, assuming the statement to be true, you won't see rain in the valley unless snow is falling on the mountain. In other words,

Snow falling on the mountain is necessary for rain in the valley.

To summarize this discussion, the statement "if p, then q" is sometimes stated in the form "p is sufficient for q" and sometimes in the form "q is necessary for p." To keep track of these two forms, which are often used in mathematics, notice that the antecedent is the sufficient part, while the consequent is the necessary part.

There are other common translations of $p \rightarrow q$. We have collected the most common ones in the box below.

COMMON TRANSLATIONS OF $p \rightarrow q$

The conditional $p \rightarrow q$ can be translated in any of the following ways.

If p, then q.	p is sufficient for q.
If p, q.	q is necessary for p.
p implies q.	All p's are q's.
p only if q.	q if p.
q when p.	

The translation of $p \rightarrow q$ into these various word forms does not in any way depend on the truth or falsity of $p \rightarrow q$.

EXAMPLE 1

Equivalent Statements

Write the following statement in eight different equivalent ways, using the common translations of $p \rightarrow q$ in the box above:

If you answer this survey, then you will be entered in the drawing.

▶Solution

If you answer this survey, you will be entered in the drawing.

Answering this survey implies that you will be entered in the drawing.

You answer this survey only if you will be entered in the drawing.

You will be entered in the drawing when you answer this survey.

Answering this survey is sufficient for you to be entered in the drawing.

Being entered in the drawing is necessary for you to answer this survey.

All who answer this survey will be entered in the drawing.

You will be entered in the drawing if you answer this survey.

EXAMPLE 2 **Equivalent Statements**

Write each statement in the form "if p, then q."

(a) Possession of a valid identification card is necessary for admission.

▸**Solution** If you are admitted, then you possess a valid identification card.

(b) You should use this door only if there is an emergency.

▸**Solution** If you should use this door, then there is an emergency.

(c) All who are weary can come and rest.

▸**Solution** If you are weary, then you can come and rest.

CAUTION Notice that "p only if q" is the opposite of "p if q." The first means $p \rightarrow q$, while the second means $q \rightarrow p$.

EXAMPLE 3 **Symbolic Statements**

Let r represent "A triangle is a right triangle" and s represent "The sum of the squares of the two sides equals the square of the hypotenuse." Write each statement in symbols.

(a) A triangle is a right triangle if the sum of the squares of the two sides equals the square of the hypotenuse.

▸**Solution**

$$s \rightarrow r$$

(b) A triangle is a right triangle only if the sum of the squares of the two sides equals the square of the hypotenuse.

▸**Solution**

$$r \rightarrow s$$

Converse, Inverse, and Contrapositive

Example 3(b) gives a statement of the Pythagorean theorem. Example 3(a) gives what is called the **converse** of the Pythagorean theorem. The converse of a statement $p \rightarrow q$ is the statement $q \rightarrow p$. For example, the converse of the statement

If today is Monday, then we have to put out the garbage

is the statement

If we have to put out the garbage, then today is Monday.

Although both the Pythagorean theorem and its converse are true, the converse of a true statement is not necessarily true, as we shall see in Example 4.

A second statement related to the conditional is the **inverse**, in which both the antecedent and consequent are negated. The inverse of the first statement above is

If today is not Monday, then we do not have to put out the garbage.

The inverse of a statement $p \rightarrow q$ is the statement $\sim p \rightarrow \sim q$. The inverse, like the converse, is not necessarily true, even if the original statement is true.

The third and final statement related to the conditional is the **contrapositive**, in which the antecedent and consequent are negated and interchanged. The contrapositive of the first statement above is

If we do not have to put out the garbage, then today is not Monday.

The contrapositive of a statement $p \rightarrow q$ is the statement $\sim q \rightarrow \sim p$. We will show in a moment that the contrapositive is logically equivalent to the original statement, so that if one is true, the other is also true. First, we summarize the three statements related to the original conditional statement below.

RELATED CONDITIONAL STATEMENTS

Original Statement	$p \rightarrow q$	(If p, then q.)
Converse	$q \rightarrow p$	(If q, then p.)
Inverse	$\sim p \rightarrow \sim q$	(If not p, then not q.)
Contrapositive	$\sim q \rightarrow \sim p$	(If not q, then not p.)

In the following truth table, we include the conditional $p \rightarrow q$ as well as the three related forms. Notice that the original conditional and the contrapositive are equivalent; that is, whenever one is true, so is the other. Notice also that the converse and the inverse are equivalent to each other. Finally, notice that the original conditional is not equivalent to the converse or the inverse; for example, in line 2 of the table, where p is true and q is false, the original conditional and the contrapositive are both false, but the inverse and converse are true.

(Equivalent: Original and Contrapositive; Equivalent: Converse and Inverse)

p	q	Original $p \rightarrow q$	Converse $q \rightarrow p$	Inverse $\sim p \rightarrow \sim q$	Contrapositive $\sim q \rightarrow \sim p$
T	T	T	T	T	T
T	F	F	T	T	F
F	T	T	F	F	T
F	F	T	T	T	T

This discussion is summarized in the following sentence.

EQUIVALENCES

The original statement and the contrapositive are equivalent, and the converse and the inverse are equivalent.

EXAMPLE 4 **Related Conditional Statements**

Consider the following statement:

If Cauchy is a cat, then Cauchy is a mammal.

Write each of the following.

(a) The converse

▶**Solution** Let c represent "Cauchy is a cat" and m represent "Cauchy is a mammal." Then the original statement is $c \rightarrow m$, and the converse is $m \rightarrow c$, or

If Cauchy is a mammal, then Cauchy is a cat.

Notice that in this case the original statement is true, while the converse is false, because Cauchy might be a mammal that is not a cat, such as a horse.

(b) The inverse

▶**Solution** The inverse of $c \rightarrow m$ is $\sim c \rightarrow \sim m$, or

If Cauchy is not a cat, then Cauchy is not a mammal.

The inverse, like the converse, is false in this case.

(c) The contrapositive

▶**Solution** The contrapositive of $c \rightarrow m$ is $\sim m \rightarrow \sim c$, or

If Cauchy is not a mammal, then Cauchy is not a cat.

In this case the contrapositive is true, as it must be if the original statement is true.

Biconditionals As we saw earlier, both the Pythagorean theorem and its converse are true. That is, letting r represent "A triangle is a right triangle" and s represent "The sum of the squares of the two sides equals the square of the hypotenuse," both $s \rightarrow r$ and $r \rightarrow s$ are true. In words, we say *s if and only if r* (often abbreviated s iff r). (Recall that *only if* is the opposite of *if . . . then*.) This is called the **biconditional** and is written $s \leftrightarrow r$. You can think of the biconditional as being defined by the equivalence

$$p \leftrightarrow q \equiv (p \rightarrow q) \wedge (q \rightarrow p)$$

or by the following truth table.

TRUTH TABLE FOR THE BICONDITIONAL p if and only if q

p	q	$p \leftrightarrow q$
T	T	T
T	F	F
F	T	F
F	F	T

EXAMPLE 5 **Biconditional Statements**

Tell whether each statement is true or false.

(a) George Washington is the first president of the United States if and only if John Adams is the second president of the United States.

▶**Solution** Since both component statements are true, the first row of the truth table tells us that this biconditional is true.

(b) Alaska is one of the original 13 states if and only if kangaroos can fly.

▶**Solution** Since both component statements are false, the last row of the truth table tells us that this biconditional is true.

(c) $2 + 2 = 4$ if and only if $7 > 10$.

▶**Solution** The first component statement is true, but the second is false. The second row of the truth table tells us that this biconditional is false.

Notice in the truth table and the previous example that when p and q have the same truth value, $p \leftrightarrow q$ is true; when p and q have different truth values, $p \leftrightarrow q$ is false.

In this and the previous two sections, truth tables have been derived for several important types of compound statements. The summary that follows describes how these truth tables may be remembered.

SUMMARY OF BASIC TRUTH TABLES

1. $\sim p$, the **negation** of p, has truth value opposite of p.
2. $p \wedge q$, the **conjunction**, is true only when both p and q are true.
3. $p \vee q$, the **disjunction**, is false only when both p and q are false.
4. $p \rightarrow q$, the **conditional**, is false only when p is true and q is false.
5. $p \leftrightarrow q$, the **biconditional**, is true only when p and q have the same truth value.

▶ 6.4 Exercises

For each given statement, write **(a)** *the converse,* **(b)** *the inverse, and* **(c)** *the contrapositive in* if . . . then *form. In some of the exercises, it may be helpful to restate the statement in* if . . . then *form.*

1. If the exit is ahead, then I don't see it.

2. If I finish reading this novel, then I'll write a review.

3. If I knew you were coming, I'd have baked a cake.

4. If I'm the bottom, you're the top.

5. Dead men don't wear plaid.

6. Beggars can't be choosers.

7. $p \rightarrow \sim q$

8. $\sim q \rightarrow \sim p$

9. $p \rightarrow (q \vee r)$ (*Hint:* Use one of De Morgan's Laws as necessary.)

10. $(r \vee \sim q) \rightarrow p$ (*Hint:* Use one of De Morgan's Laws as necessary.)

11. Discuss the equivalences that exist among the direct conditional statement, the converse, the inverse, and the contrapositive.

12. State the contrapositive of "If the square of a natural number is even, then the natural number is even." The two statements must have the same truth value. Use several examples and inductive reasoning to decide whether both are true or both are false.

Write each statement in the form "if p . . . then q."

13. Your signature implies that you accept the conditions.

14. His tardiness implies that he doesn't care.

15. You can take this course pass/fail only if you have prior permission.

16. You can purchase this stock only if you have \$1000.

17. You can skate on the pond when the temperature is below 10°.

18. The party will be stopped when more than 200 people attend.

19. Eating 10 hot dogs is sufficient to make someone sick.

20. Two hours in the desert sun is sufficient to give the typical person a sunburn.

21. A valid passport is necessary for travel to France.

22. Support from the party bosses is necessary to get the nomination.

23. For a number to have a real square root, it is necessary that it be nonnegative.

24. For a number to have a real square root, it is sufficient that it be nonnegative.

25. All brides are beautiful.

26. All passengers for Hempstead must change trains at Jamaica station.

27. A number is divisible by 3 if the sum of its digits is divisible by 3.

28. A number is even if its last digit is even.

29. One of the following statements is not equivalent to all the others. Which one is it?

 a. r only if s.

 b. r implies s.

 c. If r, then s.

 d. r is necessary for s.

30. Use the statement "Being 65 years old is sufficient for being eligible for Medicare" to explain why "p is sufficient for q" is equivalent to "if p, then q."

31. Use the statement "Being over 21 is necessary for entering this club" to explain why "p is necessary for q" is equivalent to "if q, then p."

32. Explain why the statement "Elephants can fly if and only if Africa is the smallest continent" is true.

Identify each statement as true *or* false.

33. $5 = 9 - 4$ if and only if $8 + 2 = 10$.

34. $3 + 1 \neq 6$ if and only if $8 \neq 8$.

35. $8 + 7 \neq 15$ if and only if $3 \times 5 \neq 9$.

36. $6 \times 2 = 14$ if and only if $9 + 7 \neq 16$.

37. China is in Asia if and only if Mexico is in Europe.

38. The moon is made of green cheese if and only if Hawaii is one of the United States.

Construct a truth table for each statement.

39. $(\sim p \wedge q) \leftrightarrow (p \rightarrow q)$

40. $(p \leftrightarrow \sim q) \leftrightarrow (\sim p \vee q)$

Applications

BUSINESS AND ECONOMICS

41. *Income Tax* The following excerpts appear in a guide for preparing income tax reports.* Write each of the statements in *if . . . then* form.

a. Your employer must tell you if the award qualifies for full or partial tax-free treatment.

b. Medical expenses are "qualified" only if incurred *after* the HSA has been established.

c. You can avoid this interest deduction limitation if you elect to report the market discount annually as interest income.

42. *Income Tax* The following excerpts appear in a guide for preparing income tax reports.† Write the converse, inverse, and contrapositive of each statement. Which statements are equivalent?

a. If the bond was issued before July 19, 1984, the market discount interest income rule does *not* apply.

b. If the bond was held long term, the gain is long term.

c. If the net short-term loss exceeds the net long-term gain, the excess short-term loss is deductible up to the $3000 capital loss limit.

43. *Credit Cards* The following statement appeared on a monthly statement of a Verizon Visa Card.‡ Write the converse, inverse, and contrapositive. Which statements are equivalent?

If you close your account within 30 days from the date this statement was mailed, you may avoid paying the annual fee billed on this statement.

LIFE SCIENCES

44. *Medicine* The following statement appears in a home medical reference book.§ Find the contrapositive of this statement.

If you are wheezing persistently, you should see your doctor.

45. *Polar Bears* The following statement is with regard to polar bear cubs.‖

If there are triplets, the most persistent stands to gain an extra meal and it may eat at the expense of another.

a. Use symbols to write this statement.

b. Write the contrapositive of this statement.

SOCIAL SCIENCES

46. *Law* The following statements appear in a guide to common laws.# Write the converse, inverse, and contrapositive of the statements. Which statements are equivalent?

a. If you are married, you can't get married again.

b. If your job is going to cost more than $500, your contractor is legally required to put it (a bid) in writing.

c. If your application for citizenship is denied, you can appeal in federal court.

47. *Philosophy* Aristotle once said, "If liberty and equality, as is thought by some, are chiefly to be found in democracy, they will be best attained when all persons alike share in the government to the utmost."** Write the contrapositive of this statement.

*J. K. Lasser Institute, *Your Income Tax 2006*, New York: John Wiley & Sons, pp. 58, 64, 85.
†Ibid, pp. 85, 90, 101.
‡Verizon Visa Card, Wilmington, DE.
§Goldman, David R., ed., *American College of Physicians Complete Home Medical Guide*, 2nd ed., New York: DK Publishing, 2003, p. 146.
‖Rosing, Norbert, "Bear Beginnings: New Life on the Ice," *National Geographic*, December 2000, p. 33.
#Ventura, John, *Law for Dummies*, Foster City, CA: IDG Books Worldwide, 1996.
**Bartlett, John, *Bartlett's Familiar Quotations*, 15th ed., Boston: Little, Brown and Company, 1980.

48. *Political Development* It has been argued that political development in Western Europe will increase if and only if social assimilation is increasing.*

  a. Express this statement symbolically, and construct a truth table for it.

  b. The author of the article quoted above says that it is true that political development in Western Europe is increasing, but it is false that social assimilation is increasing. What can then be said about the original statement?

49. *Libya* Referring to Libya's offer to pay $2.7 billion in compensation for the families of those killed in the 1988 crash of Pan Am flight 103, a White House official said, "This is a necessary step, but it is not sufficient [for the United States to drop sanctions against Libya]."[†] Letting d represent "the United States drops sanctions against Libya" and l represent "Libya offers to pay compensation," write the White House official's statement as a statement in symbolic logic.

50. *Education* It has been argued that "a high level of education ... comes close to being a necessary [condition for democracy]."[‡] For the purpose of this exercise, consider "comes close to being" as meaning "is," and assume the statement refers to a country. Write the statement in *if ... then* form, and then write the converse, inverse, and contrapositive of the statement. Which one of these is equivalent to the original statement?

51. *Political Alliances* According to political scientist Howard Rosenthal, "the presence of [a Modéré] incumbent can be regarded as a necessary condition for a R.P.F. alliance."[§] Write the statement in *if ... then* form, and then write the converse, inverse, and contrapositive of the statement. Which one of these is equivalent to the original statement?

52. *Test of Reasoning* A test devised by psychologist Peter Wason is designed to test how people reason.[||] As an example of this test, volunteers are given the rule, "If a card has a D on one side, then it must have a 3 on the other side." Volunteers view four cards displaying D, F, 3, and 7, respectively. They are told that each card has a letter on one side and a number on the other. Which cards do they need to turn over to determine if the rule has been violated? In Wason's experiments, fewer than one-fourth of the participants gave the correct answer.

53. *Test of Reasoning* In another example of a Wason test (see previous exercise), volunteers are given the rule, "If an employee works on the weekend, then that person gets a day off during the week." Volunteers are given four cards displaying "worked on the weekend," "did not work on the weekend," "did get a day off," and "did not get a day off." Volunteers were told that one side of the card tells whether an employee worked on the weekend, and the other side tells whether an employee got a day off. Which cards must be turned over to determine if the rule has been violated? In a set of experiments, volunteers told to take the perspective of the employees tended to give the correct answer, while volunteers told to take the perspective of employers tended to turn over the second and third card.

GENERAL INTEREST

54. *Sayings* Rewrite each of the following statements as a conditional in *if ... then* form. Then write two statements that are equivalent to the *if ... then* statements. (*Hint:* Write the contrapositive of the statement and rewrite the conditional statement using *or.*)

  a. Nothing ventured, nothing gained.

  b. The best things in life are free.

  c. Every cloud has a silver lining.

55. *Sayings* Think of some wise sayings that have been around for a long time, and state them in *if ... then* form.

56. *Yahtzee®* Statements similar to the ones below appear in the instructions for the Milton Bradley game, Yahtzee®.[#] Rewrite each statement in *if ... then* form and then write an equivalent statement using *or.*

  a. You can score in this box only if the dice include three or more of the same number.

  b. You can score in this box only if the dice show any sequence of four numbers.

  c. You can score in this box only if the dice show three of one number and two of another.

*Fisher, William E., "An Analysis of the Deutsch Sociocausal Paradigm of Political Integration," *International Organizational*, Vol. 23, No. 2, Spring 1969, pp. 254–290.

[†]*The New York Times*, May 29, 2002, p. A1.

[‡]Lipset, Seymour Martin, *Political Man*, 1960, quoted by Hildebrand, David K., James D. Laing, and Howard Rosenthal, "Prediction Analysis in Political Research," *The American Political Science Review*, Vol. 70, No. 2, June 1976, pp. 509–535.

[§]Rosenthal, Howard, "The Electoral Politics of Gaullists in the Fourth French Republic: Ideology or Constituency Interest?" *The American Political Science Review*, Vol. 63, June 1969, pp. 476–487. The R.P.F. and Modéré are the names of political parties in France.

[||]Bower, Bruce, "Roots of Reason," *Science News*, Vol. 145, Jan. 29, 1994, pp. 72–73.

[#]Milton Bradley Company, East Longmeadow, MA, 1996.

6.5 Analyzing Arguments and Proofs

? THINK ABOUT IT

If a man could be in two places at one time, then I'd be with you. I am not with you, so can I conclude that a man can't be in two places at one time?

This question will be analyzed using truth tables in Example 2 of this section.

In this section we will use the logical operators discussed in the previous four sections to analyze and construct arguments, or proofs, which use deductive reasoning. A logical argument is made up of **premises** and **conclusions**. The premises are statements that we accept for the sake of the argument. Perhaps they are obviously true or commonly accepted, or perhaps we merely consider them as true for the moment to investigate what conclusions follow. Whether they are actually true or not falls outside the realm of logic. In logic we suppose the premises to be true, and then ask what statements follow using the laws of logic. The statements that follow are the conclusions. The argument is considered *valid* if the conclusions must be true when the premises are true.

VALID AND INVALID ARGUMENTS

An argument is **valid** if the fact that all the premises are true forces the conclusion to be true. An argument that is not valid is **invalid**, or a **fallacy**.

It is very important to note that *valid* and *true* are not the same—an argument can be valid even though the conclusion is false. (See the discussion after Example 3.)

We will begin by using truth tables to determine whether certain types of arguments are valid or invalid. We will then use the results from these examples to demonstrate a more powerful method of proving that an argument is valid or invalid. As our first example, consider the following argument.

If it's after midnight, then I must go to sleep.
It's after midnight.
―――――――――――――――――
I must go to sleep.

Here we use the common method of placing one premise over another, with the conclusion below a line. Alternatively, we could indicate that the last line is a conclusion using "therefore," as in "Therefore, I must go to sleep."

To test the validity of this argument, we begin by identifying the component statements found in the argument. They are "It's after midnight" and "I must go to sleep." We will use the generic variables p and q here, rather than meaningful variable names, so the argument will have a generic form.

p represents "It's after midnight";
q represents "I must go to sleep."

Now we write the two premises and the conclusion in symbols:

Premise 1: $p \rightarrow q$
Premise 2: p
―――――――――
Conclusion: q .

To decide if this argument is valid, we must determine whether the conjunction of both premises implies the conclusion for all possible cases of truth values for p and q. Therefore, write the conjunction of the premises as the antecedent of a conditional statement, and the conclusion as the consequent.

$$\underset{\text{Premise}}{\underset{\uparrow}{[(p \rightarrow q)}} \quad \underset{\text{and}}{\underset{\uparrow}{\wedge}} \quad \underset{\text{premise}}{\underset{\uparrow}{p]}} \quad \underset{\text{implies}}{\underset{\uparrow}{\rightarrow}} \quad \underset{\text{conclusion.}}{\underset{\uparrow}{q}}$$

Finally, construct the truth table for the conditional statement, as shown below.

p	q	$p \rightarrow q$	$(p \rightarrow q) \wedge p$	$[(p \rightarrow q) \wedge p] \rightarrow q$
T	T	T	T	T
T	F	F	F	T
F	T	T	F	T
F	F	T	F	T

Since the final column, shown in color, indicates that the conditional statement that represents the argument is true for all possible truth values of p and q, the statement is a tautology. Thus, the argument is valid.

The pattern of the argument in the preceding example,

$$\begin{array}{l} p \rightarrow q \\ \underline{p\quad\quad} \\ q \end{array},$$

is a common one, and is called **Modus Ponens**, or the *law of detachment.*

In summary, to test the validity of an argument using a truth table, go through the steps in the box that follows.

TESTING THE VALIDITY OF AN ARGUMENT WITH A TRUTH TABLE

1. Assign a letter to represent each component statement in the argument.
2. Express each premise and the conclusion symbolically.
3. Form the symbolic statement of the entire argument by writing the *conjunction* of *all* the premises as the antecedent of a conditional statement, and the conclusion of the argument as the consequent.
4. Complete the truth table for the conditional statement formed in step 3. If it is a tautology, then the argument is valid; otherwise, it is invalid.

EXAMPLE 1 **Invalid Argument**

Determine whether the argument is *valid* or *invalid.*

If I win the lottery, then I'll buy a new house.
I bought a new house.
____
I won the lottery.

▸Solution Let p represent "I win the lottery" and let q represent "I'll buy a new house." Using these symbols, the argument can be written in the form

$$\begin{array}{l} p \to q \\ \underline{q \quad\quad} \\ p \quad . \end{array}$$

To test for validity, construct a truth table for the statement

$$[(p \to q) \wedge q] \to p.$$

p	q	$p \to q$	$(p \to q) \wedge q$	$[(p \to q) \wedge q] \to p$
T	T	T	T	T
T	F	F	F	T
F	T	T	T	F
F	F	T	F	T

The third row of the final column of the truth table shows F, and this is enough to conclude that the argument is invalid. Even if the premises are true, the conclusion that I won the lottery is not necessarily true. Perhaps I bought a new house with money I inherited.

If a conditional and its converse were logically equivalent, then an argument of the type found in Example 1 would be valid. Since a conditional and its converse are *not* equivalent, the argument is an example of what is sometimes called the **Fallacy of the Converse**.

EXAMPLE 2 ?

Valid Argument

Determine whether the argument is *valid* or *invalid.*

If a man could be in two places at one time, I'd be with you.
I am not with you.
____
A man can't be in two places at one time.

▸Solution If p represents "a man could be in two places at one time" and q represents "I'd be with you," the argument becomes

$$\begin{array}{l} p \to q \\ \underline{\sim q \quad\quad} \\ \sim p \quad . \end{array}$$

The symbolic statement of the entire argument is

$$[(p \to q) \wedge \sim q] \to \sim p.$$

The truth table for this argument, shown below, indicates a tautology, and the argument is valid.

p	q	$p \rightarrow q$	$\sim q$	$(p \rightarrow q) \wedge \sim q$	$\sim p$	$[(p \rightarrow q) \wedge \sim q] \rightarrow \sim p$
T	T	T	F	F	F	T
T	F	F	T	F	F	T
F	T	T	F	F	T	T
F	F	T	T	T	T	T

The pattern of reasoning of this example is called **Modus Tollens**, or the *law of contraposition.*

With reasoning similar to that used to name the fallacy of the converse, the fallacy

$$\begin{array}{l} p \rightarrow q \\ \underline{\sim p} \\ \sim q \end{array}$$

is often called the **Fallacy of the Inverse**. An example of such a fallacy is "If it rains, I get wet. It doesn't rain. Therefore, I don't get wet."

EXAMPLE 3 Valid Argument

Determine whether the argument is *valid* or *invalid*.

I'll win this race or I'll eat my hat.
I didn't win this race.
____
I'll eat my hat.

▶Solution Let p represent "I'll win this race" and let q represent "I'll eat my hat." Using these symbols, the argument can be written in the form

$$\begin{array}{l} p \vee q \\ \underline{\sim p} \\ q \end{array}$$

Set up a truth table for

$$[(p \vee q) \wedge \sim p] \rightarrow q.$$

p	q	$p \vee q$	$\sim p$	$(p \vee q) \wedge \sim p$	$[(p \vee q) \wedge \sim p] \rightarrow q$
T	T	T	F	F	T
T	F	T	F	F	T
F	T	T	T	T	T
F	F	F	T	F	T

The statement is a tautology and the argument is valid. Any argument of this form is valid by the law of **Disjunctive Syllogism**.

Suppose you notice a few months later that I didn't win the race or eat my hat. You object that the conclusion of the argument in Example 3 is false. Did that make the argument invalid? No. The problem is that I lied about the first premise, namely, that I would win the race or eat my hat. A valid argument only guarantees that *if* all the premises are true, then the conclusion must also be true. Whether the premises of an argument are actually true is a separate issue. I might have a perfectly valid argument, yet you disagree with my conclusion because you disagree with one or more of my premises.

EXAMPLE 4

Valid Argument

Determine whether the following argument is *valid* or *invalid.*

If you are wheezing persistently, you should see your doctor.*
If you should see your doctor, you better have health insurance.
____
If you are wheezing persistently, you better have health insurance.

Solution Let p represent "you are wheezing persistently," let q represent "you should see your doctor," and r represent "you better have health insurance." The argument takes on the general form

$$\begin{array}{c} p \rightarrow q \\ \underline{q \rightarrow r} \\ p \rightarrow r. \end{array}$$

Make a truth table for the following statement:

$$[(p \rightarrow q) \wedge (q \rightarrow r)] \rightarrow (p \rightarrow r).$$

It will require eight rows.

p	q	r	$p \rightarrow q$	$q \rightarrow r$	$p \rightarrow r$	$(p \rightarrow q) \wedge (q \rightarrow r)$	$[(p \rightarrow q) \wedge (q \rightarrow r)] \rightarrow (p \rightarrow r)$
T	T	T	T	T	T	T	T
T	T	F	T	F	F	F	T
T	F	T	F	T	T	F	T
T	F	F	F	T	F	F	T
F	T	T	T	T	T	T	T
F	T	F	T	F	T	F	T
F	F	T	T	T	T	T	T
F	F	F	T	T	T	T	T

This argument is valid since the final statement is a tautology. The pattern of argument shown in this example is called **Reasoning by Transitivity**, or the *law of hypothetical syllogism.*

*Goldman, David R., ed., *American College of Physicians Complete Home Medical Guide,* 2nd ed., New York: DK Publishing, 2003, p. 146.

A summary of the valid and invalid forms of argument presented so far is given below.

VALID ARGUMENT FORMS

Modus Ponens	Modus Tollens	Disjunctive Syllogism	Reasoning by Transitivity
$p \rightarrow q$	$p \rightarrow q$	$p \vee q$	$p \rightarrow q$
p	$\sim q$	$\sim p$	$q \rightarrow r$
q	$\sim p$	q	$p \rightarrow r$

INVALID ARGUMENT FORMS (FALLACIES)

Fallacy of the Converse	Fallacy of the Inverse
$p \rightarrow q$	$p \rightarrow q$
q	$\sim p$
p	$\sim q$

When an argument contains three or more premises, it will be necessary to determine the truth values of the conjunction of all of them. Remember that if *at least one* premise in a conjunction of several premises is false, then the entire conjunction is false. This will be used in the next example.

EXAMPLE 5

Invalid Argument

Determine whether the following argument is *valid* or *invalid.*

If Ed reaches the semifinals, then Liz will be happy.
If Liz is not happy, then Roxanne will bake her a cake.
Roxanne did not bake Liz a cake.
Therefore, Ed reaches the semifinals.

▸**Solution** Let e represent "Ed reaches the semifinals," l represent "Liz is happy," and r represent "Roxanne bakes Liz a cake." The symbolic form of the argument is as follows.

1. $e \rightarrow l$ Premise
2. $\sim l \rightarrow r$ Premise
3. $\sim r$ Premise
e Conclusion

Our strategy will be to apply the valid argument forms either to the premises, or to conclusions that have already been reached. Notice, for example, that we can apply Modus Tollens to lines 2 and 3 to reach $\sim(\sim l) \equiv l$. We write this as follows, giving the statements and the argument form used.

4. l 2, 3, Modus Tollens

From statements 1 and 4, can we conclude e? Only if we use the Fallacy of the Converse! Thus this argument appears to be invalid.

Someone might object that the argument is actually valid by a clever proof that we haven't discovered. We could refute this objection by creating a truth table, as we did in Examples 1 through 4. A less tedious approach is to find a truth value for all the components e, l, and r that make the premises true and the conclusion false. If we can do that, we know for certain that the conclusion does not follow from the premises. Write the premises and the conclusion, with the desired truth value next to each.

1. $e \rightarrow l$	T
2. $\sim l \rightarrow r$	T
3. $\underline{\sim r}$	T
e	F

In statement 3, to make $\sim r$ true, we know that r must be false. From the conclusion, e must be false. Statement 1 is then automatically true. To make statement 2 true, simply make l true, so $\sim l$ is false. In summary, the following assignment of truth values make all the premises true, yet the conclusion is false.

e = "Ed reaches the semifinals" = F

l = "Liz is happy" = T

r = "Roxanne bakes Liz a cake" = F

Let's summarize what we have seen.

- To show that an argument is valid, prove it using the four valid argument forms discussed in this section. We can also use any of the laws given in the Equivalent Statement box in Section 6.3.
- To show that an argument is invalid, give an assignment of truth values that makes the premises true and the conclusion false.

It's not always clear which of these two strategies you should try first, but if one doesn't work, try the other. An argument may be valid, but it may not be provable using only the rules learned so far. In the next section we will study additional rules for proving arguments. When all else fails, you can always create a truth table, but if the argument has many variables, this could be very tedious.

EXAMPLE 6

Valid Argument

Lewis Carroll* gave humorous logic puzzles in his book *Symbolic Logic*. In each puzzle, he presented several premises, and the reader was to find valid conclusions. Here is one of his puzzles. What is the valid conclusion?

Babies are illogical.

Nobody is despised who can manage a crocodile.

Illogical persons are despised.

*Lewis Carroll is the pseudonym for Charles Dodgson (1832–1898), mathematician and author of *Alice in Wonderland*. The puzzles in this book are from *The Complete Works of Lewis Carroll*, Vintage Books, 1976.

▶**Solution** First, write each premise in *if . . . then* form.

If you are a baby, then you are illogical.
If you can manage a crocodile, then you are not despised.
If you are illogical, then you are despised.

Let b represent "you are a baby," i represent "you are illogical," m represent "you can manage a crocodile," and d represent "you are despised." The statements can then be written symbolically as

1. $b \to i$ Premise
2. $m \to \sim d$ Premise
3. $i \to d$ Premise

Notice that we can combine the first and third statements using reasoning by transitivity.

4. $b \to d$ 1, 3, Transitivity

How can we combine this with statement 2, which has not yet been used? Both statements 2 and 4 have d at the end, but in statement 2 the d is negated. Use the contrapositive to get statement 2 in a more useful form.

5. $d \to \sim m$ 2, Contrapositive

Now we can use transitivity.

6. $b \to \sim m$ 4, 5, Transitivity

In words, the conclusion is "If you are a baby, then you cannot manage a crocodile," or, as Lewis Carroll put it, "Babies cannot manage crocodiles."

Often there are different ways to do a proof. In the previous example, you might first apply contrapositive to statement 2, and then combine the result with statement 3 using transitivity, and finally combine that result with statement 1 using transitivity. In the exercises in this textbook, your proofs might look different from those in the back of the book but still be correct.

EXAMPLE 7 **Valid Argument**

Determine whether the following argument is *valid* or *invalid.*

If tomorrow is Saturday and sunny, then it is a beach day.
Tomorrow is Saturday.
Tomorrow is not a beach day.
Therefore, tomorrow must not be sunny.

▶**Solution** Let t represent "tomorrow is Saturday," s represent "tomorrow is sunny," and b represent "tomorrow is a beach day." We give a proof that this argument is valid.

1. $(t \wedge s) \rightarrow b$ Premise

2. t Premise

3. $\sim b$ Premise

4. $\sim(t \wedge s)$ 1, 3, Modus Tollens

5. $\sim t \vee \sim s$ 4, De Morgan's Law

6. $\sim s$ 2, 5, Disjunctive Syllogism

How did we figure out this proof? We started by looking for a rule of logic that combines two of the three premises. Notice that statement 3 has the negation of the right side of statement 1; this is a clue that Modus Tollens might be helpful. Once we get statement 4, using one of De Morgan's Laws is an obvious choice, as it should be whenever we see a negation over $\wedge$ or $\vee$. Statement 2 still hasn't been used at this point, but notice that it is the negation of one part of the $\vee$ in statement 5, leading us to try Disjunctive Syllogism.

EXAMPLE 8

Invalid Argument

Determine whether the following argument is *valid* or *invalid.*

If your teeth are white, you use Extreme Bright toothpaste.

If your teeth are white, you will have sex appeal.

Therefore, if you use Extreme Bright toothpaste, you will have sex appeal.

►**Solution** Let w represent "your teeth are white," t represent "you use Extreme Bright toothpaste," and s represent "you will have sex appeal."

In symbolic form, this argument can be written as follows.

$$\begin{array}{c} w \rightarrow t \\ \underline{w \rightarrow s} \\ t \rightarrow s \end{array}$$

This looks like a misguided attempt at using reasoning by transitivity and so appears invalid. To show this, we need to find a way to make the premises true and the conclusion false. The only way the conclusion, $t \rightarrow s$, can be false is if t is true and s is false. Both premises will automatically be true if we make w false. Thus the argument is invalid. The following assignment of truth values make all the premises true and the conclusion false.

w = "your teeth are white" = F

t = "you use Extreme Bright toothpaste" = T

s = "you will have sex appeal" = F

6.5 Exercises

Each of the following arguments is either valid by one of the forms of valid arguments discussed in this section, or is a fallacy by one of the forms of invalid arguments discussed. (See the summary boxes.) Decide whether the argument is valid *or* invalid, *and give the form that applies.*

1. If she weighs the same as a duck, she's made of wood.
If she's made of wood, she's a witch.
If she weighs the same as a duck, she's a witch.

2. If love is out of the question, there's no point to this affection.
If there's no point to this affection, then these lyrics make no sense.
If love is out of the question, then these lyrics make no sense.

3. If I had a hammer, I'd hammer in the morning.
I have a hammer.
I hammer in the morning.

4. If I were a swan, I'd be gone.
I am a swan.
I am gone.

5. If you want to make trouble, the door is that way.
The door is that way.
You want to make trouble.

6. If you finish the test, you can leave early.
You can leave early.
You finish the test.

7. If Martin Broudeur plays, the opponent gets shut out.
The opponent does not get shut out.
Martin Broudeur does not play.

8. If you want to follow along, we're on p. 315.
We're not on p. 315.
You don't want to follow along.

9. "If we evolved a race of Isaac Newtons, that would not be progress." (quote from Aldous Huxley)
We have not evolved a race of Isaac Newtons.
That is progress.

10. "If I have seen farther than others, it is because I stood on the shoulders of giants." (quote from Sir Isaac Newton)
I have not seen farther than others.
I have not stood on the shoulders of giants.

11. Something is rotten in the state of Denmark, or my name isn't Hamlet.
My name is Hamlet.
Something is rotten in the state of Denmark.

12. We shall conquer together or we shall die together.
We shall not die together.
We shall conquer together.

Determine whether each argument is valid *or* invalid. *If it is valid, give a proof. If it is invalid, give an assignment of truth values to the variables that makes the premises true and the conclusion false.*

13. $\begin{array}{l} p \lor q \\ \underline{p\qquad} \\ \sim q \end{array}$

14. $\begin{array}{l} p \lor \sim q \\ \underline{p\qquad} \\ \sim q \end{array}$

15. $\begin{array}{l} p \rightarrow q \\ \underline{q \rightarrow p} \\ p \land q \end{array}$

16. $\begin{array}{l} \sim p \rightarrow q \\ \underline{p\qquad} \\ \sim q \end{array}$

17. $\begin{array}{l} \sim p \rightarrow \sim q \\ \underline{q\qquad} \\ p \end{array}$

18. $\begin{array}{l} p \rightarrow \sim q \\ \underline{q\qquad} \\ \sim p \end{array}$

19. $\begin{array}{l} p \rightarrow q \\ \sim q \\ \underline{\sim p \rightarrow r} \\ r \end{array}$

20. $\begin{array}{l} p \lor q \\ \sim p \\ \underline{r \rightarrow \sim q} \\ \sim r \end{array}$

21. $\begin{array}{l} p \rightarrow q \\ q \rightarrow r \\ \underline{\sim r\qquad} \\ \sim p \end{array}$

22. $\begin{array}{l} p \rightarrow q \\ \underline{r \rightarrow \sim q} \\ p \rightarrow \sim r \end{array}$

23. $\begin{array}{l} p \rightarrow q \\ q \rightarrow \sim r \\ p \\ \underline{r \lor s\qquad} \\ s \end{array}$

24. $\begin{array}{l} p \rightarrow q \\ \sim p \rightarrow r \\ \underline{s \rightarrow \sim q} \\ \sim r \rightarrow \sim s \end{array}$

Use a truth table, similar to those in Examples 1–4, to prove each rule of logic. The rules in Exercises 25–27 are known as simplification, amplification, *and* conjunction, *respectively.*

25. $\begin{array}{l} \underline{p \land q} \\ p \end{array}$

26. $\begin{array}{l} \underline{p\qquad} \\ p \lor q \end{array}$

27. $\begin{array}{l} p \\ \underline{q\qquad} \\ p \land q \end{array}$

28. Cheryl Arabie made the following observation: "If I want to determine whether an argument leading to the statement

$$[(p \rightarrow q) \land \sim q] \rightarrow \sim p$$

is valid, I only need to consider the lines of the truth table that lead to T for the column headed $(p \rightarrow q) \land \sim q$." Cheryl was very perceptive. Can you explain why her observation was correct?

Applications

GENERAL INTEREST

Determine whether each of the following arguments is valid *or* invalid. *If it is valid, give a proof. If it is invalid, give an assignment of truth values to the variables that makes the premises true and the conclusion false.*

29. *Computers* My computer is working. If the power supply is faulty, my computer does not work. If the power supply is not faulty, my stereo works. Therefore, my stereo works.

30. *Politics* If Boehme runs for senator, then Hoffman will run for governor. If Tobin runs for congressman, then Hoffman will not run for governor. Therefore, if Tobin runs for congressman, then Boehme will not run for senator.

31. *Illness* If you have strep throat, then you have a fever. You have either a serious cough or a fever. You do not have a serious cough. Therefore, you do not have strep throat.

32. *Love* If I've got you under my skin, then you are deep in the heart of me. If you are deep in the heart of me, then you are not really a part of me. You are deep in the heart of me or you are really a part of me. Therefore, if I've got you under my skin, then you are really a part of me.

33. *Baseball* The Yankees will be in the World Series or the Marlins won't be there. If the Marlins are not in the World Series, the National League cannot win. In fact, the National League wins. Therefore, the Yankees are in the World Series.

34. *Music* If Matt plays the drums, Jimmy will play the guitar. Karla will play the flute or Matt will play the drums. Karla will not play the flute. Therefore, Jimmy will play the guitar.

35. *Love* If I were your woman and you were my man, then I'd never stop loving you. I've stopped loving you. Therefore, I am not your woman or you are not my man.

36. *Transportation* If my car starts, then either the battery is strong or I got a jump. Either my car starts or I'm taking the train. The battery is not strong and I did not get a jump. Therefore, I'm taking the train.

37. *Time* Suppose that you ask someone for the time and you get the following response:

> "If I tell you the time, then we'll start chatting. If we start chatting, then you'll want to meet me at a truck stop. If we meet at a truck stop, then we'll discuss my family. If we discuss my family, then you'll find out that my daughter is available for marriage. If you find out that she is available for marriage, then you'll want to marry her. If you want to marry her, then my life will be miserable since I don't want my daughter married to some fool who can't afford a $20 watch."

Use Reasoning by Transitivity to draw a valid conclusion.

Lewis Carroll Exercises 38–43 are from problems in Lewis Carroll's book Symbolic Logic.* *Write each premise in symbols, and then in the final part, give a conclusion that yields a valid argument.*

38. Let d be "it is a duck," p be "it is my poultry," o be "one is an officer," and w be "one waltzes."

a. No ducks waltz.

b. No officers ever decline to waltz.

c. All my poultry are ducks.

d. Give a conclusion that yields a valid argument.

39. Let l be "one is able to do logic," j be "one is fit to serve on a jury," s be "one is sane," and y be "he is your son."

a. Everyone who is sane can do logic.

b. No lunatics are fit to serve on a jury.

c. None of your sons can do logic.

d. Give a conclusion that yields a valid argument.

40. Let h be "one is honest," p be "one is a pawnbroker," b be "one is a promise breaker," t be "one is trustworthy," c be "one is very communicative," and w be "one is a wine drinker."

a. Promise breakers are untrustworthy.

b. Wine drinkers are very communicative.

c. A man who keeps his promise is honest.

d. No teetotalers are pawnbrokers. (*Hint:* Assume "teetotaler" is the opposite of "wine drinker.")

e. One can always trust a very communicative person.

f. Give a conclusion that yields a valid argument.

41. Let g be "it is a guinea pig," i be "it is hopelessly ignorant of music," s be "it keeps silent while the *Moonlight Sonata* is being played," and a be "it appreciates Beethoven."

a. Nobody who really appreciates Beethoven fails to keep silent while the *Moonlight Sonata* is being played.

b. Guinea pigs are hopelessly ignorant of music.

c. No one who is hopelessly ignorant of music ever keeps silent while the *Moonlight Sonata* is being played.

d. Give a conclusion that yields a valid argument.

42. Let s be "it begins with 'Dear Sir'," c be "it is crossed," d be "it is dated," f be "it is filed," i be "it is in black ink," t be "it is in the third person," r be "I can read it," p be "it is on blue paper," o be "it is on one sheet," and b be "it is written by Brown."

a. All the dated letters in this room are written on blue paper.

b. None of them are in black ink, except those that are written in the third person.

c. I have not filed any of them that I can read.

d. None of them that are written on one sheet are undated.

e. All of them that are not crossed are in black ink.

f. All of them written by Brown begin with "Dear Sir."

g. All of them written on blue paper are filed.

**The Complete Works of Lewis Carroll,* Vintage Books, 1976.

h. None of them written on more than one sheet are crossed.

i. None of them that begin with "Dear Sir" are written in the third person.

j. Give a conclusion that yields a valid argument.

43. Let p be "he is going to a party," b be "he brushes his hair," s be "he has self-command," l be "he looks fascinating," o be "he is an opium eater," t be "he is tidy," and w be "he wears white kid gloves."

a. No one who is going to a party ever fails to brush his hair.

b. No one looks fascinating if he is untidy.

c. Opium eaters have no self-command.

d. Everyone who has brushed his hair looks fascinating.

e. No one wears white kid gloves unless he is going to a party. (*Hint:* "not a unless b" $\equiv a \rightarrow b$.)

f. A man is always untidy if he has no self-command.

g. Give a conclusion that yields a valid argument.

6.6 Analyzing Arguments with Quantifiers

? THINK ABOUT IT **If some U.S. presidents won the popular vote and George W. Bush is the U.S. president, did he win the popular vote?**

This question will be analyzed in Example 6 of this section.

There is a subtle but important point that we have ignored in some of the logic puzzles of the previous section. Consider the statement "Babies are illogical" in Example 6 in that section. We reinterpreted this statement as "If you are a baby, then you are illogical." Letting b represent "you are a baby" and i represent "you are illogical," we wrote this statement as $b \rightarrow i$. This was perfectly adequate for solving that puzzle. But now suppose we want to add two more premises: "Madeleine is a baby" and "Noa is a baby." It would seem that two valid conclusions are: "Madeleine is illogical" and "Noa is illogical." But how do we write these statements symbolically? We can't very well have b represent both "Madeleine is a baby" and "Noa is a baby." We might try letting these two statements be represented by m and n, but then it's not clear how we can combine m or n with $b \rightarrow i$.

Logicians solve this problem with the use of **quantifiers**. The words *all*, *each*, *every*, and *no(ne)* are called **universal quantifiers**, while words and phrases such as *some, there exists*, and *(for) at least one* are called **existential quantifiers**. Quantifiers are used extensively in mathematics to indicate *how many* cases of a particular situation exist.

In the previous example, we could let $b(x)$ represent "x is a baby," where x could represent any baby. Then $b(m)$ could represent "Madeleine is a baby" and $b(n)$ could represent "Noa is a baby." We want to say that $b(x) \rightarrow i(x)$ *for all x*. Logicians use the symbol $\forall$ to represent "for all." We could then represent the statement "Babies are illogical" as

$$\forall x\,[b(x) \rightarrow i(x)].$$

Suppose, instead, that we don't want to claim that all babies are illogical, but that some babies are illogical, or, equivalently, that there exists someone who is a baby and who is illogical. Logicians use the symbol $\exists$ to represent "there exists." We could then write the statement "Some babies are illogical" as

$$\exists x\,[b(x) \wedge i(x)].$$

NOTE The important point to note here is that when a statement contains the word "all," a universal quantifier ($\forall$) is usually called for, and when it contains the word "some," an existential quantifier ($\exists$) is usually called for. ■

Negation of Quantifiers We must be careful in forming the negation of a statement involving quantifiers. Suppose we wish to say that the statement "Babies are illogical" is false. This is *not* the same as saying "Babies are logical." Maybe some babies are illogical and some are logical. Let $b(x)$ represent "x is a baby" and $i(x)$ represent "x is illogical." To deny the claim that for all x, the statement $b(x) \rightarrow i(x)$ is true, is equivalent to saying that for some x, the statement $b(x) \rightarrow i(x)$ is false. Thus,

$$\sim\{\forall x\,[b(x) \rightarrow i(x)]\} \equiv \exists x\,\{\sim[b(x) \rightarrow i(x)]\}.$$

Recall from Section 6.3 that the right side of this last statement is equivalent to $\exists x\,\{[b(x) \wedge \sim i(x)]\}$. Therefore,

$$\sim\{\forall x\,[b(x) \rightarrow i(x)]\} \equiv \exists x\,\{[b(x) \wedge \sim i(x)]\}.$$

This last statement makes intuitive sense; it says that if it's false that all babies are illogical, then there exists someone who is a baby and who is not illogical. In other words, some babies are logical.

In a similar way, suppose we wish to say that the statement "Some babies are illogical" is false. To deny the existence of an x such that $b(x) \wedge i(x)$ is the same as saying that for all x, the statement $b(x) \wedge i(x)$ is false. Thus,

$$\sim\{\exists x\,[b(x) \wedge i(x)]\} \equiv \forall x\,\{\sim[b(x) \wedge i(x)]\}.$$

Recall from Section 6.3 that $\sim(p \rightarrow q) \equiv p \wedge \sim q$, so negating both sides of this equivalence and replacing q with $\sim q$ implies $\sim(p \wedge q) \equiv p \rightarrow \sim q$. We can use this fact to rewrite the previous equivalence as

$$\sim\{\exists x\,[b(x) \wedge i(x)]\} \equiv \forall x\,\{[b(x) \rightarrow \sim i(x)]\}.$$

This equivalence makes intuitive sense. It says that if it's false that some babies are illogical, then all babies are not illogical, which can be phrased more clearly as "All babies are logical."

Let's summarize the above discussion.

NEGATIONS OF QUANTIFIED STATEMENTS

Statement	Symbolic	Negation	Symbolic for Negation
$s(x)$ is true for all x.	$\forall x\,[s(x)]$	$s(x)$ is false for some x.	$\exists x\,[\sim s(x)]$
$s(x)$ is true for some x.	$\exists x\,[s(x)]$	$s(x)$ is false for all x.	$\forall x\,[\sim s(x)]$

Essentially, negation changes a $\forall$ into a $\exists$ and vice versa.

EXAMPLE 1 **Negation**

Write each statement symbolically. Then write the negation symbolically, and translate the negation back into words.

(a) Some Texans eat quiche.

▶**Solution** Let $t(x)$ represent "x is a Texan" and $q(x)$ represent "x eats quiche." The statement can be written as

$$\exists x[t(x) \wedge q(x)].$$

Its negation is

$$\forall x\{\sim[t(x) \wedge q(x)]\},$$

which is equivalent to

$$\forall x[t(x) \rightarrow \sim q(x)].$$

In words, "All Texans don't eat quiche," which can be expressed more clearly as "No Texan eats quiche."

(b) Some Texans do not eat quiche.

▶**Solution** Using the notation from part (a), the statement can be written as

$$\exists x[t(x) \wedge \sim q(x)].$$

Its negation is

$$\forall x\{\sim[t(x) \wedge \sim q(x)]\},$$

which is equivalent to

$$\forall x[t(x) \rightarrow q(x)].$$

In words, "All Texans eat quiche."

(c) No Texans each quiche.

▶**Solution** This is the same as saying "All Texans do not eat quiche," which can be written as

$$\forall x\ [t(x) \rightarrow \sim q(x)].$$

Its negation is

$$\exists x\{\ \sim[t(x) \rightarrow \sim q(x)]\},$$

which is equivalent to

$$\exists x\ [t(x) \wedge q(x)].$$

In words, "Some Texans eat quiche." Notice that the statements in parts (a) and (c) are negations of each other.

Just as we had rules for analyzing arguments without quantifiers in the last section, there are also rules for analyzing arguments with quantifiers. Instead of presenting such rules, we will instead demonstrate a visual technique based on **Euler diagrams**, illustrated in the following examples. Euler (pronounced "oiler") diagrams are named after the great Swiss mathematician Leonhard Euler (1707–1783).

EXAMPLE 2 **Valid Argument**

Represent the following argument symbolically. Is the argument valid?

All elephants have wrinkles.
Babar is an elephant.

Babar has wrinkles.

▶Solution If we let $e(x)$ represent "x is an elephant," $w(x)$ represent "x has wrinkles," and b represent "Babar," we could represent the argument symbolically as follows.

$$\begin{array}{l} \forall x\,[e(x) \rightarrow w(x)] \\ \underline{e(b)\qquad\qquad\qquad} \\ w(b) \end{array}$$

Notice that this argument resembles Modus Ponens from the previous section, but with a quantifier, so we suspect that the argument is valid. To verify our suspicion, we draw regions to represent the first premise. One is the region for "elephants." Since all elephants have wrinkles, the region for "elephants" goes inside the region for "have wrinkles," as in Figure 14(a).

The second premise, "Babar is an elephant," suggests that "Babar" should go inside the region representing "elephants." Let b represent "Babar." Figure 14(b) shows that "Babar" is also inside the region for "have wrinkles." Therefore, if both premises are true, the conclusion that "Babar has wrinkles" must also be true. Euler diagrams illustrate that this argument is valid.

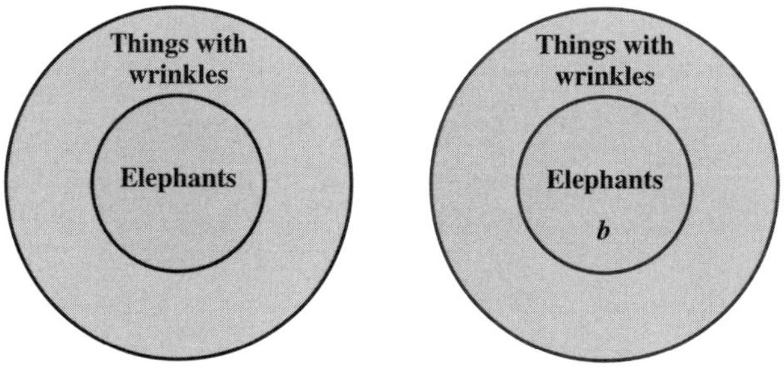

b represents Babar.

(a) **(b)**

FIGURE 14

EXAMPLE 3 **Valid Argument**

Represent the following argument symbolically. Is the argument valid?

All frogs are slimy.
Kermit is not slimy.

Kermit is not a frog.

▶**Solution** If we let $f(x)$ represent "x is a frog," $s(x)$ represent "x is slimy," and k represent "Kermit," we could represent the argument symbolically as follows.

$$\begin{array}{l} \forall x\,[f(x) \rightarrow s(x)] \\ \underline{\sim s(k) \qquad\qquad\quad} \\ \sim f(k) \end{array}$$

This argument resembles Modus Tollens with a quantifier, so it is probably valid. We will use an Euler diagram to verify this.

In Figure 15(a) the region for "frogs" is drawn entirely inside the region for "slimy." Since "Kermit is *not* slimy," place a k for "Kermit" *outside* the region for "slimy." (See Figure 15b.) Placing the k outside the region for "slimy" automatically places it outside the region for "frogs." Thus, if the first two premises are true, the conclusion that Kermit is not a frog must also be true, so the argument is valid.

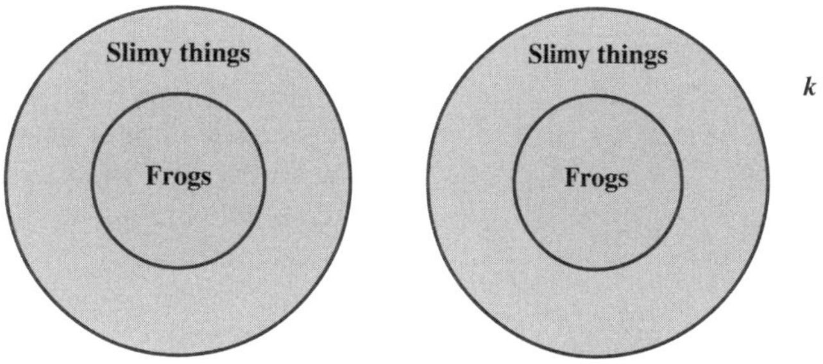

k represents Kermit.

(a) (b)

FIGURE 15

EXAMPLE 4 **Fallacy**

Represent the following argument symbolically. Is the argument valid?

All well-run businesses generate profits.
<u>Monsters, Inc., generates profits.</u>
Monsters, Inc., is a well-run business.

▶**Solution** If we let $w(x)$ represent "x is a well-run business," $p(x)$ represent "x generates profits," and m represent "Monsters, Inc.," we could represent the argument symbolically as follows.

$$\begin{array}{l} \forall x\,[w(x) \rightarrow p(x)] \\ \underline{p(m) \qquad\qquad\qquad} \\ w(m) \end{array}$$

This argument resembles the Fallacy of the Converse with a quantifier. We will use an Euler diagram to verify that the argument is invalid.

The region for "well-run business" goes entirely inside the region for "generates a profit." (See Figure 16.) It is not clear where to put the m for "Monsters, Inc." It must go inside the region for "generates profits," but it could go inside or outside the region "well-run business." Even if the premises are true, the conclusion may or may not be true, so the argument is invalid.

FIGURE 16

As we mentioned in the previous section, the validity of an argument is not the same as the truth of its conclusion. The argument in Example 4 was invalid, but the conclusion "Monsters, Inc." may or may not be true. We cannot make a valid conclusion.

EXAMPLE 5 **Valid Argument**

Represent the following argument symbolically. Is the argument valid?

All squirrels eat nuts.
All those who eat nuts are healthy.
All who are healthy avoid cigarettes.
———
All squirrels avoid cigarettes.

▶Solution If we let $s(x)$ represent "x is a squirrel," $e(x)$ represent "x eats nuts," $h(x)$ represent "x is healthy," and $a(x)$ represent "x avoids cigarettes," we could represent the argument symbolically as follows.

$$\forall x\,[s(x) \to e(x)]$$
$$\forall x\,[e(x) \to h(x)]$$
$$\underline{\forall x\,[h(x) \to a(x)]}$$
$$\forall x\,[s(x) \to a(x)]$$

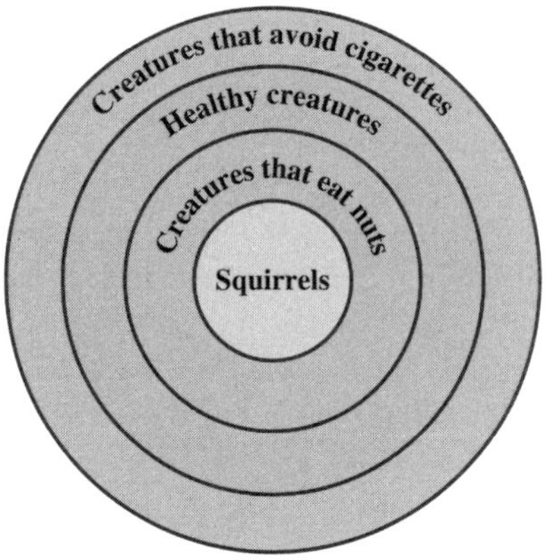

FIGURE 17

This argument should remind you of Reasoning by Transitivity. We will use the Euler diagram in Figure 17 to verify that the argument is valid. If each premise is true, then the conclusion must be true because the region for "squirrels" lies completely within the region for "avoid cigarettes." Thus, the argument is valid.

Our last example in this section illustrates an argument with the word "some," which means that an existential quantifier is needed.

EXAMPLE 6 **Invalid Argument**

Represent the following argument symbolically. Is the argument valid?

Some U.S. presidents have won the popular vote.
George W. Bush is president of the United States.
George W. Bush won the popular vote.

?

▶**Solution** If we let $p(x)$ represent "x is a U.S. president," $v(x)$ represent "x won the popular vote," and w represent "George W. Bush," we could represent the argument symbolically as follows.

$$\exists x\,[p(x) \wedge v(x)]$$
$$\underline{p(w)\qquad\qquad}$$
$$v(w)$$

This argument doesn't resemble any of those in the previous section. An Euler diagram might help us see whether this argument is valid or not. The first premise is sketched in Figure 18(a). We have indicated that some U.S. presidents have won the popular vote by putting an x in the region that belongs to both the set of presidential candidates who have won the popular vote and the set of U.S. presidents. There are two possibilities for w, as shown in Figure 18(b).

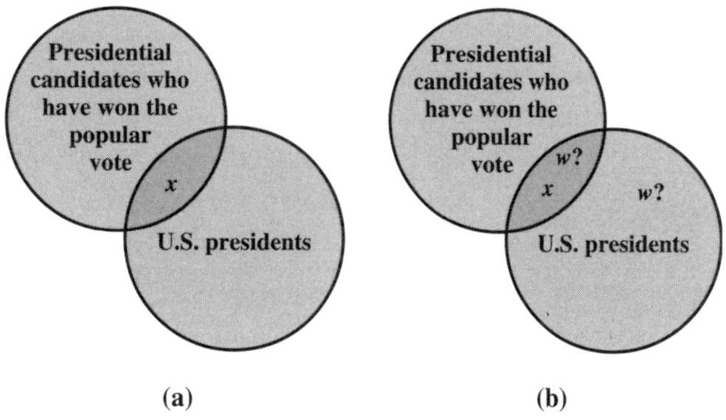

FIGURE 18

One possibility is that Bush won the popular vote; the other is that Bush did not win the popular vote. Since the truth of the premises does not force the conclusion to be true, the argument is invalid.

NOTE This argument is not valid regardless of whether George W. Bush won the popular vote. In fact, in 2000 George W. Bush was elected president even though he received 50,456,002 votes, compared with 50,999,897 for Al Gore.* ■

6.6 Exercises

For Exercises 1–6, **(a)** *write the statement symbolically,* **(b)** *write the negative of the statement in part (a), and* **(c)** *translate your answer from part (b) into words.*

1. Some books are bestsellers.

2. Every dog has his day.

3. No CEO sleeps well at night.

4. There's no business like show business.

5. All the leaves are brown.

6. Some days are better than others.

In Exercises 7–20, **(a)** *represent the argument symbolically, and* **(b)** *use an Euler diagram to determine if the argument is valid.*

7. Girls just want to have fun.
Teri Lovelace is a girl.
Teri Lovelace just wants to have fun.

8. All sophomores have earned at least 60 credits.
Liliana Molina is a sophomore.
Liliana Molina has earned at least 60 credits.

9. All professors are covered with chalk dust.
Otis Taylor is covered with chalk dust.
Otis Taylor is a professor.

10. All dinosaurs are extinct.
The dodo is extinct.
The dodo is a dinosaur.

11. All accountants use spreadsheets.
Nancy Hart does not use spreadsheets.
Nancy Hart is not an accountant.

12. All fish have gills.
Whales do not have gills.
Whales are not fish.

13. Some people who are turned down for a mortgage have a second income.
All people who are turned down for a loan need a mortgage broker.
Some people with a second income need a mortgage broker.

14. Some residents of Minnesota don't like snow.
All skiers like snow.
Some residents of Minnesota are not skiers.

15. Some who wander are lost.
Martha McDonald wanders.
Martha McDonald is lost.

16. Some old houses have root cellars.
My house has a root cellar.
My house is old.

17. Some psychologists are university professors.
Some psychologists have a private practice.
Some university professors have a private practice.

18. Someone who is responsible must pay for this.
Nola Akala is responsible.
Nola Akala must pay for this.

*www.fec.gov/pubrec/fe2000/prespop.htm.

19. Everybody is either a saint or a sinner.
Some people are not saints.
———
Some people are sinners.

20. If you're here for the first time, we want to make you feel welcome.
If you're here after being gone for a while, we want to make you feel welcome.
Some people are here for the first time or are here after being gone for a while.
———
There are some people here whom we want to make feel welcome.

21. Refer to Example 4. If the second premise and the conclusion were interchanged, would the argument then be valid?

22. Refer to Example 5. Give a different conclusion than the one given there, so that the argument is still valid.

Construct a valid argument based on the Euler diagram shown.

23.

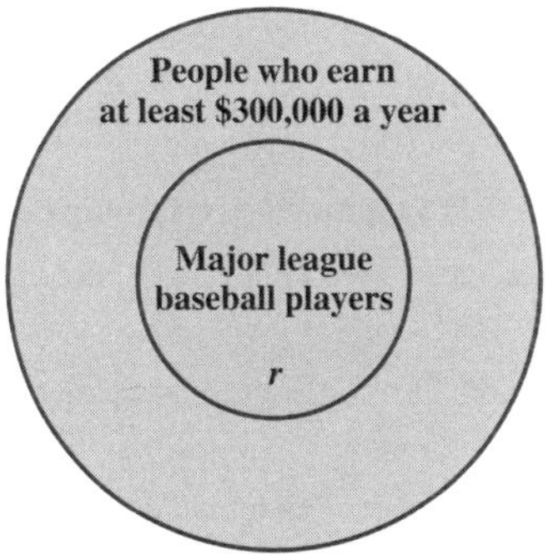

r represents Ryan Howard.

24.

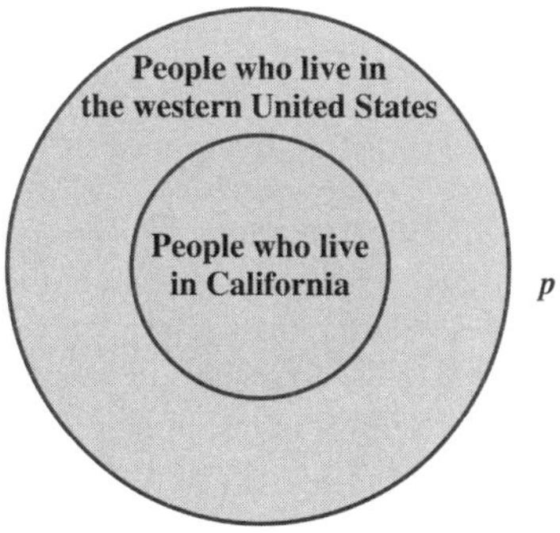

p represents Phyllis Crittenden.

As mentioned in the text, an argument can have a true conclusion yet be invalid. In these exercises, each argument has a true conclusion. Identify each argument as valid *or* invalid.

25. All houses have roofs.
All roofs have nails.
———
All houses have nails.

26. All platypuses have bills.
All ducks have bills.
———
A platypus is not a duck.

27. All mammals have fur.
All tigers have fur.
———
All tigers are mammals.

28. All mammals have fur.
All tigers are mammals.
———
All tigers have fur.

29. California is adjacent to Arizona.
Arizona is adjacent to Nevada.
———
California is adjacent to Nevada.

30. Seattle is northwest of Boise.
Seattle is northwest of Salt Lake City.
———
Boise is northwest of Salt Lake City.

31. A rectangle has four sides.
A square has four sides.
———
A square is a rectangle.

32. No integer is irrational.
The number π is irrational.
———
The number π is not an integer.

33. Explain the difference between the following statements:
All students did not pass the test.
Not all students passed the test.

34. Write the following statement using *every:* There is no one here who has not done that at one time or another.

Applications

BUSINESS AND ECONOMICS

35. *Advertising* Incorrect use of quantifiers often is heard in everyday language. Suppose you hear that a local electronics chain is having a 30% off sale, and the radio advertisement states, "All items are not available in all stores." Do you think that, literally translated, the ad really means what it says? What do you think really is meant? Explain your answer.

36. *Portfolios* Repeat Exercise 35 for the following: "All people don't have the time to devote to maintaining their financial portfolios properly."

SOCIAL SCIENCES

Constitution Each of the following exercises gives a passage from the U.S. Constitution, followed by another statement. **(a)** *Translate the passage from the Constitution into a statement in symbolic logic.* **(b)** *Find a valid conclusion using as premises the passage from the Constitution and the statement that follows it.* **(c)** *Illustrate the argument with an Euler diagram.*

37. All legislative powers herein granted shall be vested in a Congress of the United States . . . (Article 1, Section 1)

The power to collect taxes is a legislative power herein granted.

38. No person shall be a Representative who shall not have attained to the age of twenty-five years, and been seven years a citizen of the United States, and who shall not, when elected, be an inhabitant of that State in which he shall be chosen. (Article 1, Section 2)

Dennis Hastert is a Representative.

39. No bill of attainder or ex post facto law shall be passed. (Article 1, Section 9)

The law forbidding members of the Communist Party to serve as an officer or as an employee of a labor union was a bill of attainder.*

40. No State shall enter into any treaty, alliance, or confederation. (Article 1, Section 10)

Texas is a state.

41. *Tourette's Syndrome* The following diagram appeared in an article on Tourette's syndrome, an inherited disorder characterized by involuntary utterances and body movements.†

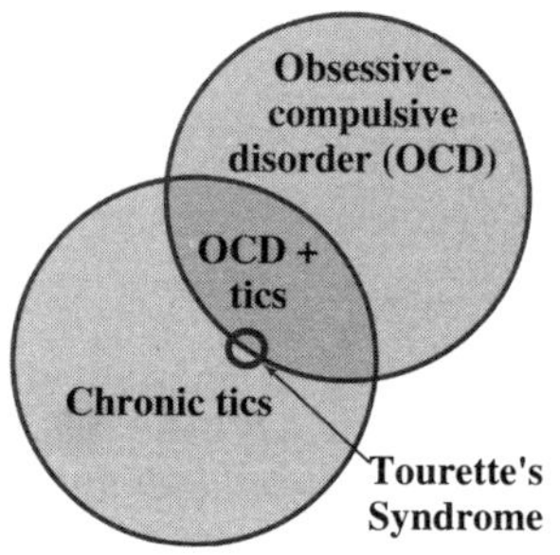

Which of the following conclusions are valid?

a. All those with Tourette's syndrome have chronic tics.

b. All those with Tourette's syndrome have obsessive-compulsive disorder.

c. Some of those with chronic tics have obsessive-compulsive disorder.

d. Some of those with obsessive-compulsive disorder do not have Tourette's syndrome.

e. All those with chronic tics have Tourette's syndrome.

GENERAL INTEREST

42. *Bible* Write the negation of each of the following quotes from the Bible.‡

a. "Everyone who hears about this will laugh with me."—Genesis 21:6

b. "Someone came to destroy your lord the king."—1 Samuel 26:15

c. "There is no one who does good."—Psalm 53:3

d. "Everyone is the friend of one who gives gifts."—Proverbs 19:6

e. "Everyone who quotes proverbs will quote this proverb about you: 'Like mother, like daughter.'"—Ezekiel 16:44

**United States v. Brown*, 381 U.S. 437 (1965).
†Source: From Olson, *SCIENCE*, Vol. 305: p. 1390–1392 (2004). Reprinted with permission of AAAS.
‡*Today's New International Version.*

Animals In Exercises 43–48, the premises marked A, B, *and* C *are followed by several possible conclusions. Take each conclusion in turn, and check whether the resulting argument is* valid *or* invalid.

A. *All kittens are cute animals.*

B. *All cute animals are admired by animal lovers.*

C. *Some dangerous animals are admired by animal lovers.*

43. Some kittens are dangerous animals.

44. Some cute animals are dangerous.

45. Some dangerous animals are cute.

46. Kittens are not dangerous animals.

47. All kittens are admired by animal lovers.

48. Some things admired by animal lovers are dangerous animals.

Chapter 6 Review

Chapter Summary

In this chapter we introduced symbolic logic, which uses letters to represent statements, and symbols for words such as

- *and* (conjunction, denoted by $\wedge$),
- *or* (disjunction, denoted by $\vee$),
- *not* (negation, denoted by $\sim$),
- *if . . . then* (conditional, denoted by $\rightarrow$), and
- *if and only if* (biconditional, denoted by $\leftrightarrow$).

Statements are declarative sentences that are either true or false, but not both simultaneously. Using logical connectives, two or more statements can be combined to form a compound statement. Truth values of various compound statements were explored using truth tables. We saw that two logical statements are equivalent (denoted as $\equiv$) if they have the same truth value. We used symbolic logic to design circuits, and then used logical equivalences to simplify the circuits. We saw that the contrapositive, a statement related to the conditional, is equivalent to the original conditional statement, but that two other related statements, the inverse and the converse, are not equivalent to the original conditional statement. We next explored how to prove valid arguments and give counterexamples to invalid arguments. Finally, we discussed the quantifiers *for all* (denoted by $\forall$) and *there exists* (denoted by $\exists$), and we used Euler diagrams to determine the validity of an argument involving quantifiers.

LOGIC SUMMARY

Truth Tables for Logical Operators

p	q	$p \wedge q$	$p \vee q$	$p \rightarrow q$	$p \leftrightarrow q$
T	T	T	T	T	T
T	F	F	T	F	F
F	T	F	T	T	F
F	F	F	F	T	T

Writing a Conditional as an *or* Statement

$p \to q \equiv \sim p \vee q$

Negation of a Conditional Statement

$\sim(p \to q) \equiv p \wedge \sim q$

Equivalent Statements

1a.	$p \vee q \equiv q \vee p$	Commutative Laws
b.	$p \wedge q \equiv q \wedge p$	
2a.	$p \vee (q \vee r) \equiv (p \vee q) \vee r$	Associative Laws
b.	$p \wedge (q \wedge r) \equiv (p \wedge q) \wedge r$	
3a.	$p \vee (q \wedge r) \equiv (p \vee q) \wedge (p \vee r)$	Distributive Laws
b.	$p \wedge (q \vee r) \equiv (p \wedge q) \vee (p \wedge r)$	
4a.	$\sim(p \wedge q) \equiv \sim p \vee \sim q$	De Morgan's Laws
b.	$\sim(p \vee q) \equiv \sim p \wedge \sim q$	
5a.	$p \vee p \equiv p$	Idempotent Laws
b.	$p \wedge p \equiv p$	
6a.	$(p \wedge q) \vee p \equiv p$	Absorption Laws
b.	$(p \vee q) \wedge p \equiv p$	
7.	$\sim(\sim p) \equiv p$	Double Negative
8.	$p \to q \equiv \sim p \vee q$	Conditional as an "or"
9.	$p \to q \equiv \sim q \to \sim p$	Contrapositive

If T represents any true statement and F represents any false statement, then

10a.	$p \vee \mathrm{T} \equiv \mathrm{T}$	Identity Laws
b.	$p \wedge \mathrm{T} \equiv p$	
c.	$p \vee \mathrm{F} \equiv p$	
d.	$p \wedge \mathrm{F} \equiv \mathrm{F}$	
11a.	$p \vee \sim p \equiv \mathrm{T}$	Negation Laws
b.	$p \wedge \sim p \equiv \mathrm{F}$	

Common Translations of $p \to q$

If p, then q.	p is sufficient for q.
If p, q.	q is necessary for p.
p implies q.	All p's are q's.
p only if q.	q if p.
q when p.	

Related Conditional Statements

Original Statement	$p \to q$
Converse	$q \to p$
Inverse	$\sim p \to \sim q$
Contrapositive	$\sim q \to \sim p$

Valid Argument Forms

Modus Ponens	Modus Tollens	Disjunctive Syllogism	Reasoning by Transitivity
$p \to q$	$p \to q$	$p \vee q$	$p \to q$
p	$\sim q$	$\sim p$	$q \to r$
q	$\sim p$	q	$p \to r$

Invalid Argument Forms (Fallacies)

Fallacy of the Converse	Fallacy of the Inverse
$p \rightarrow q$	$p \rightarrow q$
q	$\sim p$
p	$\sim q$

Key Terms

6.1 **statement**, **logical connective**, **compound statement**, **component statement**, **negation**, **truth table**, **conjunction**, **disjunction**

6.2 **equivalent**, **De Morgan's Laws**

6.3 **conditional**, **antecedent**, **consequent**, **contradiction**, **tautology**, **circuit**, **series**, **parallel**, **Principle of Duality**

6.4 **converse**, **inverse**, **contrapositive**, **biconditional**

6.5 **premise**, **conclusion**, **valid**, **invalid**, **fallacy**, **Modus Ponens**, **Fallacy of the Converse**, **Modus Tollens**, **Fallacy of the Inverse**, **Disjunctive Syllogism**, **Reasoning by Transitivity**

6.6 **quantifier**, **universal quantifier**, **existential quantifier**, **Euler diagram**

Concept Check

Determine whether each of the following statements is true or false, and explain why.

1. A compound statement is a negation, a conjunction, a disjunction, a conditional, or a biconditional.

2. A truth table with 5 variables has 10 rows.

3. A truth table can have an odd number of rows.

4. Using one of De Morgan's Laws, the negation of a disjunction may be written as a conditional.

5. The negation of a conditional statement is a disjunction.

6. Elements in a circuit that are in parallel are connected in a logic statement with an *or.*

7. A tautology might be true or false.

8. A statement might be true even though its inverse is false.

9. The conclusion of a valid argument must be true.

10. The conclusion of a fallacy must be false.

11. Euler diagrams can be used to determine whether an argument with quantifiers is valid or invalid.

12. The negation of a statement with the universal quantifier involves the existential quantifier.

Chapter 6 Review Exercises

Write the negation of each statement.

1. If she doesn't pay me, I won't have enough cash.

2. I fought the law and the law won.

Let l represent "He loses the election" *and let w represent* "He wins the hearts of the voters." *Write each statement in symbols.*

3. He loses the election, but he wins the hearts of the voters.

4. If he wins the hearts of the voters, then he doesn't lose the election.

5. He loses the election only if he doesn't win the hearts of the voters.

6. He loses the election if and only if he doesn't win the hearts of the voters.

Using the same statements as for Exercises 3–6, write each mathematical statement in words.

7. $\sim l \wedge w$

8. $\sim(l \vee \sim w)$

Assume that p is true and that q and r are false. Find the truth value of each statement.

9. $\sim q \wedge \sim r$

10. $r \vee (p \wedge \sim q)$

11. $r \rightarrow (s \vee r)$ (The truth value of the statement s is unknown.)

12. $p \leftrightarrow (p \rightarrow q)$

13. Explain in your own words why, if p is a statement, the biconditional $p \leftrightarrow \sim p$ must be false.

14. State the necessary conditions for

a. a conditional statement to be false;

b. a conjunction to be true;

c. a disjunction to be false.

Construct a truth table for each statement. Is the statement a tautology?

15. $p \wedge (\sim p \vee q)$

16. $\sim(p \wedge q) \rightarrow (\sim p \vee \sim q)$

Write each conditional statement in if . . . then *form.*

17. All mathematicians are loveable.

18. You can have dessert only if you eat your vegetables.

19. Having at least as many equations as unknowns is necessary for a system to have a unique solution.

20. Having a feasible region is sufficient for a linear programming problem to have a minimum.

For each statement, write **(a)** *the converse,* **(b)** *the inverse, and* **(c)** *the contrapositive.*

21. If the proposed regulations have been approved, then we need to change the way we do business.

22. $(p \vee q) \rightarrow \sim r$ (Use one of De Morgan's Laws to simplify.)

Write a logical statement representing each circuit. Simplify each circuit when possible.

23.

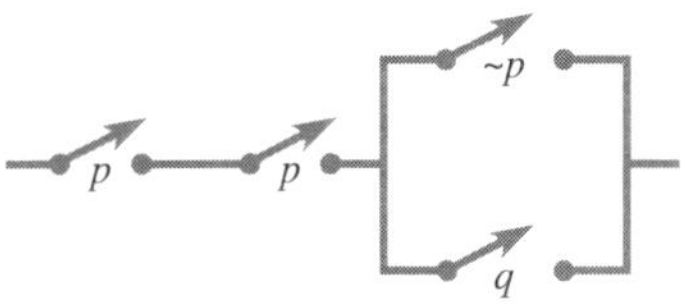

24.

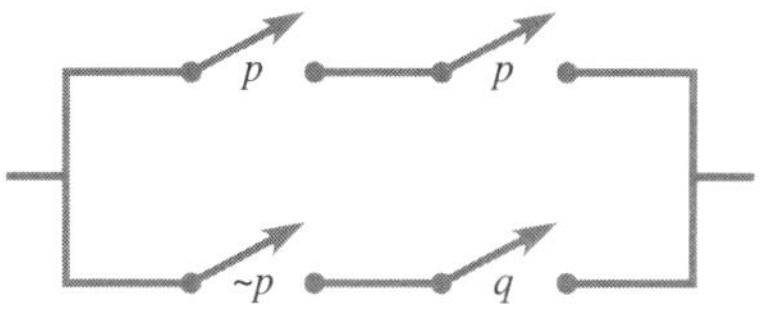

Draw the circuit representing each statement as it is given. Simplify if possible.

25. $(p \wedge q) \vee (p \wedge p)$

26. $p \wedge (p \vee q)$

For Exercises 27 and 28, consider the exclusive disjunction introduced in Exercises 35–38 of Section 6.2. In exclusive disjunction, denoted by $p \underline{\vee} q$, *either p or q is true, but not both.*

27. Use a truth table to show that $p \underline{\vee} q \equiv (p \vee q) \wedge \sim(p \wedge q)$.

28. Use a truth table to show that $p \underline{\vee} q \equiv \sim[(p \vee q) \rightarrow (p \wedge q)]$.

29. Consider the statement "If this year is 2002, then 1 + 1 = 3."

a. Is the above statement true?

b. Was the above statement true in 2002?

c. Discuss how a statement such as the one above changes its truth value over time, and how this agrees or disagrees with your intuitive notion of whether a conditional statement is true or false.

30. a. Consider the statement "If Shakespeare didn't write *Hamlet*, then someone else did." Explain why this statement is true.

b. Consider the statement "If Shakespeare hadn't written *Hamlet*, then someone else would have." Explain why this statement should be false in any reasonable logic.

c. If we let *s* represent "Shakespeare wrote *Hamlet*" and *e* represent "someone else wrote *Hamlet*," then the logic of this chapter leads us to represent either the statement of part (a) or part (b) as $\sim s \rightarrow e$. Discuss the issue of how this statement could be true in part (a) but false in part (b).*

Each of the following arguments is either valid by one of the forms of valid arguments discussed in this chapter, or is a fallacy by one of the forms of invalid arguments discussed. Decide whether the argument is valid *or* invalid, *and give the form that applies.*

31. If you're late one more time, you'll be docked.
You're late one more time.
____
You'll be docked.

32. If the company makes a profit, its stock goes up.
The company doesn't makes a profit.
____
Its stock doesn't go up.

33. The instructor is late or my watch is wrong.
My watch is not wrong.
____
The instructor is late.

34. If the parent is loving, then the child will be happy.
If the child is happy, then the teacher can teach.
____
If the parent is loving, then the teacher can teach.

*The problem of logical statements using the subjunctive mood is described in more detail in *Formal Logic: Its Scope and Limits*, 2nd ed., by Richard Jeffrey, New York: McGraw-Hill, 1981.

35. If you play that song one more time, I'm going nuts.
I'm going nuts.
___
You play that song one more time.

36. If it's after five, the store is closed.
The store is not closed.
___
It's not after five.

Determine whether each argument is valid *or* invalid. *If it is valid, give a proof. If it is invalid, give an assignment of truth values to the variables that makes the premises true and the conclusion false.*

37. If we hire a new person, then we'll spend more on training. If we rewrite the manual, then we won't spend more on training. We rewrite the manual. Therefore, we don't hire a new person.

38. It is not true that Smith or Jones received enough votes to qualify. But if Smith got enough votes to qualify, then the election was rigged. Therefore, the election was not rigged.

39. $\sim p \rightarrow \sim q$
$q \rightarrow p$
___
$p \vee q$

40. $p \rightarrow q$
$r \rightarrow \sim q$
___
$p \rightarrow \sim r$

In Exercises 41 and 42, **(a)** *write the statement symbolically,* **(b)** *write the negation of the statement in part (a), and* **(c)** *translate your answer from part (b) into words.*

41. All dogs go to heaven.

42. Some cars have manual transmissions.

In Exercises 43 and 44, **(a)** *represent the argument symbolically, and* **(b)** *use an Euler diagram to determine whether the argument is valid.*

43. All members of that fraternity do well academically.
John Cross is a member of that fraternity.
___
John Cross does well academically.

44. Some members of that fraternity don't do well academically.
Bill Hoffman does well academically.
___
Bill Hoffman is not a member of that fraternity.

45. Construct a truth table for $p \rightarrow (q \rightarrow r)$ and $(p \rightarrow q) \rightarrow r$. Are these two statements equivalent?

46. **a.** Convert the statement $p \rightarrow (q \rightarrow r)$ into an equivalent statement with an "or," but without the conditional.

b. Convert the statement $(p \rightarrow q) \rightarrow r$ into an equivalent statement with an "or," but without the conditional.

47. **a.** Construct a truth table for the statement $(p \wedge \sim p) \rightarrow q$.

b. Based on the truth table for part a, explain why any conclusion may be reached from a contradictory premise.

Applications

BUSINESS AND ECONOMICS

Income Tax *The following excerpts appear in a guide for preparing income tax reports.**

a. *If your top tax bracket is 10% or 15%, your capital gain rate is generally 5%.*

b. *The Schedule D Tax Worksheet in the Schedule D instructions is used only if you have a net 28% rate gain or unrecaptured Section 1250 gain.*

c. *This is true whether or not you are personally liable for the debt, and whether or not the buyer assumes the mortgage or merely takes the property subject to the mortgage.*

d. *See 8.10 for IRS guidelines on granting a waiver.*

48. Which of these excerpts are statements?

49. Which of these excerpts are compound statements?

50. Write the converse of statement a.

51. Write the contrapositive of statement b.

LIFE SCIENCES

52. *Medicine* The following statement appears in a home medical reference book with regard to pregnancy and rubella.† Write the contrapositive of the statement and then write a statement equivalent to the original using the word *or.*

If the angina becomes more severe despite drug treatment, your doctor may advise you to consider surgery to either widen the arteries or improve the blood flow to the heart.

SOCIAL SCIENCES

53. *Gun Control* The following statements appear in an article about gun control.‡ Negate each statement.

a. Regulations have both costs and benefits, and rules that are passed to solve a problem can sometimes make it worse.

b. Shooters overwhelmingly have problems with alcoholism and have long criminal histories, particularly arrests for violent acts.

c. They are disproportionately involved in automobile crashes and are more likely to have had their driver's license suspended or revoked.

54. *Democratization* According to one sociologist, "If, for example, democratization always occurs in the company of widening splits within ruling oligarchies (that is, such splits are active candidates for necessary conditions of democratization), a valid causal story will most likely connect democratization with such splits."§ Let w represent "widening splits occur within ruling oligarchies," d represent "democratization occurs," and v represent "a valid causal story connects democratization with splits." Represent the sociologist's statement as a statement in logic.

55. *Philosophy* In an article in *Skeptical Inquirer*, Ralph Estling said, "Positivists such as [Stephen] Hawking tell us that reality cannot be dealt with . . . [and that] we can only deal with what we can measure in some way. . . . This seems sensible . . . until we watch as our physicists slowly slide down the slippery slopes . . . [and] tell us that only the measurable is real."‖

a. Draw an Euler diagram with three circles showing real things, things we can measure, and things we can deal

*J. K. Lasser Institute, *Your Income Tax 2006,* New York: John Wiley & Sons, pp. 101, 105, 112, 157.
†Goldman, David R., ed., *American College of Physicians Complete Home Medical Guide*, 2nd ed., New York: DK Publishing, 2003, p. 409.
‡Lott, John R., "When Gun Control Costs Lives," *National Forum: The Phi Kappa Phi Journal*, Vol. 80, No. 4, Fall 2000, pp. 29–32.
§Tilly, Charles, "Processes and Mechanisms of Democratization," *Sociological Theory*, Vol. 18, No. 1, March 2000, pp. 1–16
‖Estling, Ralph, "It's a Good Thing Cows Can't Fly in Mobile," *Skeptical Inquirer*, Nov./Dec. 2002, pp. 57–58.

with. Your diagram should simultaneously illustrate the following statements:

i. Real things are not things we can deal with.

ii. Only things we can measure are things we can deal with.

iii. Only things we can measure are real things.

b. Using the Euler diagram from part a, explain why things that we can measure and deal with are not real, assuming the previous statements are true.

c. Philosopher Timothy Chambers suggests adding the plausible premise, "All things we can measure are things we can deal with."* Explain why this premise, added to the previous three, implies that nothing is real.

GENERAL INTEREST

Monopoly Junior™ The following excerpts can be found on the box to the game, Monopoly Junior.™ †

a. Join Rich Uncle Pennybags and his nieces and nephews for a thrill-filled day at the Boardwalk Amusements—the Roller Coaster, the Magic Show, the Water Slide, the Video Arcade, and more.

b. Draw a Chance card and take a ride on the Miniature Railroad, win a free Ticket Booth . . . or pay $3 to visit the Rest Rooms!

c. When someone finally runs out of money, the player with the most cash on hand wins this colorful, fast-paced junior version of the world's most popular board game.

d. Set up Ticket Booths at the Amusements and collect fees from other players who land on them.

56. Which of these excerpts are statements?

57. Which of these excerpts are compound statements?

Lewis Carroll The following exercises are from problems by Lewis Carroll.‡ Write each premise in symbols, and then in the final part, give a conclusion that yields a valid argument.

58. Let s be "the puppy lies still," g be "the puppy is grateful to be lent a skipping rope," l be "the puppy is lame," and w be "the puppy cares to do worsted work."

a. Puppies that will not lie still are always grateful for the loan of a skipping rope.

b. A lame puppy would not say "thank you" if you offered to lend it a skipping rope.

c. None but lame puppies ever care to do worsted work.

d. Give a conclusion that yields a valid argument.

59. Let o be "the bird is an ostrich," h be "the bird is at least 9 feet high," a be "the bird is in this aviary," m be "the bird belongs to me," and p be "the bird lives on mince pies."

a. No birds, except ostriches, are 9 feet high. (*Hint:* Interpret as: If a bird is at least 9 feet high, it is an ostrich.)

b. There are no birds in this aviary that belong to any one but me.

c. No ostrich lives on mince pies.

d. I have no birds less than 9 feet high.

e. Give a conclusion that yields a valid argument.

60. Let f be "the kitten loves fish," t be "the kitten is teachable," a be "the kitten has a tail," g be "the kitten will play with a gorilla," w be "the kitten has whiskers," and e be "the kitten has green eyes."

a. No kitten that loves fish is unteachable.

b. No kitten without a tail will play with a gorilla.

c. Kittens with whiskers always love fish.

d. No teachable kitten has green eyes.

e. No kittens have tails unless they have whiskers. (*Hint:* "$(\sim a)$ unless b" $\equiv a \rightarrow b$)

f. Give a conclusion that yields a valid argument.

61. Let u be "the writer understands human nature," c be "the writer is clever," p be "the writer is a true poet," r be "the writer can stir the hearts of men," s be "the writer is Shakespeare," and h be "the writer wrote *Hamlet*."

a. All writers who understand human nature are clever.

b. No one is a true poet unless he can stir the hearts of men. (*Hint:* "$(\sim a)$ unless b" $\equiv a \rightarrow b$)

c. Shakespeare wrote *Hamlet.*

d. No writer who does not understand human nature can stir the hearts of men.

e. None but a true poet could have written *Hamlet.*

f. Give a conclusion that yields a valid argument.

*Chambers, Timothy, "On Venn Diagrams," *The Mathematics Teacher*, Vol. 97, No. 1, Jan. 2004, p. 3.

†Monopoly® is Hasbro, Inc.'s trademark for its real estate trading board game equipment.

‡*The Complete Works of Lewis Carroll*, Vintage Books, 1976.

Logic Puzzles

Some people find that logic puzzles, which appear in periodicals such as *World-Class Logic Problems* (Penny Press) and *Logic Puzzles* (Dell), provide hours of enjoyment. They are based on deductive reasoning, and players answer questions based on clues given. The following explanation on solving such problems appeared in the Autumn 2003 issue of *World-Class Logic Problems*.

HOW TO SOLVE LOGIC PROBLEMS

Solving logic problems is entertaining and challenging. All the information you need to solve a logic problem is given in the introduction and clues, and in illustrations, when provided. If you've never solved a logic problem before, our sample should help you get started. Fill in the Sample Solving Chart in Figure 19 as you follow our explanation. We use a "•" to signify "Yes" and an "×" to signify "No."

Five couples were married last week, each on a different weekday. From the information provided, determine the woman (one is Cathy) and man (one is Paul) who make up each couple, as well as the day on which each couple was married.

1. Anne was married on Monday, but not to Wally.
2. Stan's wedding was on Wednesday. Rob was married on Friday, but not to Ida.
3. Vern (who married Fran) was married the day after Eve.

SAMPLE SOLVING CHART:

	PAUL	ROB	STAN	VERN	WALLY	MONDAY	TUESDAY	WEDNESDAY	THURSDAY	FRIDAY
ANNE										
CATHY										
EVE										
FRAN										
IDA										
MONDAY										
TUESDAY										
WEDNESDAY										
THURSDAY										
FRIDAY										

FIGURE 19

Anne was married Monday (1), so put a "•" at the intersection of Anne and Monday. Put "×" in all the other days in Anne's row and all the other names in the Monday column. (Whenever you establish a relationship, as we did here, be sure to place "×" at the intersections of all relationships that become impossible as a result.) Anne wasn't married to Wally (1), so put an "×" at the intersection of Anne and Wally. Stan's wedding was Wednesday (2), so put a "•" at the intersection of Stan and Wednesday (don't forget the "×"s). Stan didn't marry Anne, who was married Monday, so put an "×" at the intersection of Anne and Stan. Rob was married Friday, but not to Ida (2), so put a "•" at the intersection of Rob and Friday, and "×" at the intersections of Rob and Ida and Ida and Friday. Rob also didn't marry Anne, who was married Monday, so put an "×" at the intersection of Anne and Rob. Now your chart should look like Figure 20.

	PAUL	ROB	STAN	VERN	WALLY	MONDAY	TUESDAY	WEDNESDAY	THURSDAY	FRIDAY
ANNE		×	×		×	•	×	×	×	×
CATHY						×				
EVE						×				
FRAN						×				
IDA		×				×				×
MONDAY		×	×							
TUESDAY		×	×							
WEDNESDAY	×	×	•	×	×					
THURSDAY		×	×							
FRIDAY	×	•	×	×	×					

FIGURE 20

Vern married Fran (3), so put a "•" at the intersection of Vern and Fran. This leaves Anne's only possible husband as Paul, so put a "•" at the intersection of Anne and Paul and Paul and Monday. Vern and Fran's wedding was the day after Eve's (3), which wasn't Monday [Anne], so Vern's wasn't Tuesday. It must have been Thursday [see chart], so Eve's was Wednesday (3). Put "•" at the intersections of Vern and Thursday, Fran and Thursday, and Eve and Wednesday. Now your chart should look like Figure 21.

	PAUL	ROB	STAN	VERN	WALLY	MONDAY	TUESDAY	WEDNESDAY	THURSDAY	FRIDAY
ANNE	•	×	×	×	×	•	×	×	×	×
CATHY	×			×		×		×	×	
EVE	×			×		×	×	•	×	×
FRAN	×	×	×	•	×	×	×	×	•	×
IDA	×	×		×		×		×	×	×
MONDAY	•	×	×	×	×					
TUESDAY	×	×	×	×						
WEDNESDAY	×	×	•	×	×					
THURSDAY	×	×	×	•	×					
FRIDAY	×	•	×	×	×					

FIGURE 21

The chart shows that Cathy was married Friday, Ida was married Tuesday, and Wally was married Tuesday. Ida married Wally, and Cathy's wedding was Friday, so she married Rob. After this information is filled in, Eve could only have married Stan. You've completed the puzzle, and your chart should now look like Figure 22 on the next page.

	PAUL	ROB	STAN	VERN	WALLY	MONDAY	TUESDAY	WEDNESDAY	THURSDAY	FRIDAY
ANNE	●	×	×	×	×	●	×	×	×	×
CATHY	×	●	×	×	×	×	×	×	×	●
EVE	×	×	●	×	×	×	×	●	×	×
FRAN	×	×	×	●	×	×	×	×	●	×
IDA	×	×	×	×	●	×	●	×	×	×
MONDAY	●	×	×	×	×					
TUESDAY	×	×	×	×	●					
WEDNESDAY	×	×	●	×	×					
THURSDAY	×	×	×	●	×					
FRIDAY	×	●	×	×	×					

FIGURE 22

In summary: Anne and Paul, Monday; Cathy and Rob, Friday; Eve and Stan, Wednesday; Fran and Vern, Thursday; Ida and Wally, Tuesday.

In some problems, it may be necessary to make a logical guess based on facts you've established. When you do, always look for clues or other facts that disprove it. If you find that your guess is incorrect, eliminate it as a possibility.

EXERCISES

1. ***Water, Water, Everywhere**** After an invigorating workout, five fitness-conscious friends know that nothing is more refreshing than a tall, cool glass of mineral water! Each person (including Annie) has a different, favorite form of daily exercise (one likes to rollerblade), and each drinks a different brand of mineral water (one is Crystal Spring). From the information provided, determine the type of exercise and brand of water each person prefers.
   - **a.** The one who bicycles in pursuit of fitness drinks Bevé.
   - **b.** Tim enjoys aerobicizing every morning before work. Ben is neither the one who drinks Sparkling Creek nor the one who imbibes Bevé.
   - **c.** Page (who is neither the one who jogs nor the one who walks to keep in shape) drinks Purity. Meg drinks Mountain Clear, but not after jogging.

		EXERCISE					WATER				
		AEROBICIZING	BICYCLING	JOGGING	ROLLERBLADING	WALKING	BEVÉ	CRYSTAL SPRING	MOUNTAIN CLEAR	PURITY	SPARKLING CREEK
PERSON	ANNIE										
	BEN										
	MEG										
	PAGE										
	TIM										
WATER	BEVÉ										
	CRYSTAL SPRING										
	MOUNTAIN CLEAR										
	PURITY										
	SPARKLING CREEK										

2. ***Let's Get Physical***† The Anytown Community Center, in conjunction with the Board of Education's adult-outreach program, has scheduled a week-long series of lectures this fall on topics in physics. The goals are to increase awareness of the physical sciences and to attract renowned scientists (including Dr. Denton) to the community. Each of the five lectures will be held on a different weekday, and each will feature a different physicist lecturing on a different topic (one is magnetism). So far, the community has shown great interest in their upcoming physical training! From the information provided, determine the physicist who will speak on each weekday and the topic of his or her lecture.
   - **a.** Dr. Hoo, who is from Yale, will not be lecturing on Thursday. Dr. Zhivago's lecture will be exactly three days after the lecture on chaos theory.
   - **b.** If Dr. Jay is lecturing on Thursday, then the person giving the kinetic-energy lecture will appear on Tuesday; otherwise, Dr. Jay will speak on Tuesday, and kinetic energy will be the topic of Monday's lecture.
   - **c.** Dr. Know (who is not giving the lecture on quantum mechanics) is not the Harvard physicist who will speak on Monday. The photonics lecture will not be given on either Wednesday or Thursday.

		PHYSICIST					TOPIC				
		DR. DENTON	DR. HOO	DR. JAY	DR. KNOW	DR. ZHIVAGO	CHAOS THEORY	KINETIC ENERGY	MAGNETISM	PHOTONICS	QUANTUM MECHANICS
WEEKDAY	MONDAY										
	TUESDAY										
	WEDNESDAY										
	THURSDAY										
	FRIDAY										
TOPIC	CHAOS THEORY										
	KINETIC ENERGY										
	MAGNETISM										
	PHOTONICS										
	QUANTUM MECHANICS										

**World-Class Logic Problems Special*, Summer 2003, p. 6.
†*World-Class Logic Problems Special*, October 2003, p. 23.

3. ***What's in Store?**** I had a day off from work yesterday, so I figured it was the perfect time to do some shopping. I hit the road shortly after breakfast and visited five stores (one was Bullseye). At each shop, I had intended to buy a different one of five items (a pair of andirons, a Crock-pot, pruning shears, a pair of sneakers, or a toaster oven). Unfortunately, no store had the item I was looking for in stock. The trips weren't a total loss, however, as I purchased a different item (a CD, a fondue pot, a garden gnome, spark plugs, or a winter coat) that had caught my eye in each store. Despite my failure to acquire any of the things I had sought, there were a couple of positive outcomes. I now have some nifty new things that I know I'll enjoy, and I have a shopping list written and ready to go for my next day off! From the information provided, determine the order in which I visited the five stores, as well as the item I sought and the item I bought at each store.

a. I went to the store where I bought a CD (which isn't where I sought pruning shears) immediately after I visited PJ Nickle but immediately before I went to the shop where I intended to buy a Crock-pot.

b. I didn't purchase the fondue pot at Costington's. I went into one store intending to buy a Crock-pot, but came out with a garden gnome instead. I didn't go to PJ Nickle for a pair of andirons.

c. The store at which I sought a toaster oven (which wasn't the third one I visited) isn't the place where I eventually bought a winter coat. Neither Lacy's nor S-Mart was the fourth shop I visited.

d. I went to Costington's immediately after I visited the shop where I sought pruning shears (which wasn't S-Mart) but immediately before I went to the store where I purchased a set of spark plugs.

		STORE					ITEM SOUGHT					ITEM BOUGHT				
		BULLSEYE	COSTINGTON'S	LACY'S	PJ NICKLE	S-MART	ANDIRONS	CROCK-POT	PRUNING SHEARS	SNEAKERS	TOASTER OVEN	CD	FONDUE POT	GARDEN GNOME	SPARK PLUGS	WINTER COAT
ORDER	FIRST															
	SECOND															
	THIRD															
	FOURTH															
	FIFTH															
ITEM BOUGHT	CD															
	FONDUE POT															
	GARDEN GNOME															
	SPARK PLUGS															
	WINTER COAT															
ITEM SOUGHT	ANDIRONS															
	CROCK-POT															
	PRUNING SHEARS															
	SNEAKERS															
	TOASTER OVEN															

4. ***High Five***† Otis Lifter is the elevator operator at Schwarzenbach Tower, downtown Brownsville's tallest building. Since he works in such a towering edifice, Otis gets a chance to chat with his passengers on the way to their destinations. Five people who always have a friendly word for Otis work on the Schwarzenbach's top floors. Each person works on a different floor, which is home to a different company (one is the Watershed Co.). Each company is in a different business (one is a real-estate agency). Otis is content with his job, but he'll be the first to tell you that, like any profession, it has its ups and downs! From the information provided, can you determine the floor (41st through 45th) to which Otis took each person, as well as the name of his or her company and the type of business it conducts?

a. Edwina's company is exactly 1 floor above Nelson & Leopold but exactly 1 floor below the accounting firm. Brierwood Ltd. is on the 44th floor.

b. Zed's company is exactly 1 floor above Ogden's. Keith's business is on the 45th floor. Trish works at the public-relations firm (which is exactly 2 floors above Glyptic).

c. Glyptic and the Thebes Group are the literary agency and the Web-design firm, in some order. The Web-design firm is not on the 41st floor.

		FLOOR					COMPANY					BUSINESS				
		41st	42nd	43rd	44th	45th	BRIERWOOD LTD.	GLYPTIC	NELSON & LEOPOLD	THEBES GROUP	WATERSHED CO.	ACCOUNTING	LITERARY AGENCY	PUBLIC RELATIONS	REAL ESTATE	WEB DESIGN
PERSON	EDWINA															
	KEITH															
	OGDEN															
	TRISH															
	ZED															
BUSINESS	ACCOUNTING															
	LITERARY AGENCY															
	PUBLIC RELATIONS															
	REAL ESTATE															
	WEB DESIGN															
COMPANY	BRIERWOOD LTD.															
	GLYPTIC															
	NELSON & LEOPOLD															
	THEBES GROUP															
	WATERSHED CO.															

**World-Class Logic Problems Special,* Autumn 2003, p. 20.
†*World-Class Logic Problems Special,* Autumn 2003, p. 6.

5. ***First Ratings**** At long last, Macrocosm Industries has released Q Sphere, its new video-game console. To fully demonstrate the Q Sphere's capabilities, each of Macrocosm's five Q Sphere games (including Idle Hands) is a different genre. Anxious to be the first publication to spotlight this new gaming system, *All Game* magazine featured reviews of the Q Sphere games in its latest issue. Each game was played extensively by a different *All Game* staff reviewer (including Chadwick) and given a different rating (from lowest to highest, "don't bother," "just okay," "pretty cool," "almost perfect," or "totally awesome"). In the end, though, true video-game aficionados will want to try all of the Q Sphere games for themselves, despite the ratings! From the information provided, determine the genre of the game reviewed by each *All Game* staff member (identified by first and last names—one surname is Ploof), as well as the rating given to each game.

   a. At least one game was given a lower rating than the sports game (which is called Pitching Duel). The person surnamed Corley reviewed the action game.

   b. The puzzle game's rating was "just okay," which was higher than the rating Darren Castles gave.

   c. King of the Road is the racing game. The person surnamed Munoz (who isn't Milton) gave the "totally awesome" rating.

   d. The person surnamed Gilligan reviewed the simulation game, which isn't Hypnotic Trace (which was rated "almost perfect").

   e. Alise gave one game a "pretty cool" rating. Kourtney spent many hours playing Fiji in order to write her review.

		LAST NAME					GAME					GENRE					RATING				
		CASTLES	CORLEY	GILLIGAN	MUNOZ	PLOOF	FIJI	HYPNOTIC TRACE	IDLE HANDS	KING OF THE ROAD	PITCHING DUEL	ACTION	PUZZLE	RACING	SIMULATION	SPORTS	DON'T BOTHER	JUST OKAY	PRETTY COOL	ALMOST PERFECT	TOTALLY AWESOME
FIRST NAME	ALISE																				
	CHADWICK																				
	DARREN																				
	KOURTNEY																				
	MILTON																				
RATING	DON'T BOTHER																				
	JUST OKAY																				
	PRETTY COOL																				
	ALMOST PERFECT																				
	TOTALLY AWESOME																				
GENRE	ACTION																				
	PUZZLE																				
	RACING																				
	SIMULATION																				
	SPORTS																				
GAME	FIJI																				
	HYPNOTIC																				
	IDLE HANDS																				
	KING OF THE ROAD																				
	PITCHING DUEL																				

DIRECTIONS FOR GROUP PROJECT

Construct your own logic puzzle.

***World-Class Logic Problems Special,* Autumn 2003, p. 7.

7

Sets and Probability

The study of probability begins with counting. An exercise in Section 2 of this chapter counts trucks carrying different combinations of early, late, and extra late peaches from the orchard to canning facilities. You'll see trees in another context in Section 5, where we use branching tree diagrams to calculate conditional probabilities.

In this chapter and the next, we introduce the basic ideas of probability theory, a branch of mathematics that has become increasingly important in management and in the biological and social sciences. Probability theory is valuable because it provides a way to deal with uncertainty. Since the language of sets and set operations is used in the study of probability, we begin there.

7.1 Sets

? THINK ABOUT IT

In how many ways can two candidates win the 50 states plus the District of Columbia in a U.S. presidential election?

Using knowledge of sets, we will answer this question in one of the exercises.

Think of a **set** as a well-defined collection of objects in which it is possible to determine if a given object is included in the collection. A set of coins might include one of each type of coin now put out by the U.S. government. Another set might be made up of all the students in your English class. By contrast, a collection of young adults does not constitute a set unless the designation "young adult" is clearly defined. For example, this set might be defined as those aged 18 to 29.

In mathematics, sets are often made up of numbers. The set consisting of the numbers 3, 4, and 5 is written

$$\{3, 4, 5\},$$

with set braces, $\{\ \}$, enclosing the numbers belonging to the set. The numbers 3, 4, and 5 are called the **elements** or **members** of this set. To show that 4 is an element of the set $\{3, 4, 5\}$, we use the symbol $\in$ and write

$$4 \in \{3, 4, 5\},$$

read "4 is an element of the set containing 3, 4, and 5." Also, $5 \in \{3, 4, 5\}$.

To show that 8 is *not* an element of this set, place a slash through the symbol:

$$8 \notin \{3, 4, 5\}.$$

Sets often are named with capital letters, so that if

$$B = \{5, 6, 7\},$$

then, for example, $6 \in B$ and $10 \notin B$.

It is possible to have a set with no elements. Some examples are the set of counting numbers less than one, the set of foreign-born presidents of the United States, and the set of men more than 10 feet tall. A set with no elements is called the **empty set** and is written $\emptyset$.

CAUTION Be careful to distinguish between the symbols 0, $\emptyset$, $\{0\}$, and $\{\emptyset\}$. The symbol 0 represents a *number*; $\emptyset$ represents a *set* with 0 elements; $\{0\}$ represents a set with one element, 0; and $\{\emptyset\}$ represents a set with one element, $\emptyset$. ■

We use the symbol $n(A)$ to indicate the *number* of elements in a finite set A. For example, if $A = \{a, b, c, d, e\}$, then $n(A) = 5$. Using this notation, we can write the information in the Caution on the previous page as $n(\emptyset) = 0$ and $n(\{0\}) = n(\{\emptyset\}) = 1$.

Two sets are *equal* if they contain the same elements. The sets $\{5, 6, 7\}$, $\{7, 6, 5\}$, and $\{6, 5, 7\}$ all contain exactly the same elements and are equal. In symbols,

$$\{5, 6, 7\} = \{7, 6, 5\} = \{6, 5, 7\}.$$

This means that the ordering of the elements in a set is unimportant. Note that each element of the set is only listed once. Sets that do not contain exactly the same elements are *not equal*. For example, the sets $\{5, 6, 7\}$ and $\{7, 8, 9\}$ do not contain exactly the same elements and thus are not equal. To indicate that these sets are not equal, we write

$$\{5, 6, 7\} \neq \{7, 8, 9\}.$$

Sometimes we are interested in a common property of the elements in a set, rather than a list of the elements. This common property can be expressed by using **set-builder notation**, for example,

$$\{x \mid x \text{ has property } P\}$$

(read "the set of all elements x such that x has property P") represents the set of all elements x having some stated property P.

EXAMPLE 1 **Sets**

Write the elements belonging to each set.

(a) $\{x \mid x \text{ is a natural number less than } 5\}$

▶**Solution** The natural numbers less than 5 make up the set $\{1, 2, 3, 4\}$.

(b) $\{x \mid x \text{ is a state that borders Florida}\}$

▶**Solution** The states that border Florida make up the set {Alabama, Georgia}.

The **universal set** for a particular discussion is a set that includes all the objects being discussed. In elementary school arithmetic, for instance, the set of whole numbers might be the universal set, while in a college algebra class the universal set might be the set of real numbers. The universal set will be specified when necessary, or it will be clearly understandable from the context of the problem.

Subsets

Sometimes every element of one set also belongs to another set. For example, if

$$A = \{3, 4, 5, 6\}$$

and

$$B = \{2, 3, 4, 5, 6, 7, 8\},$$

then every element of A is also an element of B. This is an example of the following definition.

SUBSET

Set A is a **subset** of set B (written $A \subseteq B$) if every element of A is also an element of B. Set A is a *proper subset* (written $A \subset B$) if $A \subseteq B$ and $A \neq B$.

To indicate that A is *not* a subset of B, we write $A \not\subseteq B$.

EXAMPLE 2 **Sets**

Decide whether the following statements are *true* or *false*.

(a) $\{3, 4, 5, 6\} = \{4, 6, 3, 5\}$

▶**Solution** Both sets contain exactly the same elements, so the sets are equal and the given statement is true. (The fact that the elements are listed in a different order does not matter.)

(b) $\{5, 6, 9, 12\} \subseteq \{5, 6, 7, 8, 9, 10, 11\}$

▶**Solution** The first set is not a subset of the second because it contains an element, 12, that does not belong to the second set. Therefore, the statement is false.

By the definition of subset, the empty set (which contains no elements) is a subset of every set. That is, if A is any set, and the symbol $\emptyset$ represents the empty set, then $\emptyset \subseteq A$. Also, the definition of subset can be used to show that every set is a subset of itself; that is, if A is any set, then $A \subseteq A$.

For any set A,

$$\emptyset \subseteq A \qquad \textbf{and} \qquad A \subseteq A.$$

EXAMPLE 3 **Subsets**

List all possible subsets for each set.

(a) $\{7, 8\}$

▶**Solution** There are 4 subsets of $\{7, 8\}$:

$$\emptyset, \quad \{7\}, \quad \{8\}, \quad \text{and} \quad \{7, 8\}.$$

(b) $\{\text{a, b, c}\}$

▶**Solution** There are 8 subsets of $\{\text{a, b, c}\}$:

$$\emptyset, \quad \{a\}, \quad \{b\}, \quad \{c\}, \quad \{a, b\}, \quad \{a, c\}, \quad \{b, c\}, \quad \text{and} \quad \{a, b, c\}.$$

A good way to find the subsets of $\{7, 8\}$ and the subsets of $\{\text{a, b, c}\}$ in Example 3 is to use a **tree diagram**—a systematic way of listing all the subsets of a given set. Figure 1 shows tree diagrams for finding the subsets of $\{7, 8\}$ and $\{\text{a, b, c}\}$.

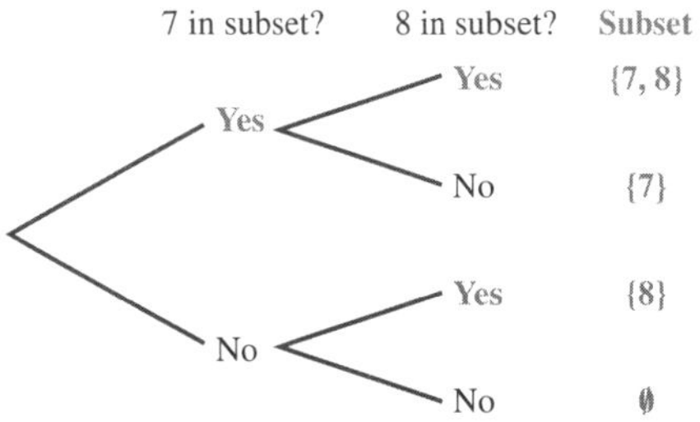

(a) Find the subsets of {7, 8}.

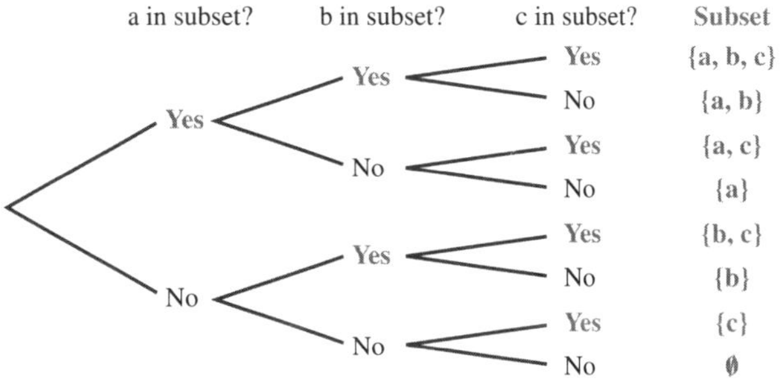

(b) Find the subsets of {a, b, c}.

FIGURE 1

As Figure 1 shows, there are two possibilities for each element (either it's in the subset or it's not), so a set with 2 elements has $2 \cdot 2 = 2^2 = 4$ subsets, and a set with 3 elements has $2^3 = 8$ subsets. This idea can be extended to a set with any finite number of elements, which leads to the following conclusion.

> A set of k distinct elements has 2^k subsets.

In other words, if $n(A) = k$, then $n(\text{the set of all subsets of } A) = 2^k$.

EXAMPLE 4 Subsets

Find the number of subsets for each set.

(a) {3, 4, 5, 6, 7}

▶**Solution** This set has 5 elements; thus, it has 2^5 or 32 subsets.

(b) $\{x \mid x \text{ is a day of the week}\}$

▶**Solution** This set has 7 elements and therefore has $2^7 = 128$ subsets.

(c) ∅

▶**Solution** Since the empty set has 0 elements, it has $2^0 = 1$ subset—itself.

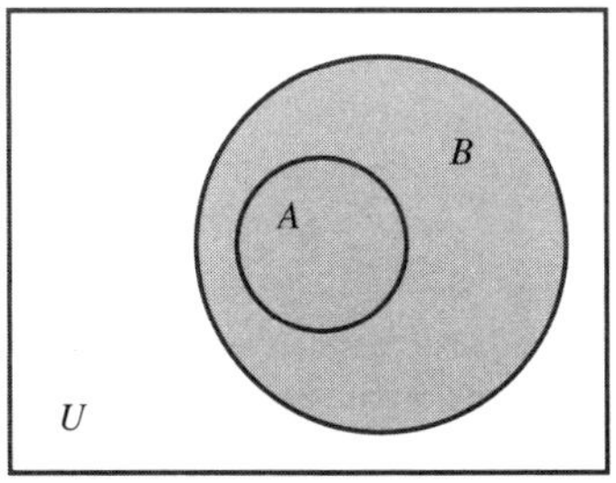

$A \subseteq B$

FIGURE 2

Figure 2 shows a set A that is a subset of set B. The rectangle represents the universal set, U. Such diagrams, called **Venn diagrams**—after the English logician John Venn (1834–1923), who invented them in 1876—are used to help illustrate relationships among sets. Venn diagrams are very similar to Euler diagrams, described in Section 6.6. Euler diagrams are used in logic to denote variables having a certain property or not, while Venn diagrams are used in the context of sets to denote something being an element of a set or not.

Set Operations It is possible to form new sets by combining or manipulating one or more existing sets. Given a set A and a universal set U, the set of all elements of U that do *not* belong to A is called the *complement* of set A. For example, if set A is the set of all the female students in a class, and U is the set of all students in the class, then the complement of A would be the set of all male students in the class. The complement of set A is written A', read "A-prime."

COMPLEMENT OF A SET

Let A be any set, with U representing the universal set. Then the **complement** of A, colored pink in the figure, is

$$A' = \{x \mid x \notin A \text{ and } x \in U\}.$$

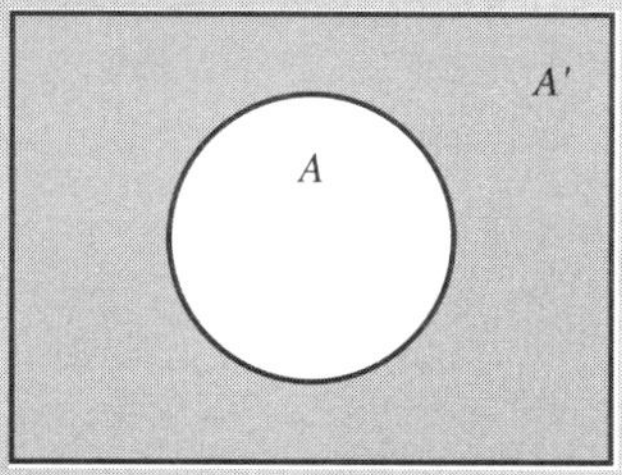

EXAMPLE 5 **Set Operations**

Let $U = \{1, 2, 3, 4, 5, 6, 7, 8, 9, 10, 11\}$, $A = \{1, 2, 4, 5, 7\}$, and $B = \{2, 4, 5, 7, 9, 11\}$. Find each set.

(a) A'

▶**Solution** Set A' contains the elements of U that are not in A.

$$A' = \{3, 6, 8, 9, 10, 11\}$$

(b) $B' = \{1, 3, 6, 8, 10\}$

(c) $\emptyset' = U$ and $U' = \emptyset$

(d) $(A')' = A$

Given two sets A and B, the set of all elements belonging to *both* set A and set B is called the *intersection* of the two sets, written $A \cap B$. For example, the elements that belong to both set $A = \{1, 2, 4, 5, 7\}$ and set $B = \{2, 4, 5, 7, 9, 11\}$ are 2, 4, 5, and 7, so that

$$A \cap B = \{1, 2, 4, 5, 7\} \cap \{2, 4, 5, 7, 9, 11\} = \{2, 4, 5, 7\}.$$

INTERSECTION OF TWO SETS

The **intersection** of sets A and B, shown in green in the figure, is

$$\boldsymbol{A \cap B = \{x \mid x \in A \text{ and } x \in B\}}.$$

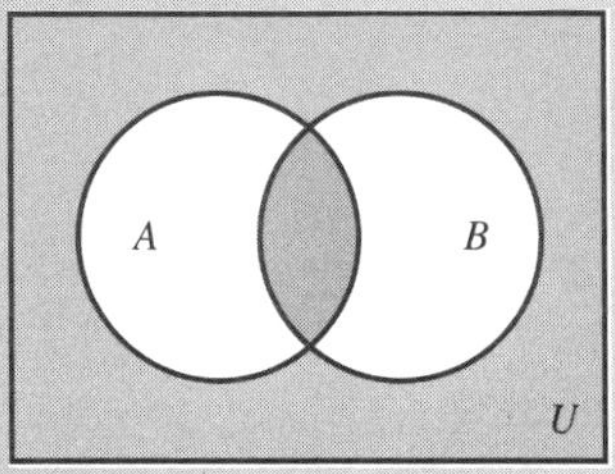

EXAMPLE 6 **Set Operations**

Let $A = \{3, 6, 9\}$, $B = \{2, 4, 6, 8\}$, and the universal set $U = \{0, 1, 2, \ldots, 10\}$. Find each set.

(a) $A \cap B$

►Solution

$$A \cap B = \{3, 6, 9\} \cap \{2, 4, 6, 8\} = \{6\}$$

(b) $A \cap B'$

►Solution

$$A \cap B' = \{3, 6, 9\} \cap \{0, 1, 3, 5, 7, 9, 10\} = \{3, 9\}$$

Two sets that have no elements in common are called *disjoint sets*. For example, there are no elements common to both $\{50, 51, 54\}$ and $\{52, 53, 55, 56\}$, so these two sets are disjoint, and

$$\{50, 51, 54\} \cap \{52, 53, 55, 56\} = \emptyset.$$

This result can be generalized as follows.

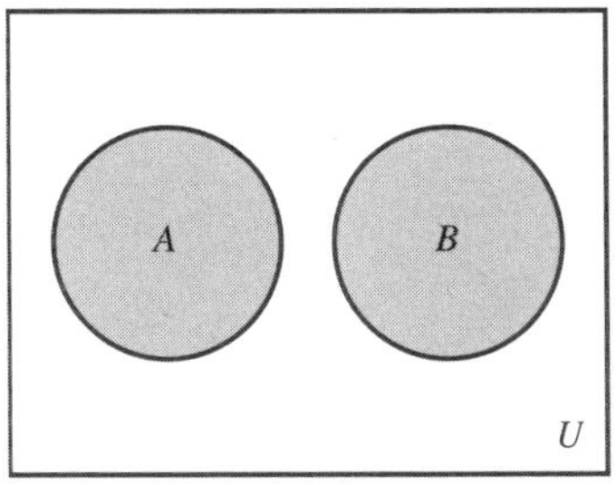

A and *B* are disjoint sets.

FIGURE 3

DISJOINT SETS

For any sets A and B, if A and B are **disjoint sets**, then $A \cap B = \emptyset$.

Figure 3 shows a pair of disjoint sets.

The set of all elements belonging to set A, to set B, or to both sets is called the *union* of the two sets, written $A \cup B$. For example,

$$\{1, 3, 5\} \cup \{3, 5, 7, 9\} = \{1, 3, 5, 7, 9\}.$$

UNION OF TWO SETS

The **union** of sets A and B, shown in blue in the figure, is

$$\mathbf{A \cup B = \{x \mid x \in A \text{ or } x \in B\}}.$$

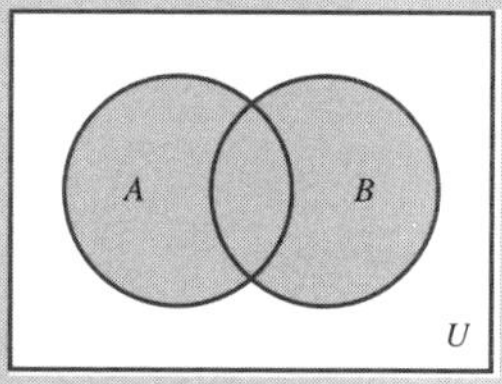

EXAMPLE 7

Union of Sets

Let $A = \{1, 3, 5, 7, 9, 11\}$, $B = \{3, 6, 9, 12\}$, $C = \{1, 2, 3, 4, 5\}$, and the universal set $U = \{0, 1, 2, \ldots, 12\}$. Find each set.

(a) $A \cup B$

▶**Solution** Begin by listing the elements of the first set, $\{1, 3, 5, 7, 9, 11\}$. Then include any elements from the second set *that are not already listed*. Doing this gives

$$A \cup B = \{1, 3, 5, 7, 9, 11\} \cup \{3, 6, 9, 12\} = \{1, 3, 5, 7, 9, 11, 6, 12\}$$
$$= \{1, 3, 5, 6, 7, 9, 11, 12\}.$$

(b) $(A \cup B) \cap C'$

▶**Solution** Begin with the expression in parentheses, which we calculated in part (a), and then intersect this with C'.

$$(A \cup B) \cap C' = \{1, 3, 5, 6, 7, 9, 11, 12\} \cap \{0, 6, 7, 8, 9, 10, 11, 12\}$$
$$= \{6, 7, 9, 11, 12\}$$

NOTE

1. As Example 7 shows, when forming sets, do not list the same element more than once. In our final answer, we listed the elements in numerical order to make it easier to see what elements are in the set, but the set is the same, regardless of the order of the elements.
2. As shown in the definitions, an element is in the *intersection* of sets A and B if it is in A *and* B. On the other hand, an element is in the *union* of sets A and B if it is in A *or* B (or both).

EXAMPLE 8

Stocks

The following table gives the 52-week high and low prices, the closing price, and the change from the previous day for six stocks in the Standard & Poor's 100 on April 11, 2006.*

Stock	Low	High	Close	Change
AT&T	21.75	28.82	25.78	−0.25
CocaCola	39.36	45.26	41.29	−0.74
Dell Inc.	28.60	41.99	29.66	−0.06
Disney	22.89	28.85	27.77	−0.02
Office Max Inc.	24.20	33.95	32.09	+0.02
Texas Instruments	22.51	34.95	32.89	+0.13

**The New York Times,* April 12, 2006, p. C8.

Let the universal set U consist of the six stocks listed in the table. Let A contain all stocks with a high price greater than \$34, B all stocks with a closing price between \$26 and \$30, and C all stocks with a positive price change. Find the following.

(a) A'

►**Solution** Set A' contains all the listed stocks that are not in set A, or those with a high price less than or equal to \$34, so

$$A' = \{\text{AT\&T, Disney, Office Max Inc.}\}.$$

(b) $A \cap C$

►**Solution** The intersection of A and C will contain those stocks that are in both sets A and C, or those with a high price greater than \$34 and a positive price change.

$$A \cap C = \{\text{Texas Instruments}\}$$

(c) $A \cup B$

►**Solution** The union of A and B contains all stocks with a high price greater than \$34 and a closing price between \$26 and \$30.

$$A \cup B = \{\text{CocaCola, Dell Inc., Disney, Texas Instruments}\}$$

EXAMPLE 9 **Employment**

A department store classifies credit applicants by gender, marital status, and employment status. Let the universal set be the set of all applicants, M be the set of male applicants, S be the set of single applicants, and E be the set of employed applicants. Describe each set in words.

(a) $M \cap E$

►**Solution** The set $M \cap E$ includes all applicants who are both male *and* employed; that is, employed male applicants.

(b) $M' \cup S$

►**Solution** This set includes all applicants who are female (not male) *or* single. *All* female applicants and *all* single applicants are in this set.

(c) $M' \cap S'$

►**Solution** These applicants are female *and* married (not single); thus, $M' \cap S'$ is the set of all married female applicants.

(d) $M \cup E'$

►**Solution** $M \cup E'$ is the set of applicants that are male *or* unemployed. The set includes *all* male applicants and *all* unemployed applicants.

7.1 Exercises

In Exercises 1–9, write true or false for each statement.

1. $3 \in \{2, 5, 7, 9, 10\}$

2. $6 \in \{-2, 6, 9, 5\}$

3. $9 \notin \{2, 1, 5, 8\}$

4. $3 \notin \{7, 6, 5, 4\}$

5. $\{2, 5, 8, 9\} = \{2, 5, 9, 8\}$

6. $\{3, 7, 12, 14\} = \{3, 7, 12, 14, 0\}$

7. $\{\text{all whole numbers greater than 7 and less than 10}\} = \{8, 9\}$

8. $\{x \mid x \text{ is an odd integer; } 6 \leq x \leq 18\} = \{7, 9, 11, 15, 17\}$

9. $0 \in \emptyset$

10. What is set-builder notation? Give an example.

Let $A = \{2, 4, 6, 8, 10, 12\}$, $B = \{2, 4, 8, 10\}$, $C = \{4, 8, 12\}$, $D = \{2, 10\}$, $E = \{6\}$, *and* $U = \{2, 4, 6, 8, 10, 12, 14\}$. *Insert* $\subseteq$ *or* $\not\subseteq$ *to make the statement true.*

11. A __ U

12. E __ A

13. A __ E

14. B __ C

15. $\emptyset$ __ A

16. $\{0, 2\}$ __ D

17. D __ B

18. A __ C

19. Repeat Exercises 11–18 except insert $\subset$ or $\not\subset$ to make the statement true.

Insert a number in each blank to make the statement true, using the sets for Exercises 11–18.

20. There are exactly __ subsets of A.

21. There are exactly __ subsets of B.

22. There are exactly __ subsets of C.

23. There are exactly __ subsets of D.

24. Describe the intersection and union of sets. How do they differ?

Insert $\cap$ *or* $\cup$ *to make each statement true.*

25. $\{5, 7, 9, 19\}$ __ $\{7, 9, 11, 15\} = \{7, 9\}$

26. $\{8, 11, 15\}$ __ $\{8, 11, 19, 20\} = \{8, 11\}$

27. $\{2, 1, 7\}$ __ $\{1, 5, 9\} = \{1\}$

28. $\{6, 12, 14, 16\}$ __ $\{6, 14, 19\} = \{6, 14\}$

29. $\{3, 5, 9, 10\}$ __ $\emptyset = \emptyset$

30. $\{3, 5, 9, 10\}$ __ $\emptyset = \{3, 5, 9, 10\}$

31. $\{1, 2, 4\}$ __ $\{1, 2, 4\} = \{1, 2, 4\}$

32. Is it possible for two nonempty sets to have the same intersection and union? If so, give an example.

Let $U = \{1, 2, 3, 4, 5, 6, 7, 8, 9\}$, $X = \{2, 4, 6, 8\}$, $Y = \{2, 3, 4, 5, 6\}$, *and* $Z = \{1, 2, 3, 8, 9\}$. *List the members of each set, using set braces.*

33. $X \cap Y$

34. $X \cup Y$

35. X'

36. Y'

37. $X' \cap Y'$

38. $X' \cap Z$

39. $Y \cap (X \cup Z)$

40. $X' \cap (Y' \cup Z)$

41. $(X \cap Y') \cup Z'$

42. **a.** In Example 6, what set do you get when you calculate $(A \cap B) \cup (A \cap B')$?

b. Explain in words why $(A \cap B) \cup (A \cap B') = A$.

Let $U = \{$all students in this school$\}$, $M = \{$all students taking this course$\}$, $N = \{$all students taking accounting$\}$, *and* $P = \{$all students taking zoology$\}$. *Describe each set in words.*

43. M'

44. $M \cup N$

45. $N \cap P$

46. $N' \cap P'$

47. Refer to the sets listed for Exercises 11–18. Which pairs of sets are disjoint?

48. Refer to the sets listed for Exercises 33–41. Which pairs are disjoint?

Refer to Example 8 in the text. Describe each set in Exercises 49–52 in words; then list the elements of each set.

49. B'

50. $A \cap B$

51. $(A \cap B)'$

52. $(A \cup C)'$

53. Let $A = \{1, 2, 3, \{3\}, \{1, 4, 7\}\}$. Answer each of the following as *true* or *false*.

a. $1 \in A$ **b.** $\{3\} \in A$ **c.** $\{2\} \in A$ **d.** $4 \in A$

e. $\{\{3\}\} \subset A$ **f.** $\{1, 4, 7\} \in A$ **g.** $\{1, 4, 7\} \subseteq A$

54. Let $B = \{a, b, c, \{d\}, \{e, f\}\}$. Answer each of the following as *true* or *false*.

a. $a \in B$ **b.** $\{b, c, d\} \subset B$ **c.** $\{d\} \in B$ **d.** $\{d\} \subseteq B$

e. $\{e, f\} \in B$ **f.** $\{a, \{e, f\}\} \subset B$ **g.** $\{e, f\} \subset B$

Applications

BUSINESS AND ECONOMICS

*Mutual Funds The table below shows the top five holdings of four major mutual funds on January 27, 2006.**

Vanguard 500	Janus Fund	Fidelity Magellan	T. Rowe Price Blue Chip Growth Fund
General Electric Co.	Boeing Co.	Nokia Corp.	UnitedHealth Group
ExxonMobil Corp.	Procter & Gamble	UnitedHealth Group	General Electric Co.
Citigroup, Inc.	Yahoo!, Inc.	Schlumberger Ltd.	Microsoft Corp.
Microsoft Corp.	UnitedHealth Group	Google Inc.	Citigroup, Inc.
Procter & Gamble	Microsoft Corp.	General Electric Co.	American International Group

Let U be the smallest possible set that includes all the corporations listed, and V, J, F, and T be the set of top-five holdings for each mutual fund, respectively. Find each set.

55. $V \cap J$

56. $V \cap (F \cup T)$

57. $(J \cup F)'$

58. $J' \cap T'$

LIFE SCIENCES

Health The following table shows some symptoms of an overactive thyroid and an underactive thyroid.†

Underactive Thyroid	Overactive Thyroid
Sleepiness, s	Insomnia, i
Dry hands, d	Moist hands, m
Intolerance of cold, c	Intolerance of heat, h
Goiter, g	Goiter, g

Let U be the smallest possible set that includes all the symptoms listed, N be the set of symptoms for an underactive thyroid, and O be the set of symptoms for an overactive thyroid. Find each set.

59. O'

60. N'

61. $N \cap O$

62. $N \cup O$

63. $N \cap O'$

SOCIAL SCIENCES

?

64. *Electoral College* U.S. presidential elections are decided by the Electoral College, in which each of the 50 states, plus the District of Columbia, gives all of its votes to a candidate.‡ Ignoring the number of votes each state has in the Electoral College, but including all possible combinations of states that could be won by either candidate, how many outcomes are possible in the Electoral College if there are two candidates? (*Hint:* The states that can be won by a candidate form a subset of all the states.)

GENERAL INTEREST

65. *Musicians* A concert featured a cellist, a flutist, a harpist, and a vocalist. Throughout the concert, different subsets of

*Top five holdings found for each fund at www.vanguard.com, ww4.janus.com, www.fidelity.com, www.troweprice.com, respectively.

†*The Merck Manual of Diagnosis and Therapy,* 16th ed., Merck Research Laboratories, 1992, pp. 1075 and 1080.

‡The exceptions are Maine and Nebraska, which allocate their electoral college votes according to the winner in each congressional district.

the four musicians performed together, with at least two musicians playing each piece. How many subsets of at least two are possible?

66. *Cat Food* Suppose 9 flavors of cat food are available in a store. Euclid, the mathematical cat, could like all 9 flavors, or none, or any combination of selected flavors. How many possibilities are there for the set of flavors that Euclid likes? (*Hint:* Each set of flavors is a subset of the original 9 flavors.)

Pay-Cable Services The following table lists the top five pay-cable services for 2004–2005. Use this information for Exercises 67–72.*

Network	Subscribers (millions)	Content
The Disney Channel	85.0	Movies, cartoons
Showtime	39.5	Movies, variety, comedy, sports
HBO	39.0	Movies, variety, sports, documentaries
Encore	24.5	Movies
Starz	14.0	Movies

List the elements of the following sets.

67. F, the set of networks with more than 20 million subscribers

68. G, the set of networks that feature sports

69. H, the set of networks that feature only movies

70. $F \cap H$ **71.** $G \cup H$ **72.** G'

73. *Games* In David Gale's game of Subset Takeaway, the object is for each player, at his or her turn, to pick a nonempty proper subset of a given set subject to the condition that no subset chosen earlier by either player can be a subset of the newly chosen set.† The winner is the last person who can make a legal move. Consider the set $A = \{1, 2, 3\}$. Suppose Joe and Dorothy are playing the game and Dorothy goes first. If she chooses the proper subset $\{1\}$, then Joe cannot choose any subset that includes the element 1. Joe can, however, choose $\{2\}$ or $\{3\}$ or $\{2, 3\}$. Develop a strategy for Joe so that he can always win the game if Dorothy goes first.

States In the following list of states, let A = {states whose name contains the letter e}, *let* B = {states with a population of more than 4,000,000}, *and* C = {states with an area greater than 40,000 square miles}.‡

State	Population (1000s)	Area (sq. mi.)
Alabama	4447	52,419
Alaska	627	663,267
Colorado	4301	104,094
Florida	15,982	65,755
Hawaii	1212	10,931
Indiana	6080	36,418
Kentucky	4042	40,409
Maine	1275	35,385
Nebraska	1711	77,354
New Jersey	8414	8721

74. **a.** Describe in words the set $A \cup (B \cap C)'$.

b. List all elements in the set $A \cup (B \cap C)'$.

75. **a.** Describe in words the set $(A \cup B)' \cap C$.

b. List all elements in the set $(A \cup B)' \cap C$.

**The New York Times 2006 Almanac,* p. 401.

†Stewart, Ian, "Mathematical Recreations: A Strategy for Subsets," *Scientific American,* Mar. 2000, pp. 96–98.

‡*The New York Times 2006 Almanac,* pp. 177–211.

7.2 Applications of Venn Diagrams

? THINK ABOUT IT **The responses to a survey of 100 households show that 21 have a DVD player, 56 have a videocassette recorder, and 12 have both. How many have neither a DVD player nor a videocassette recorder?**

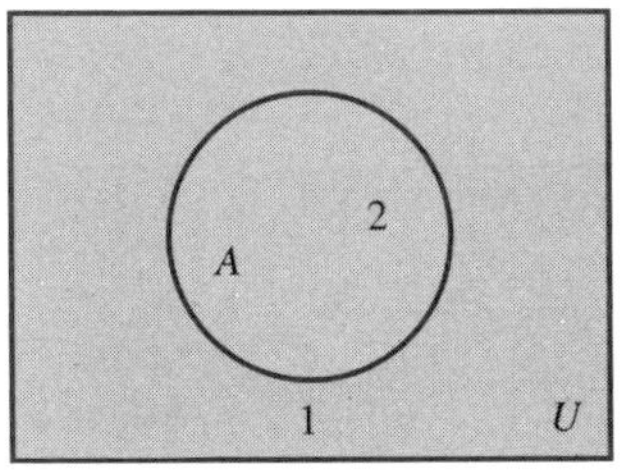

One set leads to 2 regions (numbering is arbitrary).

FIGURE 4

It is difficult to answer this question from the given information. In this section we show how a Venn diagram can be used to sort out such information, and later in this section, we are able to answer this question.

Venn diagrams were used in the previous section to illustrate set union and intersection. The rectangular region of a Venn diagram represents the universal set U. Including only a single set A inside the universal set, as in Figure 4, divides U into two regions. Region 1 represents those elements of U outside set A (that is, the elements in A'), and region 2 represents those elements belonging to set A. (The numbering of these regions is arbitrary.)

The Venn diagram in Figure 5(a) shows two sets inside U. These two sets divide the universal set into four regions. As labeled in Figure 5(a), region 1 represents the set whose elements are outside both set A and set B. Region 2 shows the set whose elements belong to A and not to B. Region 3 represents the set whose elements belong to both A and B. Which set is represented by region 4? (Again, the labeling is arbitrary.)

Two other situations can arise when representing sets by Venn diagrams. If it is known that $A \cap B = \emptyset$, then the Venn diagram is drawn as in Figure 5(b). If it is known that $A \subseteq B$, then the Venn diagram is drawn as in Figure 5(c). For the material presented throughout this chapter we will only refer to Venn diagrams like the one in Figure 5(a), and note that some of the regions of the Venn diagram may be equal to the null set.

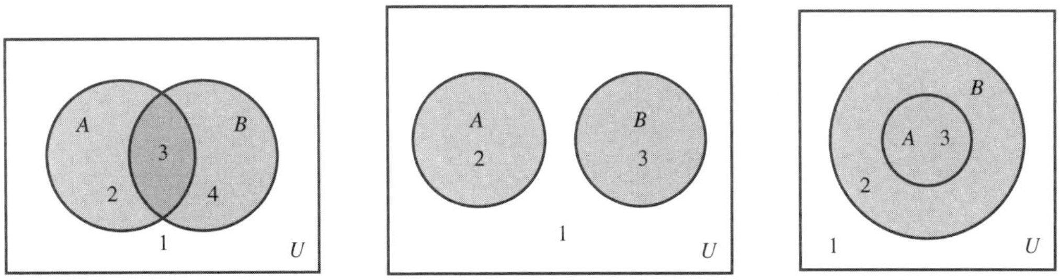

Two sets lead to 4 regions (numbering is arbitrary).

(a)

Two sets lead to 3 regions (numbering is arbitrary).

(b)

Two sets lead to 3 regions (numbering is arbitrary).

(c)

FIGURE 5

EXAMPLE 1 **Venn Diagrams**

Draw Venn diagrams similar to Figure 5(a) and shade the regions representing each set.

(a) $A' \cap B$

▶**Solution** Set A' contains all the elements outside set A. As labeled in Figure 5(a), A' is represented by regions 1 and 4. Set B is represented by regions 3 and 4. The intersection of sets A' and B, the set $A' \cap B$, is given by the region common to the combined regions 1 and 4 and the combined regions 3 and 4. The result is the set represented by region 4, which is blue in Figure 6. When looking for the intersection, remember to choose the area that is in one region *and* the other region.

In addition to the fact that region 4 in Figure 6 is $A' \cap B$, notice that region 1 is $A' \cap B'$, region 2 is $A \cap B'$, and region 3 is $A \cap B$.

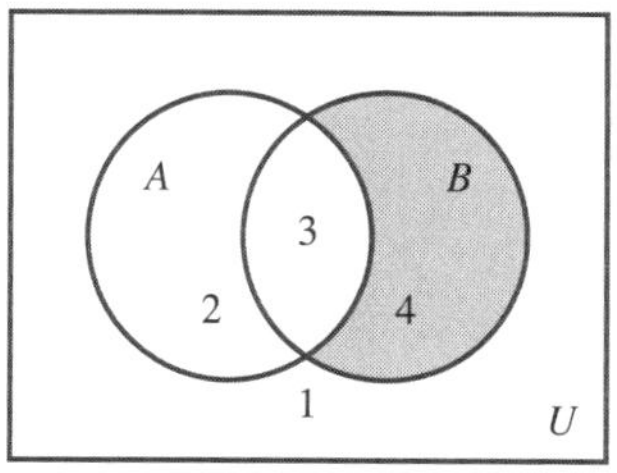

FIGURE 6

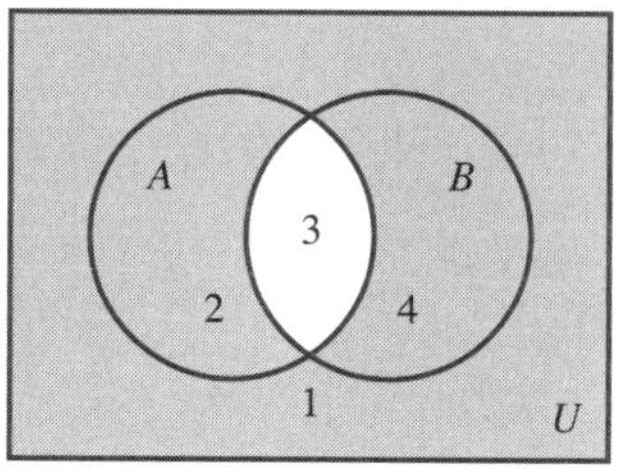

FIGURE 7

(b) $A' \cup B'$

▶**Solution** Again, set A' is represented by regions 1 and 4, and set B' by regions 1 and 2. To find $A' \cup B'$, identify the region that represents the set of all elements in A', B', or both. The result, which is blue in Figure 7, includes regions 1, 2, and 4. When looking for the union, remember to choose the area that is in one region *or* the other region (or both).

Venn diagrams also can be drawn with three sets inside U. These three sets divide the universal set into eight regions, which can be numbered (arbitrarily) as in Figure 8.

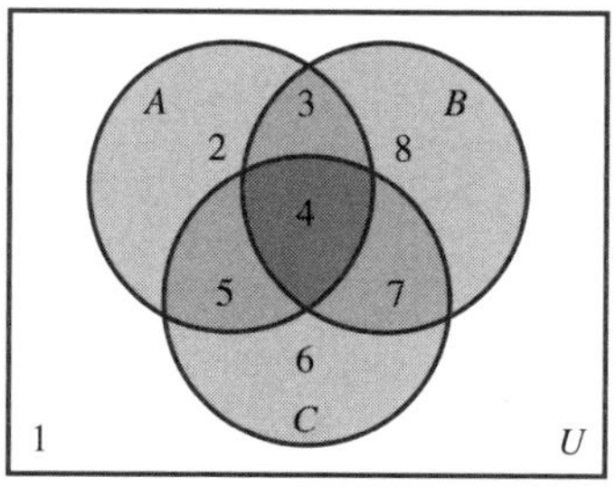

Three sets lead to 8 regions.

FIGURE 8

EXAMPLE 2 **Venn Diagram**

In a Venn diagram, shade the region that represents $A' \cup (B \cap C')$.

▶**Solution** First find $B \cap C'$. Set B is represented by regions 3, 4, 7, and 8, and set C' by regions 1, 2, 3, and 8. The overlap of these regions (regions 3 and 8)

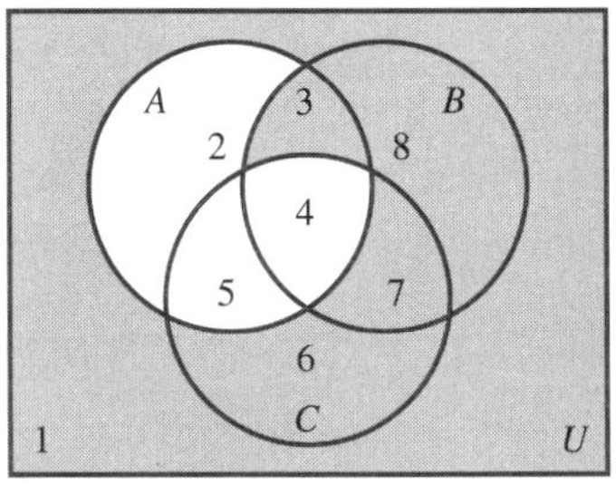

FIGURE 9

represents the set $B \cap C'$. Set A' is represented by regions 1, 6, 7, and 8. The union of the set represented by regions 3 and 8 and the set represented by regions 1, 6, 7, and 8 is the set represented by regions 1, 3, 6, 7, and 8, which are blue in Figure 9.

Applications We can now use a Venn diagram to answer the question posed at the beginning of this section. A researcher collecting data on 100 households finds that

21 have a DVD player;

56 have a videocassette recorder (VCR); and

12 have both.

The researcher wants to answer the following questions.

?

(a) How many do not have a VCR?

(b) How many have neither a DVD player nor a VCR?

(c) How many have a DVD player but not a VCR?

►**Solution** A Venn diagram like the one in Figure 10 will help sort out the information. In Figure 10(a), we put the number 12 in the region common to both a VCR and a DVD player, because 12 households have both. Of the 21 with a DVD player, $21 - 12 = 9$ have no VCR, so in Figure 10(b) we put 9 in the region for a DVD but no VCR. Similarly, $56 - 12 = 44$ households have a VCR but not a DVD player, so we put 44 in that region. Finally, the diagram shows that $100 - 44 - 12 - 9 = 35$ households have neither a VCR nor a DVD player. Now we can answer the questions:

(a) $35 + 9 = 44$ do not have a VCR.

(b) 35 have neither.

(c) 9 have a DVD player but not a VCR.

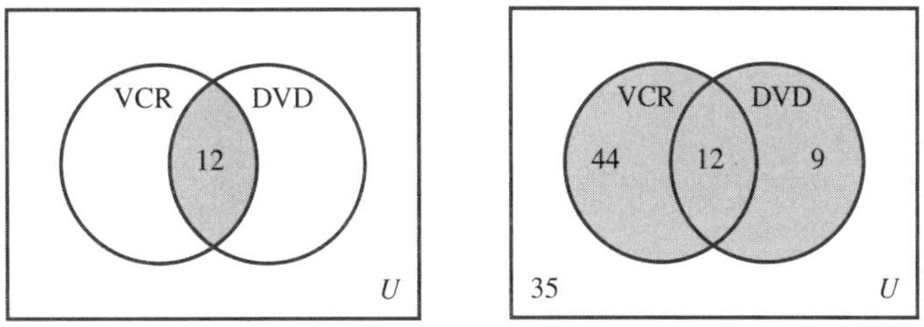

(a) (b)

FIGURE 10

EXAMPLE 3 **Magazines**

A survey of 77 freshman business students at a large university produced the following results.

25 of the students read *Business Week*;

19 read *The Wall Street Journal*;

27 do not read *Fortune*;

11 read *Business Week* but not *The Wall Street Journal*;

11 read *The Wall Street Journal* and *Fortune*;

13 read *Business Week* and *Fortune*;

9 read all three.

Use this information to answer the following questions.

(a) How many students read none of the publications?

(b) How many read only *Fortune*?

(c) How many read *Business Week* and *The Wall Street Journal*, but not *Fortune*?

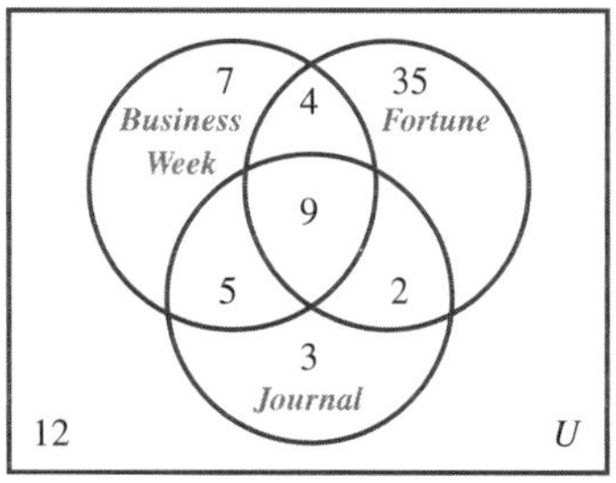

FIGURE 11

▶**Solution** Since 9 students read all three publications, begin by placing 9 in the area that belongs to all three regions, as shown in Figure 11. Of the 13 students who read *Business Week* and *Fortune*, 9 also read *The Wall Street Journal*. Therefore, only $13 - 9 = 4$ read just *Business Week* and *Fortune*. Place the number 4 in the area of Figure 11 common only to *Business Week* and *Fortune* readers.

In the same way, place $11 - 9 = 2$ in the region common only to *Fortune* and *The Wall Street Journal*. Of the 11 students who read *Business Week* but not *The Wall Street Journal*, 4 read *Fortune*, so place $11 - 4 = 7$ in the region for those who read only *Business Week*.

The data show that 25 students read *Business Week*. However, $7 + 4 + 9 = 20$ readers have already been placed in the region representing *Business Week*. The balance of this region will contain only $25 - 20 = 5$ students. These students read *Business Week* and *The Wall Street Journal*, but not *Fortune*. In the same way, $19 - (5 + 9 + 2) = 3$ students read only *The Wall Street Journal*.

Using the fact that 27 of the 77 students do not read *Fortune*, we know that 50 do read *Fortune*. We already have $4 + 9 + 2 = 15$ students in the region representing *Fortune*, leaving $50 - 15 = 35$ who read only *Fortune*.

A total of $7 + 4 + 35 + 5 + 9 + 2 + 3 = 65$ students are placed in the three circles in Figure 11. Since 77 students were surveyed, $77 - 65 = 12$ students read none of the three publications, and 12 is placed outside all three regions.

Now Figure 11 can be used to answer the questions asked above.

(a) There are 12 students who read none of the three publications.

(b) There are 35 students who read only *Fortune*.

(c) The overlap of the regions representing readers of *Business Week* and *The Wall Street Journal* shows that 5 students read *Business Week* and *The Wall Street Journal* but not *Fortune*.

CAUTION A common error in solving problems of this type is to make a circle represent one set and another circle represent its complement. In Example 3, with one circle representing those who read *Business Week*, we did not draw another for those who do not read *Business Week*. An additional circle is not only unnecessary (because those not in one set are automatically in the other) but very confusing, because the region outside or inside both circles must be empty. Similarly, if a problem involves men and women, do not draw one circle for men and another for women. Draw one circle; if you label it "women," for example, then men are automatically those outside the circle. ■

EXAMPLE 4 **Utility Maintenance**

Jeff Friedman is a section chief for an electric utility company. The employees in his section cut down trees, climb poles, and splice wire. Friedman reported the following information to the management of the utility.

"Of the 100 employees in my section,
45 can cut trees;
50 can climb poles;
57 can splice wire;
22 can climb poles but can't cut trees;
20 can climb poles and splice wire;
25 can cut trees and splice wire;
14 can cut trees and splice wire but can't climb poles;
9 can't do any of the three (management trainees)."

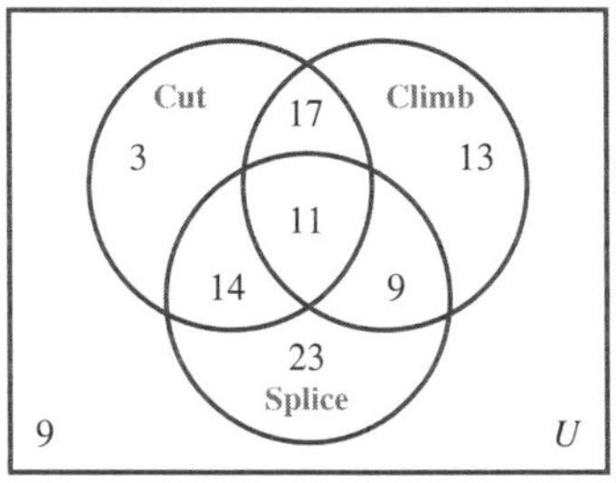

FIGURE 12

The data supplied by Friedman lead to the numbers shown in Figure 12. Add the numbers from all of the regions to get the total number of employees:

$$9 + 3 + 14 + 23 + 11 + 9 + 17 + 13 = 99.$$

Friedman claimed to have 100 employees, but his data indicate only 99. Management decided that Friedman didn't qualify as a section chief, and he was reassigned as a night-shift meter reader in Guam. (*Moral:* He should have taken this course.)

NOTE In all the examples above, we started with a piece of information specifying the relationship with all the categories. This is usually the best way to begin solving problems of this type.

As we saw in the previous section, we use the symbol $n(A)$ to indicate the *number* of elements in a finite set A. The following statement about the number of elements in the union of two sets will be used later in our study of probability.

UNION RULE FOR SETS

$$n(A \cup B) = n(A) + n(B) - n(A \cap B)$$

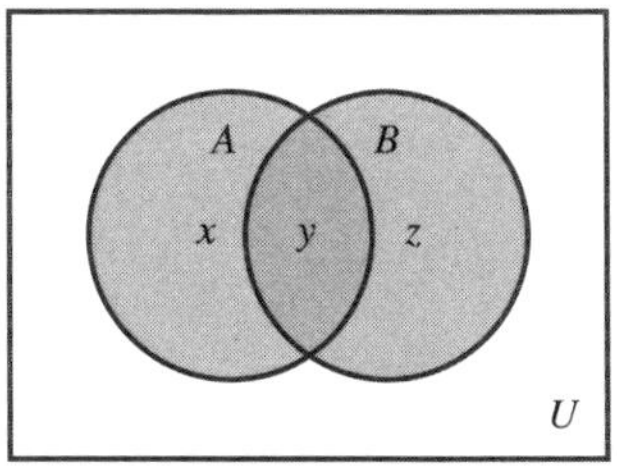

FIGURE 13

To prove this statement, let $x + y$ represent $n(A)$, y represent $n(A \cap B)$, and $y + z$ represent $n(B)$, as shown in Figure 13. Then

$$n(A \cup B) = x + y + z,$$

$$n(A) + n(B) - n(A \cap B) = (x + y) + (y + z) - y = x + y + z,$$

so

$$n(A \cup B) = n(A) + n(B) - n(A \cap B).$$

EXAMPLE 5 **School Activities**

A group of 10 students meet to plan a school function. All are majoring in accounting or economics or both. Five of the students are economics majors and 7 are majors in accounting. How many major in both subjects?

▶**Solution** Let A represent the set of accounting majors and B represent the set of economics majors. Use the union rule, with $n(A) = 5$, $n(B) = 7$, and $n(A \cup B) = 10$. Find $n(A \cap B)$.

$$n(A \cup B) = n(A) + n(B) - n(A \cap B)$$
$$10 = 5 + 7 - n(A \cap B),$$
so
$$n(A \cap B) = 5 + 7 - 10 = 2.$$

When A and B are disjoint, then $n(A \cap B) = 0$, so the union rule simplifies to $n(A \cup B) = n(A) + n(B)$.

CAUTION The rule $n(A \cup B) = n(A) + n(B)$ is *only* valid when A and B are disjoint. When A and B are *not* disjoint, use the rule $n(A \cup B) = n(A) + n(B) - n(A \cap B)$.

EXAMPLE 6 **Endangered Species**

The following table gives the number of threatened and endangered animal species in the world as of April 2006.*

	Endangered (*E*)	Threatened (*T*)	Totals
Amphibians and reptiles (*A*)	99	50	149
Arachnids and insects (*I*)	52	9	61
Birds (*B*)	252	21	273
Clams, crustaceans, and snails (*C*)	108	23	131
Fishes (*F*)	85	47	132
Mammals (*M*)	323	33	356
Totals	919	183	1102

Using the letters given in the table to denote each set, find the number of species in each of the following sets.

(a) $E \cap B$

▶**Solution** The set $E \cap B$ consists of all species that are endangered *and* are birds. From the table, we see that there are 252 such species.

(b) $E \cup B$

▶**Solution** The set $E \cup B$ consists of all species that are endangered *or* are birds. We include all 919 endangered species, plus the 21 bird species who are threatened but not endangered, for a total of 940. Alternatively, we could use the formula $n(E \cup B) = n(E) + n(B) - n(E \cap B) = 919 + 273 - 252 = 940$.

*U.S. Fish and Wildlife Service, http://ecos.fws.gov.

(c) $(F \cup M) \cap T'$

▶Solution Begin with the set $F \cup M$, which is all species that are fish or mammals. This consists of the four categories with 85, 47, 323, and 33 species. Of this set, take those that are *not* threatened, for a total of $85 + 323 = 408$ species. This is the number of species of fish and mammals that are not threatened.

EXAMPLE 7

Chat Rooms

Suppose that a group of 150 students have joined at least one of three chat rooms: one on auto-racing, one on bicycling, and one for college students. For simplicity, we will call these rooms A, B, and C. In addition,

90 students joined room A;

50 students joined room B;

70 students joined room C;

15 students joined rooms A and C;

12 students joined rooms B and C;

10 students joined all three rooms.

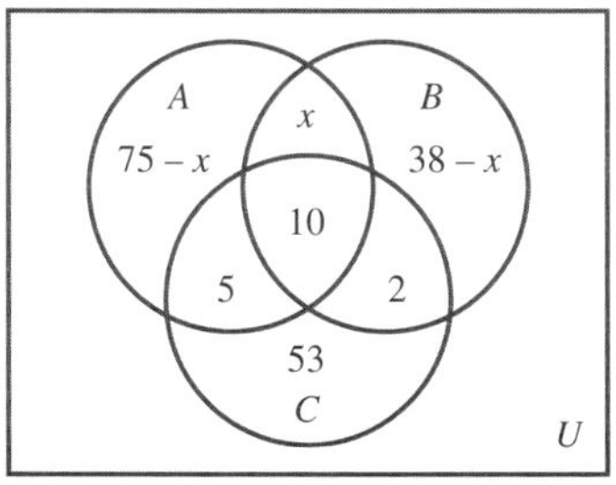

FIGURE 14

Determine how many students joined both chat rooms A and B.

▶Solution Since 10 students joined all three rooms, begin by placing 10 in the area that belongs to all three regions, as shown in Figure 14. Of the 15 students who joined rooms A and C, 10 also joined room B. Thus, only $15 - 10 = 5$ students were in the area of Figure 14 common only to rooms A and C. Likewise, there are $12 - 10 = 2$ students who joined only rooms B and C. Since there are already $5 + 10 + 2 = 17$ students in room C, there are $70 - 17 = 53$ students who joined only room C.

We cannot use the information about room A, since there are two regions in A for which we have no information. Similarly, we cannot use the information about room B. In such cases, we label a region with the variable x. Here we place x in the region common only to A and B, as shown in Figure 14.

Of the 90 students in room A, the number who joined only room A must be $90 - x - 10 - 5 = 75 - x$, and this expression is placed in the appropriate region in Figure 14. Similarly, the number who only joined room B is $50 - x - 10 - 2 = 38 - x$. Notice that because all 150 students joined at least one room, there are no elements in the region outside the three circles.

Now that the diagram is filled out, we can determine the value of x by recalling that the total number of students who joined at least one chat room was 150. Thus,

$$(75 - x) + 5 + x + 10 + (38 - x) + 2 + 53 = 150.$$

Simplifying, we have $183 - x = 150$, implying that $x = 33$. The number of students who joined both chat rooms A and B is

$$33 + 10 = 43.$$

7.2 Exercises

Sketch a Venn diagram like the one in the figure, and use shading to show each set.

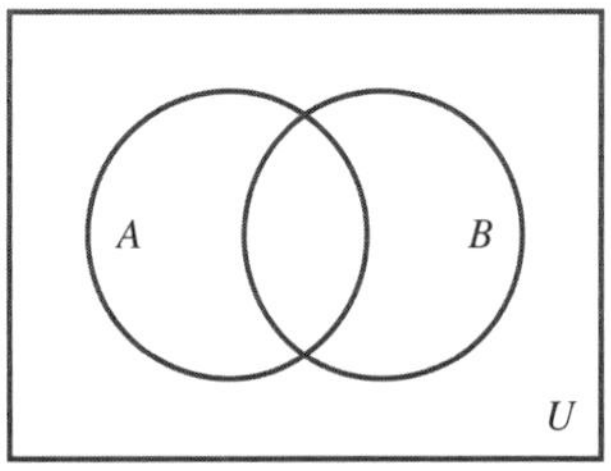

1. $B \cap A'$
2. $A \cup B'$
3. $A' \cup B$
4. $A' \cap B'$
5. $B' \cup (A' \cap B')$
6. $(A \cap B) \cup B'$
7. U'
8. $\emptyset'$
9. Three sets divide the universal set into at most ____ regions.
10. What does the notation $n(A)$ represent?

Sketch a Venn diagram like the one shown, and use shading to show each set.

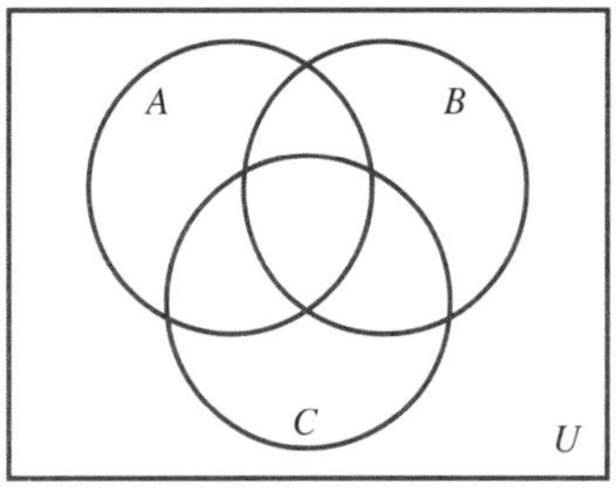

11. $(A \cap B) \cap C$
12. $(A \cap C') \cup B$
13. $A \cap (B \cup C')$
14. $A' \cap (B \cap C)$
15. $(A' \cap B') \cap C$
16. $(A \cap B') \cap C$
17. $(A \cap B') \cup C$
18. $A' \cap (B' \cup C)$
19. $(A \cup B') \cap C$
20. $A \cup (B' \cap C)$

Use the union rule to answer the following questions.

21. If $n(A) = 5$, $n(B) = 12$, and $n(A \cap B) = 4$, what is $n(A \cup B)$?
22. If $n(A) = 15$, $n(B) = 30$, and $n(A \cup B) = 33$, what is $n(A \cap B)$?
23. Suppose $n(B) = 9$, $n(A \cap B) = 5$, and $n(A \cup B) = 22$. What is $n(A)$?
24. Suppose $n(A \cap B) = 5$, $n(A \cup B) = 38$, and $n(A) = 13$. What is $n(B)$?

Draw a Venn diagram and use the given information to fill in the number of elements for each region.

25. $n(U) = 41$, $n(A) = 16$, $n(A \cap B) = 12$, $n(B') = 20$
26. $n(A) = 28$, $n(B) = 12$, $n(A \cup B) = 32$, $n(A') = 19$
27. $n(A \cup B) = 24$, $n(A \cap B) = 6$, $n(A) = 11$, $n(A' \cup B') = 25$
28. $n(A') = 31$, $n(B) = 25$, $n(A' \cup B') = 46$, $n(A \cap B) = 12$
29. $n(A) = 28$, $n(B) = 34$, $n(C) = 25$, $n(A \cap B) = 14$, $n(B \cap C) = 15$, $n(A \cap C) = 11$, $n(A \cap B \cap C) = 9$, $n(U) = 59$

30. $n(A) = 54, n(A \cap B) = 22, n(A \cup B) = 85, n(A \cap B \cap C) = 4,$
$n(A \cap C) = 15, n(B \cap C) = 16, n(C) = 44, n(B') = 63$

31. $n(A \cap B) = 6, n(A \cap B \cap C) = 4, n(A \cap C) = 7, n(B \cap C) = 4,$
$n(A \cap C') = 11, n(B \cap C') = 8, n(C) = 15, n(A' \cap B' \cap C') = 5$

32. $n(A) = 13, n(A \cap B \cap C) = 4, n(A \cap C) = 6, n(A \cap B') = 6, n(B \cap C) = 6,$
$n(B \cap C') = 11, n(B \cup C) = 22, n(A' \cap B' \cap C') = 5$

*In Exercises 33–36, show that the statement is true by drawing Venn diagrams and shading the regions representing the sets on each side of the equals sign.**

33. $(A \cup B)' = A' \cap B'$

34. $(A \cap B)' = A' \cup B'$

35. $A \cap (B \cup C) = (A \cap B) \cup (A \cap C)$

36. $A \cup (B \cap C) = (A \cup B) \cap (A \cup C)$

37. Use the union rule of sets to prove that $n(A \cup B \cup C) = n(A) + n(B) + n(C) - n(A \cap B) - n(A \cap C) - n(B \cap C) + n(A \cap B \cap C)$. (*Hint:* Write $A \cup B \cup C$ as $A \cup (B \cup C)$ and use the formula from Exercise 35.)

Applications

BUSINESS AND ECONOMICS

Use Venn diagrams to answer the following questions.

38. *Cooking Preferences* Jeff Friedman, of Example 4 in the text, was again reassigned, this time to the home economics department of the electric utility. He interviewed 140 people in a suburban shopping center to discover some of their cooking habits. He obtained the following results:

58 use microwave ovens;
63 use electric ranges;
58 use gas ranges;
19 use microwave ovens and electric ranges;
17 use microwave ovens and gas ranges;
4 use both gas and electric ranges;
1 uses all three;
2 use none of the three.

Should he be reassigned one more time? Why or why not?

39. *Harvesting Fruit* Toward the middle of the harvesting season, peaches for canning come in three types, early, late, and extra late, depending on the expected date of ripening. During a certain week, the following data were recorded at a fruit delivery station:

34 trucks went out carrying early peaches;
61 carried late peaches;
50 carried extra late;
25 carried early and late;
30 carried late and extra late;
8 carried early and extra late;
6 carried all three;
9 carried only figs (no peaches at all).

a. How many trucks carried only late variety peaches?

b. How many carried only extra late?

c. How many carried only one type of peach?

d. How many trucks (in all) went out during the week?

40. *Cola Consumption* Market research showed that the adult residents of a certain small town in Georgia fit the following categories of cola consumption. (We assume here that no one drinks both regular cola and diet cola.)

Age	Drink Regular Cola (*R*)	Drink Diet Cola (*D*)	Drink No Cola (*N*)	Totals
21–25 (*Y*)	40	15	15	70
26–35 (*M*)	30	30	20	80
Over 35 (*O*)	10	50	10	70
Totals	80	95	45	220

*The statements in Exercises 33 and 34 are known as De Morgan's Laws. They are named for the English mathematician Augustus De Morgan (1806–1871). They are analogous to De Morgan's Laws for logic seen in the previous chapter.

Using the letters given in the table, find the number of people in each set.

a. $Y \cap R$ **b.** $M \cap D$

c. $M \cup (D \cap Y)$ **d.** $Y' \cap (D \cup N)$

e. $O' \cup N$ **f.** $M' \cap (R' \cap N')$

g. Describe the set $M \cup (D \cap Y)$ in words.

41. *Investment Habits* The following table shows the results of a survey taken by a bank in a medium-sized town in Tennessee. The survey asked questions about the investment habits of bank customers. (We assume here that no one invests in more than one of type of investment.)

Age	Stocks (S)	Bonds (B)	Savings Accounts (A)	Totals
18–29 (Y)	6	2	15	23
30–49 (M)	14	5	14	33
50 or over (O)	32	20	12	64
Totals	52	27	41	120

Using the letters given in the table, find the number of people in each set.

a. $Y \cap B$ **b.** $M \cup A$ **c.** $Y \cap (S \cup B)$

d. $O' \cup (S \cup A)$ **e.** $(M' \cup O') \cap B$

f. Describe the set $Y \cap (S \cup B)$ in words.

42. *Investment Survey* Most mathematics professors love to invest their hard-earned money. A recent survey of 150 math professors revealed that

111 invested in stocks;
98 invested in bonds;
100 invested in certificates of deposit;
80 invested in stocks and bonds;
83 invested in bonds and certificates of deposit;
85 invested in stocks and certificates of deposit;
9 did not invest in any of the three.

How many mathematics professors invested in stocks and bonds and certificates of deposit?

LIFE SCIENCES

43. *Genetics* After a genetics experiment on 50 pea plants, the number of plants having certain characteristics was tallied, with the following results.

22 were tall;
25 had green peas;
39 had smooth peas;
9 were tall and had green peas;
20 had green peas and smooth peas;
6 had all three characteristics;
4 had none of the characteristics.

a. Find the number of plants that were tall and had smooth peas.

b. How many plants were tall and had peas that were neither smooth nor green?

c. How many plants were not tall but had peas that were smooth and green?

44. *Blood Antigens* Human blood can contain the A antigen, the B antigen, both the A and B antigens, or neither antigen. A third antigen, called the Rh antigen, is important in human reproduction, and again may or may not be present in an individual. Blood is called type A-positive if the individual has the A and Rh, but not the B antigen. A person having only the A and B antigens is said to have type AB-negative blood. A person having only the Rh antigen has type O-positive blood. Other blood types are defined in a similar manner. Identify the blood types of the individuals in regions (a)–(h) below.

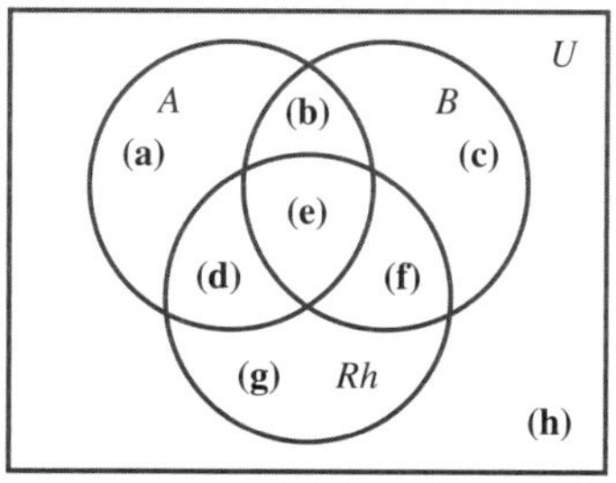

45. *Blood Antigens* (Use the diagram from Exercise 44.) In a certain hospital, the following data were recorded.

25 patients had the A antigen;
8 had the A and not the B antigen;
27 had the B antigen;
22 had the B and Rh antigens;
30 had the Rh antigen;
12 had none of the antigens;
16 had the A and Rh antigens;
15 had all three antigens.

How many patients

a. were represented? **b.** had exactly one antigen?

c. had exactly two antigens? **d.** had O-positive blood?

e. had AB-positive blood? **f.** had B-negative blood?

g. had O-negative blood? **h.** had A-positive blood?

46. *Mortality* The table on the next page lists the number of deaths in the United States during 2003 according to race and gender.* Use this information and the letters given to find the number of people in each set.

*_National Vital Statistics Reports_, Vol. 54, No. 13, April 19, 2006, p. 17.

	White (*W*)	Black (*B*)	American Indian (*I*)	Asian or Pacific Islander (*A*)
Female (F)	1,078,064	143,278	6041	18,941
Male (M)	1,025,650	148,022	7106	21,186

a. F **b.** $F \cap (I \cup A)$

c. $M \cup B$ **d.** $W' \cup I' \cup A'$

e. In words, describe the set in part b.

47. *Hockey* The table to the right lists the number of head and neck injuries for 319 ice hockey players wearing either a full shield or half shield in the Canadian Inter-University Athletics Union during one season.* Using the letters given in the table, find the number of injuries in each set.

	Half Shield (*H*)	Full Shield (*F*)
Head and Face Injuries (A)	95	34
Concussions (B)	41	38
Neck Injuries (C)	9	7
Other Injuries (D)	202	150

a. $A \cap F$ **b.** $C \cap (H \cup F)$

c. $D \cup F$ **d.** $B' \cap C'$

SOCIAL SCIENCES

48. *Military* The number of female military personnel in 2004 is given in the following table.† Use this information and the letters given to find the number of female military personnel in each set.

	Army (*A*)	Air Force (*B*)	Navy (*C*)	Marines (*D*)	Totals
Officers (O)	12,309	13,596	8111	1096	35,112
Enlisted (E)	60,395	59,436	45,415	9683	174,929
Cadets & Midshipmen (M)	639	754	722	0	2115
Totals	73,343	73,786	54,248	10,779	212,156

a. $A \cup B$ **b.** $E \cup (C \cup D)$ **c.** $O' \cap M'$

U.S. Population The projected U.S. population in 2020 (in millions) by age and race or ethnicity is given in the following table. Use this information in Exercises 49–54.‡

	Non-Hispanic White (*A*)	Hispanic (*B*)	Black (*C*)	Asian (*D*)	American Indian (*E*)	Totals
Under 45 (F)	110.6	37.6	30.2	13.1	2.2	193.7
45–64 (G)	55.3	10.3	9.9	4.3	0.6	80.4
65 and over (H)	41.4	4.7	5.0	2.2	0.3	53.6
Totals	207.3	52.6	45.1	19.6	3.1	327.7

*Benson, Brian, Nicholas Nohtaki, M. Sarah Rose, Willem Meeuwisse, "Head and Neck Injuries Among Ice Hockey Players Wearing Full Face Shields vs. Half Face Shields," *JAMA*, Vol. 282, No. 24, Dec. 22/29, 1999, pp. 2328–2332.

†*Selected Manpower Statistics, Fiscal Year 2004*, U.S. Department of Defense, p. 31.

‡*Population Projections of the United States by Age, Sex, Race, and Hispanic Origin: 1995 to 2050*, U.S. Bureau of the Census, Feb. 1996, pp. 16–17.

Using the letters given in the table on the previous page, find the number of people in each set.

49. $A \cap F$

50. $G \cup B$

51. $G \cup (C \cap H)$

52. $F \cap (B \cup H)$

53. $H \cup D$

54. $G' \cap (A' \cap C')$

Marital Status *The following table gives the population breakdown (in thousands) of the U.S. population in 2003 based on marital status and race or ethnic origin.**

	White (W)	Black (B)	Hispanic (H)	Asian or Pacific Islander (A)
Never married (N)	49,101	11,783	9862	3181
Married (M)	104,689	9279	14,239	5594
Widowed (I)	11,754	1730	918	418
Divorced/ separated (D)	21,372	4250	2917	607

Find the number of people in each set.

55. $N \cap (B \cup H)$

56. $(M \cup I) \cap A$

57. $(D \cup W) \cap A'$

58. $M' \cap (B \cup A)$

GENERAL INTEREST

59. *Chinese New Year* A survey of people attending a Lunar New Year celebration in Chinatown yielded the following results:

120 were women;
150 spoke Cantonese;
170 lit firecrackers;
108 of the men spoke Cantonese;
100 of the men did not light firecrackers;
18 of the non-Cantonese-speaking women lit firecrackers;
78 non-Cantonese-speaking men did not light firecrackers;
30 of the women who spoke Cantonese lit firecrackers.

a. How many attended?

b. How many of those who attended did not speak Cantonese?

c. How many women did not light firecrackers?

d. How many of those who lit firecrackers were Cantonese-speaking men?

The New York Times 2006 Almanac, p. 288.

60. *Native American Ceremonies* At a pow-wow in Arizona 75 Native American families from all over the Southwest came to participate in the ceremonies. A coordinator of the pow-wow took a survey and found that

15 families brought food, costumes, and crafts;
25 families brought food and crafts;
42 families brought food;
6 families brought costumes and crafts, but not food;
4 families brought crafts, but neither food nor costumes;
10 families brought none of the three items;
18 families brought costumes but not crafts.

a. How many families brought costumes and food?

b. How many families brought costumes?

c. How many families brought crafts, but not costumes?

d. How many families did not bring crafts?

e. How many families brought food or costumes?

61. *Poultry Analysis* A chicken farmer surveyed his flock with the following results. The farmer had

9 fat red roosters;
13 thin brown hens;
15 red roosters;
11 thin red chickens (hens and roosters);
17 red hens;
56 fat chickens (hens and roosters);
41 roosters;
48 hens.

Assume all chickens are thin or fat, red or brown, and hens (female) or roosters (male). How many chickens were

a. fat?

b. red?

c. fat roosters?

d. fat hens?

e. thin and brown?

f. red and fat?

7.3 Introduction to Probability

? THINK ABOUT IT **What is the probability that a randomly selected person in the United States is Hispanic or Black?**

After introducing probability, we will answer this question in one of the exercises.

If you go to a supermarket and buy 5 pounds of peaches at 99 cents per pound, you can easily find the *exact* price of your purchase: \$4.95. On the other hand, the produce manager of the market is faced with the problem of ordering peaches. The manager may have a good estimate of the number of pounds of peaches that will be sold during the day, but it is impossible to predict the *exact* amount. The number of pounds that customers will purchase during a day is *random*: the quantity cannot be predicted exactly. A great many problems that come up in applications of mathematics involve random phenomena—those for which exact prediction is impossible. The best that we can do is determine the *probability* of the possible outcomes.

Sample Spaces

In probability, an **experiment** is an activity or occurrence with an observable result. Each repetition of an experiment is called a **trial**. The possible results of each trial are called **outcomes**. The set of all possible outcomes for an experiment is the **sample space** for that experiment. A sample space for the experiment of tossing a coin is made up of the outcomes heads (h) and tails (t). If S represents this sample space, then

$$S = \{h, t\}.$$

EXAMPLE 1 **Sample Spaces**

Give the sample space for each experiment.

(a) A spinner like the one in Figure 15 is spun.

FIGURE 15

▶**Solution** The three outcomes are 1, 2, or 3, so the sample space is

$$\{1, 2, 3\}.$$

(b) For the purposes of a public opinion poll, respondents are classified as young, middle-aged, or senior, and as male or female.

▶**Solution** A sample space for this poll could be written as a set of ordered pairs:

$$\{(\text{young, male}), (\text{young, female}), (\text{middle-aged, male}),$$
$$(\text{middle-aged, female}), (\text{senior, male}), (\text{senior, female})\}.$$

(c) An experiment consists of studying the numbers of boys and girls in families with exactly 3 children. Let b represent *boy* and g represent *girl*.

▶**Solution** A three-child family can have 3 boys, written bbb, 3 girls, ggg, or various combinations, such as bgg. A sample space with four outcomes (not equally likely) is

$$S_1 = \{\text{3 boys, 2 boys and 1 girl, 1 boy and 2 girls, 3 girls}\}.$$

Notice that a family with 3 boys or 3 girls can occur in just one way, but a family of 2 boys and 1 girl or 1 boy and 2 girls can occur in more than one way. If the *order* of the births is considered, so that *bgg* is different from *gbg* or *ggb*, for example, another sample space is

$$S_2 = \{bbb, bbg, bgb, gbb, bgg, gbg, ggb, ggg\}.$$

The second sample space, S_2, has equally likely outcomes if we assume that boys and girls are equally likely. This assumption, while not quite true, is approximately true, so we will use it throughout this book. The outcomes in S_1 are not equally likely, since there is more than one way to get a family with 2 boys and 1 girl (*bbg*, *bgb*, or *gbb*) or a family with 2 girls and 1 boy (*ggb*, *gbg*, or *bgg*), but only one way to get 3 boys (*bbb*) or 3 girls (*ggg*).

CAUTION An experiment may have more than one sample space, as shown in Example 1(c). The most convenient sample spaces have equally likely outcomes, but it is not always possible to choose such a sample space. ■

Events

An **event** is a subset of a sample space. If the sample space for tossing a coin is $S = \{h, t\}$, then one event is $E = \{h\}$, which represents the outcome "heads."

An ordinary die is a cube whose six different faces show the following numbers of dots: 1, 2, 3, 4, 5, and 6. If the die is fair (not "loaded" to favor certain faces over others), then any one of the faces is equally likely to come up when the die is rolled. The sample space for the experiment of rolling a single fair die is $S = \{1, 2, 3, 4, 5, 6\}$. Some possible events are listed below.

The die shows an even number: $E_1 = \{2, 4, 6\}$.

The die shows a 1: $E_2 = \{1\}$.

The die shows a number less than 5: $E_3 = \{1, 2, 3, 4\}$.

The die shows a multiple of 3: $E_4 = \{3, 6\}$.

Using the notation introduced earlier in this chapter, notice that $n(S) = 6$, $n(E_1) = 3$, $n(E_2) = 1$, $n(E_3) = 4$, and $n(E_4) = 2$.

EXAMPLE 2 Events

For the sample space S_2 in Example 1(c), write the following events.

(a) Event H: the family has exactly two girls

▶**Solution** Families with three children can have exactly two girls with either *bgg*, *gbg*, or *ggb*, so event H is

$$H = \{bgg, gbg, ggb\}.$$

(b) Event K: the three children are the same sex

▶**Solution** Two outcomes satisfy this condition: all boys or all girls.

$$K = \{bbb, ggg\}$$

(c) Event J: the family has three girls

▶**Solution** Only ggg satisfies this condition, so

$$J = \{ggg\}.$$

In Example 2(c), event J had only one possible outcome, ggg. Such an event, with only one possible outcome, is a **simple event**. If event E equals the sample space S, then E is called a **certain event**. If event $E = \emptyset$, then E is called an **impossible event**.

EXAMPLE 3

Events

Suppose a coin is flipped until both a head and a tail appear, or until the coin has been flipped four times, whichever comes first. Write each of the following events in set notation.

(a) The coin is flipped exactly three times.

▶**Solution** This means that the first two flips of the coin did not include both a head and a tail, so they must both be heads or both be tails. Because the third flip is the last one, it must show the side of the coin not yet seen. Thus the event is

$$\{hht, tth\}.$$

(b) The coin is flipped at least three times.

▶**Solution** In addition to the outcomes listed in part a, there is also the possibility that the coin is flipped four times, which only happens when the first three flips are all heads or all tails. Thus the event is

$$\{hht, tth, hhhh, hhht, tttt, ttth\}.$$

(c) The coin is flipped at least two times.

▶**Solution** This event consists of the entire sample space:

$$S = \{ht, th, hht, tth, hhhh, hhht, tttt, ttth\}.$$

This is an example of a certain event.

(d) The coin is flipped fewer than two times.

▶**Solution** The coin cannot be flipped fewer than two times under the rules described, so the event is the empty set $\emptyset$. This is an example of an impossible event.

Since events are sets, we can use set operations to find unions, intersections, and complements of events. A summary of the set operations for events is given below.

SET OPERATIONS FOR EVENTS

Let E and F be events for a sample space S.

$E \cap F$ occurs when both E **and** F occur;

$E \cup F$ occurs when E **or** F **or both** occur;

E' occurs when E does **not** occur.

EXAMPLE 4 **Minimum-Wage Workers**

A study of workers earning the minimum wage grouped such workers into various categories, which can be interpreted as events when a worker is selected at random.* Consider the following events:

E: worker is under 20;

F: worker is white;

G: worker is female.

Describe the following events in words.

(a) E'

▶**Solution** E' is the event that the worker is 20 or over.

(b) $F \cap G'$

▶**Solution** $F \cap G'$ is the event that the worker is white and not a female, that is, the worker is a white male.

(c) $E \cup G$

▶**Solution** $E \cup G$ is the event that the worker is under 20 or is female. Note that this event includes all workers under 20, both male and female, and all female workers of any age.

Two events that cannot both occur at the same time, such as rolling an even number and an odd number with a single roll of a die, are called *mutually exclusive events.*

MUTUALLY EXCLUSIVE EVENTS

Events E and F are **mutually exclusive events** if $E \cap F = \emptyset$.

Any event E and its complement E' are mutually exclusive. By definition, mutually exclusive events are disjoint sets.

EXAMPLE 5 **Mutually Exclusive Events**

Let $S = \{1, 2, 3, 4, 5, 6\}$, the sample space for tossing a single die. Let $E = \{4, 5, 6\}$, and let $G = \{1, 2\}$. Then E and G are mutually exclusive events since they have no outcomes in common: $E \cap G = \emptyset$. See Figure 16.

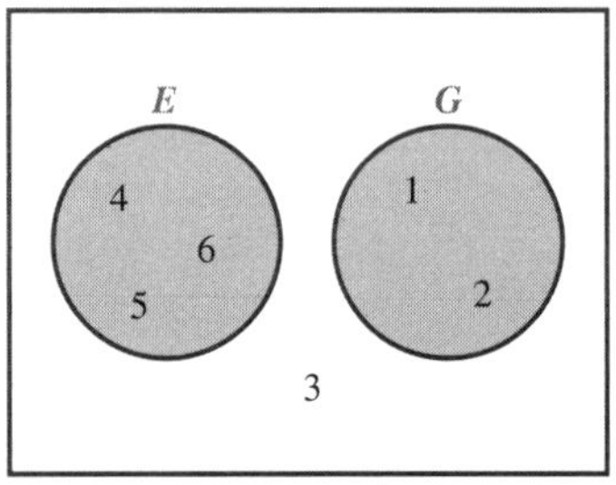

$E \cap G = \emptyset$

FIGURE 16

Probability

For sample spaces with *equally likely* outcomes, the probability of an event is defined as follows.

*http://www.epinet.org/Issuebriefs/IB133.pdf.

BASIC PROBABILITY PRINCIPLE

Let S be a sample space of equally likely outcomes, and let event E be a subset of S. Then the **probability** that event E occurs is

$$P(E) = \frac{n(E)}{n(S)}.$$

By this definition, the probability of an event is a number that indicates the relative likelihood of the event.

CAUTION The basic probability principle only applies when the outcomes are equally likely. ■

EXAMPLE 6

Basic Probabilities

Suppose a single fair die is rolled. Use the sample space $S = \{1, 2, 3, 4, 5, 6\}$ and give the probability of each event.

(a) E: the die shows an even number

▶**Solution** Here, $E = \{2, 4, 6\}$, a set with three elements. Since S contains six elements,

$$P(E) = \frac{3}{6} = \frac{1}{2}.$$

(b) F: the die shows a number less than 10

▶**Solution** Event F is a certain event, with

$$F = \{1, 2, 3, 4, 5, 6\},$$

so that

$$P(F) = \frac{6}{6} = 1.$$

(c) G: the die shows an 8

▶**Solution** This event is impossible, so

$$P(G) = 0.$$

A standard deck of 52 cards has four suits: hearts (♥), clubs (♣), diamonds (♦), and spades (♠), with 13 cards in each suit. The hearts and diamonds are red, and the spades and clubs are black. Each suit has an ace (A), a king (K), a queen (Q), a jack (J), and cards numbered from 2 to 10. The jack, queen, and king are called *face cards* and for many purposes can be thought of as having values 11, 12, and 13, respectively. The ace can be thought of as the low card (value 1) or the high card (value 14). See Figure 17 on the next page. We will refer to this standard deck of cards often in our discussion of probability.

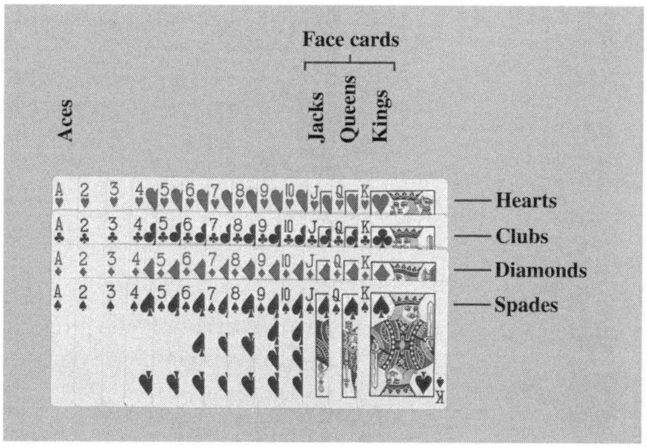

FIGURE 17

EXAMPLE 7 **Playing Cards**

If a single playing card is drawn at random from a standard 52-card deck, find the probability of each event.

(a) Drawing an ace

▶**Solution** There are 4 aces in the deck. The event "drawing an ace" is

$$\{\text{heart ace, diamond ace, club ace, spade ace}\}.$$

Therefore,

$$P(\text{ace}) = \frac{4}{52} = \frac{1}{13}.$$

(b) Drawing a face card

▶**Solution** Since there are 12 face cards (three in each of the four suits),

$$P(\text{face card}) = \frac{12}{52} = \frac{3}{13}.$$

(c) Drawing a spade

▶**Solution** The deck contains 13 spades, so

$$P(\text{spade}) = \frac{13}{52} = \frac{1}{4}.$$

(d) Drawing a spade or a heart

▶**Solution** Besides the 13 spades, the deck contains 13 hearts, so

$$P(\text{spade or heart}) = \frac{26}{52} = \frac{1}{2}.$$

In the preceding examples, the probability of each event was a number between 0 and 1. The same thing is true in general. Any event E is a subset of the sample space S, so $0 \leq n(E) \leq n(S)$. Since $P(E) = n(E)/n(S)$, it follows that $0 \leq P(E) \leq 1$.

For any event E, $\mathbf{0 \leq P(E) \leq 1}$.

EXAMPLE 8 **Congressional Service**

The following table gives the number of years of service of senators in the 109th Congress of the United States of America, which convened on January 4, 2005.*

Years of Service	Number of Senators
0–9	50
10–19	28
20–29	15
30–39	4
40 or more	3

Find the probability that a randomly selected senator of the 109th Congress served 20–29 years when Congress convened.

▶**Solution** This probability is found by dividing the number of senators who served 20–29 years by the total number of senators. Thus,

$$P(20\text{–}29 \text{ years}) = \frac{15}{100} = 0.15.$$

7.3 Exercises

1. What is meant by a "fair" coin or die?

2. What is the sample space for an experiment?

Write sample spaces for the experiments in Exercises 3–10.

3. A month of the year is chosen for a wedding.

4. A day in April is selected for a bicycle race.

5. A student is asked how many points she earned on a recent 80-point test.

6. A person is asked the number of hours (to the nearest hour) he watched television yesterday.

7. The management of an oil company must decide whether to go ahead with a new oil shale plant or to cancel it.

8. A record is kept each day for three days about whether a particular stock goes up or down.

9. A coin is tossed, and a die is rolled.

10. A box contains five balls, numbered 1, 2, 3, 4, and 5. A ball is drawn at random, the number on it recorded, and the ball replaced. The box is shaken, a second ball is drawn, and its number is recorded.

**Time Almanac 2006*, pp. 46–47.

11. Define an event.

12. What is a simple event?

For the experiments in Exercises 13–18, write out the sample space S and give the value of n(S). Then tell whether the outcomes in S are equally likely. Finally, write the indicated events in set notation.

13. A committee of 2 people is selected from 5 executives: Alam, Bartolini, Chinn, Dickson, and Ellsberg.

    a. Chinn is on the committee.

    b. Dickson and Ellsberg are not both on the committee.

    c. Both Alam and Chinn are on the committee.

14. Five states are being considered as the location for three new high-energy physics laboratories: California (CA), Colorado (CO), New Jersey (NJ), New York (NY), and Utah (UT). Three states will be chosen. Write elements of the sample space in the form (CA, CO, NJ).

    a. All three states border an ocean.

    b. Exactly two of the three states border an ocean.

    c. Exactly one of the three states is west of the Mississippi River.

15. Slips of paper marked with the numbers 1, 2, 3, 4, and 5 are placed in a box. After being mixed, two slips are drawn simultaneously.

    a. Both slips are marked with even numbers.

    b. One slip is marked with an odd number and the other is marked with an even number.

    c. Both slips are marked with the same number.

16. An unprepared student takes a three-question, true/false quiz in which he guesses the answers to all three questions, so each answer is equally likely to be correct or wrong.

    a. The student gets three answers wrong.

    b. The student gets exactly two answers correct.

    c. The student gets only the first answer correct.

17. A coin is flipped until two heads appear, up to a maximum of four flips. (If three tails are flipped, the coin is still tossed a fourth time to complete the experiment).

    a. The coin is tossed four times.

    b. Exactly two heads are tossed.

    c. No heads are tossed.

18. One jar contains four balls, labeled 1, 2, 3, and 4. A second jar contains five balls, labeled 1, 2, 3, 4, and 5. An experiment consists of taking one ball from the first jar, and then taking a ball from the second jar.

    a. The number on the first ball is even.

    b. The number on the second ball is even.

    c. The sum of the numbers on the two balls is 5.

    d. The sum of the numbers on the two balls is 1.

A single fair die is rolled. Find the probabilities of each event.

19. Getting a 2

20. Getting an odd number

21. Getting a number less than 5

22. Getting a number greater than 2

23. Getting a 3 or a 4

24. Getting any number except 3

A card is drawn from a well-shuffled deck of 52 cards. Find the probability of drawing the following.

25. A 9

26. A black card

27. A black 9

28. A heart

29. The 9 of hearts

30. A face card

31. A 2 or a queen

32. A black 7 or a red 8

33. A red card or a 10

34. A spade or a king

A jar contains 3 white, 4 orange, 5 yellow, and 8 black marbles. If a marble is drawn at random, find the probability that it is the following.

35. White

36. Orange

37. Yellow

38. Black

39. Not black

40. Orange or yellow

41. The student sitting next to you in class concludes that the probability of the ceiling falling down on both of you before class ends is $1/2$, because there are two possible outcomes—the ceiling will fall or not fall. What is wrong with this reasoning?

42. The following puzzler was given on the *Car Talk* radio program on February 24, 2001.*

"Three different numbers are chosen at random, and one is written on each of three slips of paper. The slips are then placed face down on the table. The objective is to choose the slip upon which is written the largest number. Here are the rules: You can turn over any slip of paper and look at the amount written on it. If for any reason you think this is the largest, you're done; you keep it. Otherwise you discard it and turn over a second slip. Again, if you think this is the one with the biggest number, you keep that one and the game is over. If you don't, you discard that one too. . . . The chance of getting the highest number is one in three. Or is it? Is there a strategy by which you can improve the odds?"

The answer to the puzzler is that you can indeed improve the probability of getting the highest number by the following strategy. Pick one of the slips of paper, and after looking at the number, throw it away. Then pick a second slip; if it has a larger number than the first slip, stop. If not, pick the third slip. Find the probability of winning with this strategy.

Applications

BUSINESS AND ECONOMICS

43. *Survey of Workers* The management of a firm wishes to check on the opinions of its assembly line workers. Before the workers are interviewed, they are divided into various categories. Define events E, F, and G as follows.

E: worker is female

F: worker has worked less than 5 years

G: worker contributes to a voluntary retirement plan

Describe each event in words.

a. E' **b.** $E \cap F$ **c.** $E \cup G'$

d. F' **e.** $F \cup G$ **f.** $F' \cap G'$

*Source: Car Talk, Feb. 24, 2001. http://www.cartalk.com/content/puzzler/transcripts/200107/. Cartalk.com is a production of Dewey, Cheetham and Howe. Contents © 2007, Dewey, Cheetham and Howe. This is a special case of the famous Googol problem. For more details, see "Recognizing the Maximum of a Sequence" by John P. Gilbert and Frederick Mosteller, *Journal of the American Statistical Association*, Vol. 61, No. 313, March 1966, pp. 35–73.

44. *Research Funding* In 2001, funding for university research in the United States totaled $32.723 billion. Support came from various sources, as shown in the following table.*

Source	Amount (in billions of dollars)
Federal government	19.191
State and local government	2.315
Industry	2.234
Academic institutions	6.553
Other	2.430

Find the probability that funds for a particular project came from each source.

a. Federal government

b. Industry

c. Academic institutions

45. *Investment Survey* Exercise 42 of the previous section presented a survey of 150 mathematics professors. Use the information given in that exercise to find each probability.

a. A randomly chosen professor invested in stocks and bonds.

b. A randomly chosen professor invested in stocks and bonds and certificates of deposit.

LIFE SCIENCES

46. *Medical Survey* For a medical experiment, people are classified as to whether they smoke, have a family history of heart disease, or are overweight. Define events E, F, and G as follows.

E: person smokes
F: person has a family history of heart disease
G: person is overweight

Describe each event in words.

a. G' **b.** $F \cap G$ **c.** $E \cup G'$

47. *Medical Survey* Refer to Exercise 46. Describe each event in words.

a. $E \cup F$ **b.** $E' \cap F$ **c.** $F' \cup G'$

48. *Causes of Death* There were 2,447,864 U.S. deaths in 2002. They are listed according to cause in the following table.† If a randomly selected person died in 2002, use this information to find the following probabilities.

Cause	Number of Deaths
Heart disease	695,754
Cancer	558,847
Cerebrovascular disease	163,010
Chronic lower respiratory disease	125,500
Accidents	102,303
Diabetes mellitus	73,119
Pneumonia and influenza	65,984
All other causes	663,347

a. The probability that the cause of death was heart disease

b. The probability that the cause of death was cancer or heart disease

c. The probability that the cause of death was not an accident and was not diabetes mellitus

SOCIAL SCIENCES

49. *U.S. Population* The projected U.S. population (in thousands) by race in 2020 and 2050 is given in the following table.‡

Race	2020	2050
White	207,393	207,901
Hispanic	52,652	96,508
Black	41,538	53,555
Asian and Pacific Islander	18,557	32,432
Other	2602	3535

Find the probability that a randomly selected person in the given year is of the race specified.

a. Hispanic in 2020 **b.** Hispanic in 2050

c. Black in 2020 **d.** Black in 2050

**NSF Science and Engineering Indicators 2004*, http://www.nsf.gov/statistics/seind04/c5/c5s1.htm#p4.

†*The New York Times 2006 Almanac*, p. 384.

‡*Population Projections of the United States by Age, Sex, Race, and Hispanic Origin: 1995 to 2050*, Bureau of the Census, Feb. 1996, p. 12.

50. *Civil War* Estimates of the Union Army's strength and losses for the battle of Gettysburg are given in the following table, where *strength* is the number of soldiers immediately preceding the battle and *loss* indicates a soldier who was killed, wounded, captured, or missing.*

Unit	Strength	Loss
I Corps (Reynolds)	12,222	6059
II Corps (Hancock)	11,347	4369
III Corps (Sickles)	10,675	4211
V Corps (Sykes)	10,907	2187
VI Corps (Sedgwick)	13,596	242
XI Corps (Howard)	9188	3801
XII Corps (Slocum)	9788	1082
Cavalry (Pleasonton)	11,851	610
Artillery (Tyler)	2376	242
Total	91,950	22,803

a. Find the probability that a randomly selected union soldier was from the XI Corps.

b. Find the probability that a soldier was lost in the battle.

c. Find the probability that a I Corps soldier was lost in the battle.

d. Which group had the highest probability of not being lost in the battle?

e. Which group had the highest probability of loss?

f. Explain why these probabilities vary.

51. *Civil War* Estimates of the Confederate Army's strength and losses for the battle of Gettysburg are given in the following table, where *strength* is the number of soldiers immediately preceding the battle and *loss* indicates a soldier who was killed, wounded, captured, or missing.*

Unit	Strength	Loss
I Corps (Longstreet)	20,706	7661
II Corps (Ewell)	20,666	6603
III Corps (Hill)	22,083	8007
Cavalry (Stuart)	6621	286
Total	70,076	22,557

a. Find the probability that a randomly selected confederate soldier was from the III Corps.

b. Find the probability that a confederate soldier was lost in the battle.

c. Find the probability that a I Corps soldier was lost in the battle.

d. Which group had the highest probability of not being lost in the battle?

e. Which group had the highest probability of loss?

GENERAL INTEREST

52. *Native American Ceremonies* Exercise 60 of the previous section presented a survey of families participating in a pow-wow in Arizona. Use the information given in that exercise to find each probability.

a. A randomly chosen family brought costumes and food.

b. A randomly chosen family brought crafts, but neither food nor costumes.

c. A randomly chosen family brought food or costumes.

53. *Chinese New Year* Exercise 59 of the previous section presented a survey of people attending a Lunar New Year celebration in Chinatown. Use the information given in that exercise to find each of the following probabilities.

a. A randomly chosen attendee speaks Cantonese.

b. A randomly chosen attendee does not speak Cantonese.

c. A randomly chosen attendee was a woman that did not light a firecracker.

*Busey, John and David Martin, *Regimental Strengths and Losses at Gettysburg*, Hightstown, N.J., Longstreet House, 1986, p. 270.

7.4 Basic Concepts of Probability

? THINK ABOUT IT

What is the probability that a dollar of advertising in the United States is spent on broadcast television or newspapers?

We determine the probability of this and other events in this section. But first we need to develop additional rules for calculating probability, beginning with the probability of a union of two events.

To determine the probability of the union of two events E and F in a sample space S, use the union rule for sets,

$$n(E \cup F) = n(E) + n(F) - n(E \cap F),$$

which was proved in Section 7.2. Assuming that the events in the sample space S are equally likely, divide both sides by $n(S)$, so that

$$\frac{n(E \cup F)}{n(S)} = \frac{n(E)}{n(S)} + \frac{n(F)}{n(S)} - \frac{n(E \cap F)}{n(S)}$$

$$P(E \cup F) = P(E) + P(F) - P(E \cap F).$$

Although our derivation is valid only for sample spaces with equally likely events, the result is valid for any events E and F from any sample space, and is called the **union rule for probability**.

UNION RULE FOR PROBABILITY

For any events E and F from a sample space S,

$$P(E \cup F) = P(E) + P(F) - P(E \cap F).$$

EXAMPLE 1 **Probabilities with Playing Cards**

If a single card is drawn from an ordinary deck of cards, find the probability that it will be a red or a face card.

▶**Solution** Let R represent the event "red card" and F the event "face card." There are 26 red cards in the deck, so $P(R) = 26/52$. There are 12 face cards in the deck, so $P(F) = 12/52$. Since there are 6 red face cards in the deck, $P(R \cap F) = 6/52$. By the union rule, the probability of the card being red or a face card is

$$\begin{aligned} P(R \cup F) &= P(R) + P(F) - P(R \cap F) \\ &= \frac{26}{52} + \frac{12}{52} - \frac{6}{52} = \frac{32}{52} = \frac{8}{13}. \end{aligned}$$

EXAMPLE 2 **Probabilities with Dice**

Suppose two fair dice are rolled. Find each probability.

(a) The first die shows a 2, or the sum of the results is 6 or 7.

▶**Solution** The sample space for the throw of two dice is shown in Figure 18, where 1-1 represents the event "the first die shows a 1 and the second die shows a 1," 1-2 represents "the first die shows a 1 and the second die shows a 2," and so on. Let A represent the event "the first die shows a 2," and B represent the event "the sum of the results is 6 or 7." These events are indicated in Figure 18. From the diagram, event A has 6 elements, B has 11 elements, the intersection of A and B has 2 elements, and the sample space has 36 elements. Thus,

$$P(A) = \frac{6}{36}, \quad P(B) = \frac{11}{36}, \quad \text{and} \quad P(A \cap B) = \frac{2}{36}.$$

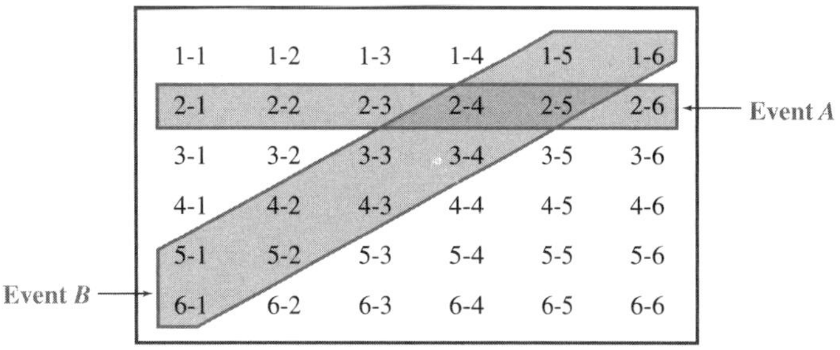

FIGURE 18

By the union rule,

$$P(A \cup B) = P(A) + P(B) - P(A \cap B)$$

$$P(A \cup B) = \frac{6}{36} + \frac{11}{36} - \frac{2}{36} = \frac{15}{36} = \frac{5}{12}.$$

(b) The sum of the results is 11, or the second die shows a 5.

▶**Solution** $P(\text{sum is } 11) = 2/36$, $P(\text{second die shows a } 5) = 6/36$, and $P(\text{sum is 11 and second die shows a } 5) = 1/36$, so

$$P(\text{sum is 11 or second die shows a } 5) = \frac{2}{36} + \frac{6}{36} - \frac{1}{36} = \frac{7}{36}.$$

CAUTION You may wonder why we did not use $S = \{2, 3, 4, 5, \ldots, 12\}$ as the sample space in Example 2. Remember, we prefer to use a sample space with equally likely outcomes. The outcomes in set S above are not equally likely—a sum of 2 can occur in just one way, a sum of 3 in two ways, a sum of 4 in three ways, and so on, as shown in Figure 18. ■

If events E and F are mutually exclusive, then $E \cap F = \emptyset$ by definition; hence, $P(E \cap F) = 0$. In this case the union rule simplifies to $P(E \cup F) = P(E) + P(F)$.

CAUTION The rule $P(E \cup F) = P(E) + P(F)$ is *only* valid when E and F are mutually exclusive. When E and F are *not* mutually exclusive, use the rule $P(E \cup F) = P(E) + P(F) - P(E \cap F)$. ■

By the definition of E', for any event E from a sample space S,

$$E \cup E' = S \quad \text{and} \quad E \cap E' = \emptyset.$$

Since $E \cap E' = \emptyset$, events E and E' are mutually exclusive, so that

$$P(E \cup E') = P(E) + P(E').$$

However, $E \cup E' = S$, the sample space, and $P(S) = 1$. Thus

$$P(E \cup E') = P(E) + P(E') = 1.$$

Rearranging these terms gives the following useful rule for complements.

COMPLEMENT RULE

$$P(E) = 1 - P(E') \quad \text{and} \quad P(E') = 1 - P(E).$$

EXAMPLE 3 **Complement Rule**

If a fair die is rolled, what is the probability that any number but 5 will come up?

▶**Solution** If E is the event that 5 comes up, then E' is the event that any number but 5 comes up. Since $P(E) = 1/6$, we have $P(E') = 1 - 1/6 = 5/6$.

EXAMPLE 4 **Complement Rule**

In Example 2, find the probability that the sum of the numbers rolled is greater than 3.

▶**Solution** To calculate this probability directly, we must find the probabilities that the sum is 4, 5, 6, 7, 8, 9, 10, 11, or 12 and then add them. It is much simpler to first find the probability of the complement, the event that the sum is less than or equal to 3.

$$\begin{aligned} P(\text{sum} \le 3) &= P(\text{sum is } 2) + P(\text{sum is } 3) \\ &= \frac{1}{36} + \frac{2}{36} \\ &= \frac{3}{36} = \frac{1}{12} \end{aligned}$$

Now use the fact that $P(E) = 1 - P(E')$ to get

$$\begin{aligned} P(\text{sum} > 3) &= 1 - P(\text{sum} \le 3) \\ &= 1 - \frac{1}{12} = \frac{11}{12}. \end{aligned}$$

Odds Sometimes probability statements are given in terms of **odds**, a comparison of $P(E)$ with $P(E')$. For example, suppose $P(E) = 4/5$. Then $P(E') = 1 - 4/5 = 1/5$. These probabilities predict that E will occur 4 out of 5 times and E' will occur 1 out of 5 times. Then we say the *odds in favor* of E are 4 to 1.

ODDS

The **odds in favor** of an event E are defined as the ratio of $P(E)$ to $P(E')$, or

$$\frac{P(E)}{P(E')}, \quad P(E') \neq 0.$$

EXAMPLE 5

Odds in Favor of Rain

Suppose the weather forecaster says that the probability of rain tomorrow is $1/3$. Find the odds in favor of rain tomorrow.

▶**Solution** Let E be the event "rain tomorrow." Then E' is the event "no rain tomorrow." Since $P(E) = 1/3$, $P(E') = 2/3$. By the definition of odds, the odds in favor of rain are

$$\frac{1/3}{2/3} = \frac{1}{2}, \quad \text{written} \quad 1 \text{ to } 2, \text{ or } 1:2.$$

On the other hand, the odds that it will *not* rain, or the *odds against* rain, are

$$\frac{2/3}{1/3} = \frac{2}{1}, \quad \text{written} \quad 2 \text{ to } 1, \text{ or } 2:1.$$

If the odds in favor of an event are, say, 3 to 5, then the probability of the event is $3/8$, while the probability of the complement of the event is $5/8$. (Odds of 3 to 5 indicate 3 outcomes in favor of the event out of a total of 8 possible outcomes.) This example suggests the following generalization.

If the odds favoring event E are m to n, then

$$P(E) = \frac{m}{m+n} \quad \text{and} \quad P(E') = \frac{n}{m+n}.$$

EXAMPLE 6

Winning Bids

The odds that a particular bid will be the low bid are 4 to 5.

(a) Find the probability that the bid will be the low bid.

▶**Solution** Odds of 4 to 5 show 4 favorable chances out of $4 + 5 = 9$ chances altogether:

$$P(\text{bid will be low bid}) = \frac{4}{4+5} = \frac{4}{9}.$$

(b) Find the odds against that bid being the low bid.

▶**Solution** There is a $5/9$ chance that the bid will not be the low bid, so the odds against a low bid are

$$\frac{P(\text{bid will not be low})}{P(\text{bid will be low})} = \frac{5/9}{4/9} = \frac{5}{4},$$

or $5:4$.

EXAMPLE 7 **Odds in Horse Racing**

If the odds in favor of a particular horse's winning a race are 5 to 7, what is the probability that the horse will win the race?

▶**Solution** The odds indicate chances of 5 out of 12 ($5 + 7 = 12$) that the horse will win, so

$$P(\text{winning}) = \frac{5}{12}.$$

Race tracks generally give odds *against* a horse winning. In this case, the track would give the odds as 7 to 5. Of course, race tracks, casinos, and other gambling establishments need to give odds that are more favorable to the house than those representing the actual probabilities, because they need to make a profit.

Empirical Probability

In many real-life problems, it is not possible to establish exact probabilities for events. Instead, useful approximations are often found by drawing on past experience. The next example shows one approach to such **empirical probabilities**.

EXAMPLE 8 **Advertising Volume**

The following table lists U.S. advertising volume in millions of dollars by medium in 2004.*

Medium	Expenditures
Direct mail	52,191
Newspapers	46,614
Broadcast TV	46,264
Cable TV	21,527
Radio	19,581
Yellow pages	14,002
Magazines	12,247
Other	51,340

Find the empirical probability that a dollar of advertising is spent on each medium.

▶**Solution** We could first find the total spent and then divide the amount spent on each medium by the total. Verify that the amounts in the table sum to 263,766. The probability that a dollar is spent on newspapers, for example, is $P(\text{newspapers}) = 46{,}614/263{,}766 \approx 0.1767$. Similarly, we could divide each amount by 263,766, with the results (rounded to four decimal places) shown in the following table.

**The New York Times 2006 Almanac*, p. 357.

Medium	Probabilities
Direct mail	0.1979
Newspapers	0.1767
Broadcast TV	0.1754
Cable TV	0.0816
Radio	0.0742
Yellow pages	0.0531
Magazines	0.0464
Other	0.1946

The numbers in this table sum to 0.9999. In theory, they should total 1.0000, but this does not always occur when the individual numbers are rounded.

?

The categories in the table are mutually exclusive simple events. Thus, to find the probability that an advertising dollar is spent on newspapers or broadcast TV, we use the union rule to calculate

$$P(\text{newspapers or broadcast TV}) = 0.1767 + 0.1754 = 0.3521.$$

We could get this same result by summing the amount spent on newspapers and broadcast television, and dividing the total by 263,766.

Thus, more than a third of all advertising dollars are spent on these two media, a figure that should be of interest to both advertisers and the owners of the various media.

A table of probabilities, as in Example 8, sets up a **probability distribution**; that is, for each possible outcome of an experiment, a number, called the probability of that outcome, is assigned. This assignment may be done in any reasonable way (on an empirical basis, as in Example 8, or by theoretical reasoning, as in Section 7.3), provided that it satisfies the following conditions.

PROPERTIES OF PROBABILITY

Let S be a sample space consisting of n distinct outcomes, $s_1, s_2, \ldots, s_n$. An acceptable probability assignment consists of assigning to each outcome s_i a number p_i (the probability of s_i) according to these rules.

1. The probability of each outcome is a number between 0 and 1.

$$0 \le p_1 \le 1, \quad 0 \le p_2 \le 1, \ldots, \quad 0 \le p_n \le 1$$

2. The sum of the probabilities of all possible outcomes is 1.

$$p_1 + p_2 + p_3 + \cdots + p_n = 1$$

Probability distributions are discussed further in the next chapter.

EXAMPLE 9 **Clothing**

Susan is a college student who receives heavy sweaters from her aunt at the first sign of cold weather. Susan has determined that the probability that a sweater is

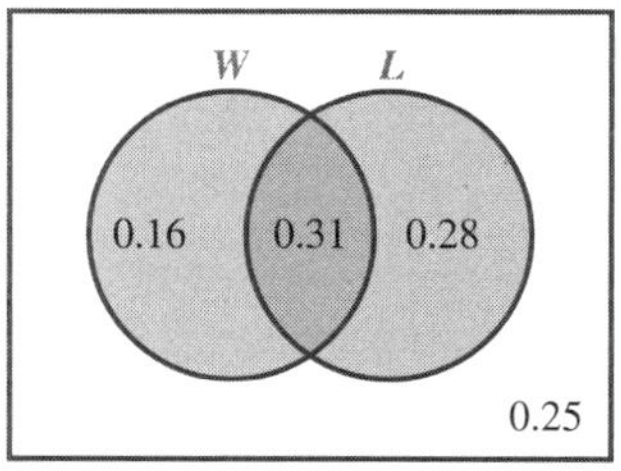

FIGURE 19

the wrong size is 0.47, the probability that it is a loud color is 0.59, and the probability that it is both the wrong size and a loud color is 0.31.

(a) Find the probability that the sweater is the correct size and not a loud color.

▶**Solution** Let W represent the event "wrong size," and L represent "loud color." Place the given information on a Venn diagram, starting with 0.31 in the intersection of the regions W and L (See Figure 19). As stated earlier, event W has probability 0.47. Since 0.31 has already been placed inside the intersection of W and L,

$$0.47 - 0.31 = 0.16$$

goes inside region W, but outside the intersection of W and L, that is, in the region $W \cap L'$. In the same way,

$$0.59 - 0.31 = 0.28$$

goes inside the region for L, and outside the overlap, that is, in the region $L \cap W'$.

Using regions W and L, the event we want is $W' \cap L'$. From the Venn diagram in Figure 19, the labeled regions have a total probability of

$$0.16 + 0.31 + 0.28 = 0.75.$$

Since the entire region of the Venn diagram must have probability 1, the region outside W and L, or $W' \cap L'$, has probability

$$1 - 0.75 = 0.25.$$

The probability is 0.25 that the sweater is the correct size and not a loud color.

(b) Find the probability that the sweater is the correct size or is not loud.

▶**Solution** The corresponding region, $W' \cup L'$, has probability

$$0.25 + 0.16 + 0.28 = 0.69.$$

7.4 Exercises

1. Define mutually exclusive events in your own words.

Decide whether the events in Exercises 2–7 are mutually exclusive.

2. Owning a dog and owning an MP3 player

3. Being a business major and being from Texas

4. Being retired and being 70 years old

5. Being a teenager and being 70 years old

6. Being one of the ten tallest people in the United States and being under 4 feet tall

7. Being male and being a nurse

Two dice are rolled. Find the probabilities of rolling the given sums.

8. a. 2 **b.** 4 **c.** 5 **d.** 6

9. a. 8 **b.** 9 **c.** 10 **d.** 13

10. **a.** 9 or more **b.** Less than 7 **c.** Between 5 and 8 (exclusive)

11. **a.** Not more than 5 **b.** Not less than 8 **c.** Between 3 and 7 (exclusive)

Two dice are rolled. Find the probabilities of the following events.

12. The first die is 3 or the sum is 8.

13. The second die is 5 or the sum is 10.

14. Three unusual dice, A, B, and C, are constructed such that die A has the numbers 3, 3, 4, 4, 8, 8; die B has the numbers 1, 1, 5, 5, 9, 9; and die C has the numbers 2, 2, 6, 6, 7, 7.

a. If dice A and B are rolled, find the probability that B beats A, that is, the number that appears on die B is greater than the number that appears on die A.

b. If dice B and C are rolled, find the probability that C beats B.

c. If dice A and C are rolled, find the probability that A beats C.

d. Which die is better? Explain.

One card is drawn from an ordinary deck of 52 cards. Find the probabilities of drawing the following cards.

15 **a.** A 9 or 10

b. A red card or a 3

c. A 9 or a black 10

d. A heart or a black card

e. A face card or a diamond

16. **a.** Less than a 4 (count aces as ones)

b. A diamond or a 7

c. A black card or an ace

d. A heart or a jack

e. A red card or a face card

Pam Snow invites 13 relatives to a party: her mother, 2 aunts, 3 uncles, 2 brothers, 1 male cousin, and 4 female cousins. If the chances of any one guest arriving first are equally likely, find the probabilities that the first guest to arrive is as follows.

17. **a.** A brother or an uncle **b.** A brother or a cousin **c.** A brother or her mother

18. **a.** An uncle or a cousin **b.** A male or a cousin **c.** A female or a cousin

The numbers 1, 2, 3, 4, and 5 are written on slips of paper, and 2 slips are drawn at random one at a time without replacement. Find the probabilities in Exercises 19 and 20.

19. **a.** The sum of the numbers is 9.

b. The sum of the numbers is 5 or less.

c. The first number is 2 or the sum is 6.

20. **a.** Both numbers are even.

b. One of the numbers is even or greater than 3.

c. The sum is 5 or the second number is 2.

Use Venn diagrams to work Exercises 21 and 22.

21. Suppose $P(E) = 0.26$, $P(F) = 0.41$, and $P(E \cap F) = 0.16$. Find the following.

a. $P(E \cup F)$ **b.** $P(E' \cap F)$ **c.** $P(E \cap F')$ **d.** $P(E' \cup F')$

22. Let $P(Z) = 0.42$, $P(Y) = 0.35$, and $P(Z \cup Y) = 0.59$. Find each probability.

a. $P(Z' \cap Y')$ **b.** $P(Z' \cup Y')$ **c.** $P(Z' \cup Y)$ **d.** $P(Z \cap Y')$

23. Define what is meant by odds.

A single fair die is rolled. Find the odds in favor of getting the results in Exercises 24–27.

24. 3 **25.** 4, 5, or 6 **26.** 2, 3, 4, or 5 **27.** Some number less than 6

28. A marble is drawn from a box containing 3 yellow, 4 white, and 11 blue marbles. Find the odds in favor of drawing the following.

a. A yellow marble **b.** A blue marble **c.** A white marble

29. Find the odds of *not* drawing a white marble in Exercise 28.

30. Two dice are rolled. Find the odds of rolling a 7 or 11.

31. In the "Ask Marilyn" column of *Parade* magazine, a reader wrote about the following game: You and I each roll a die. If your die is higher than mine, you win. Otherwise, I win. The reader thought that the probability that each player wins is 1/2. Is this correct? If not, what is the probability that each player wins?*

32. On page 134 of Roger Staubach's autobiography, *First Down, Lifetime to Go*, Staubach makes the following statement regarding his experience in Vietnam:†

> "Odds against a direct hit are very low but when your life is in danger, you don't worry too much about the odds."

Is this wording consistent with our definition of odds, for and against? How could it have been said so as to be technically correct?

33. The following table gives the odds that a particular event will occur.‡ Convert each odd to the probability that the event will occur.

Event	Odds for the Event
You will eat out today.	1 to 2
The next bottled water you buy will be nothing more than tap water.	1 to 4
The Earth will be struck by a huge meteor during your lifetime.	1 to 9000
You will go to Disney World this year.	1 to 9
You'll regain weight you lost by dieting.	9 to 10

Which of Exercises 34–41 are examples of empirical probability?

34. The probability of heads on 5 consecutive tosses of a coin

35. The probability that a freshman entering college will graduate with a degree

36. The probability that a person is allergic to penicillin

37. The probability of drawing an ace from a standard deck of 52 cards

38. The probability that a person will get lung cancer from smoking cigarettes

39. A weather forecast that predicts a 70% chance of rain tomorrow

40. A gambler's claim that on a roll of a fair die, $P(\text{even}) = 1/2$

41. A surgeon's prediction that a patient has a 90% chance of a full recovery

42. What is a probability distribution?

An experiment is conducted for which the sample space is $S = \{s_1, s_2, s_3, s_4, s_5\}$. *Which of the probability assignments in Exercises 43–48 is possible for this experiment? If an assignment is not possible, tell why.*

43.

Outcomes	s_1	s_2	s_3	s_4	s_5
Probabilities	0.09	0.32	0.21	0.25	0.13

*Source: *Parade* magazine, Nov. 6, 1994, p. 11. © 1994 Marilyn vos Savant. Initially published in *Parade* magazine. All rights reserved.

†Staubach, Roger, *First Down, Lifetime to Go*, Word Incorporated, Dallas, 1976.

‡The Forum for Investor Advice; Krantz, Les, *What the Odds Are*, Harper Perennial, 1992; and Laudan, Larry, *Danger Ahead: The Risks You Really Face on Life's Highway*, John Wiley & Sons, New York, 1997.

44.

Outcomes	s_1	s_2	s_3	s_4	s_5
Probabilities	0.92	0.03	0	0.02	0.03

45.

Outcomes	s_1	s_2	s_3	s_4	s_5
Probabilities	1/3	1/4	1/6	1/8	1/10

46.

Outcomes	s_1	s_2	s_3	s_4	s_5
Probabilities	1/5	1/3	1/4	1/5	1/10

47.

Outcomes	s_1	s_2	s_3	s_4	s_5
Probabilities	0.64	−0.08	0.30	0.12	0.02

48.

Outcomes	s_1	s_2	s_3	s_4	s_5
Probabilities	0.05	0.35	0.5	0.2	−0.3

One way to solve a probability problem is to repeat the experiment many times, keeping track of the results. Then the probability can be approximated using the basic definition of the probability of an event E: $P(E) = n(E)/n(S)$, *where E occurs* $n(E)$ *times out of* $n(S)$ *trials of an experiment. This is called the Monte Carlo method of finding probabilities. If physically repeating the experiment is too tedious, it may be simulated using a random-number generator, available on most computers and scientific or graphing calculators. To simulate a coin toss or the roll of a die on the TI-83/84 Plus, change the setting to fixed decimal mode with* 0 *digits displayed, and enter* `rand` *or* `rand*6+.5`, *respectively. For a coin toss, interpret* 0 *as a head and* 1 *as a tail. In either case, the* `ENTER` *key can be pressed repeatedly to perform multiple simulations.*

49. Suppose two dice are rolled. Use the Monte Carlo method with at least 50 repetitions to approximate the following probabilities. Compare with the results of Exercise 11.

a. *P*(the sum is not more than 5) **b.** *P*(the sum is not less than 8)

50. Suppose two dice are rolled. Use the Monte Carlo method with at least 50 repetitions to approximate the following probabilities. Compare with the results of Exercise 10.

a. *P*(the sum is 9 or more) **b.** *P*(the sum is less than 7)

51. Suppose three dice are rolled. Use the Monte Carlo method with at least 100 repetitions to approximate the following probabilities.

a. *P*(the sum is 5 or less) **b.** *P*(neither a 1 nor a 6 is rolled)

52. Suppose a coin is tossed 5 times. Use the Monte Carlo method with at least 50 repetitions to approximate the following probabilities.

a. *P*(exactly 4 heads) **b.** *P*(2 heads and 3 tails)

53. The following description of the classic "Linda problem" appeared in the *New Yorker**: "In this experiment, subjects are told, 'Linda is thirty-one years old, single, outspoken, and very bright. She majored in philosophy. As a student, she was deeply concerned with issues of discrimination and social justice and also participated in

*Menand, Louis, "Everybody's an Expert: Putting Predictions to the Test," *New Yorker*, Dec. 5, 2005, pp. 98–101.

antinuclear demonstrations.' They are then asked to rank the probability of several possible descriptions of Linda today. Two of them are 'bank teller' and 'bank teller and active in the feminist movement.'" Many people rank the second event as more likely. Explain why this violates basic concepts of probability.

54. You are given $P(A \cup B) = 0.7$ and $P(A \cup B') = 0.9$. Determine $P(A)$.* (Choose one of the following.)

a. 0.2 **b.** 0.3 **c.** 0.4 **d.** 0.6 **e.** 0.8

Applications

BUSINESS AND ECONOMICS

55. *Defective Merchandise* Suppose that 8% of a certain batch of calculators have a defective case, and that 11% have defective batteries. Also, 3% have both a defective case and defective batteries. A calculator is selected from the batch at random. Find the probability that the calculator has a good case and good batteries.

56. *Credit Charges* The table shows the probabilities of a person accumulating specific amounts of credit card charges over a 12-month period. Find the probabilities that a person's total charges during the period are the following.

a. \$500 or more **b.** Less than \$1000

c. \$500 to \$2999 **d.** \$3000 or more

Charges	Probability
Under \$100	0.21
\$100–\$499	0.17
\$500–\$999	0.16
\$1000–\$1999	0.15
\$2000–\$2999	0.12
\$3000–\$4999	0.08
\$5000–\$9999	0.07
\$10,000 or more	0.04

Customer Purchases The following table shows the probability that a customer of a department store will make a purchase in the indicated range.

Amount Spent	Probability
Below \$10	0.02
\$10–\$24.99	0.05
\$25–\$49.99	0.11
\$50–\$74.99	0.13
\$75–\$99.99	0.14
\$100–\$199.99	0.22
\$200–\$349.99	0.18
\$350–\$499.99	0.12
\$500 or more	0.03

Find the probabilities that a customer makes a purchase in the following ranges.

57. a. Less than \$25 **b.** More than \$24.99

c. \$50 to \$199.99

58. a. Less than \$350 **b.** \$75 or more

c. \$200 or more

59. *Profit* The probability that a company will make a profit this year is 0.74. Find the odds against the company making a profit.

LIFE SCIENCES

60. *Body Types* A study on body types gave the following results: 45% were short; 25% were short and overweight; and 24% were tall and not overweight. Find the probabilities that a person is the following.

a. Overweight

b. Short, but not overweight

c. Tall and overweight

*Problem 3 from the 2005 Sample Exam P of the *Education and Examination Committee of the Society of Actuaries.* Reprinted by permission of the Society of Actuaries.

61. *Color Blindness* Color blindness is an inherited characteristic that is more common in males than in females. If M represents male and C represents red-green color blindness, we use the relative frequencies of the incidences of males and red-green color blindness as probabilities to get

$$P(C) = 0.039, P(M \cap C) = 0.035, P(M \cup C) = 0.491.^*$$

Find the following probabilities.

a. $P(C')$ **b.** $P(M)$ **c.** $P(M')$

d. $P(M' \cap C')$ **e.** $P(C \cap M')$ **f.** $P(C \cup M')$

62. *Genetics* Gregor Mendel, an Austrian monk, was the first to use probability in the study of genetics. In an effort to understand the mechanism of character transmittal from one generation to the next in plants, he counted the number of occurrences of various characteristics. Mendel found that the flower color in certain pea plants obeyed this scheme:

Pure red crossed with pure white produces red.

From its parents, the red offspring received genes for both red (R) and white (W), but in this case red is *dominant* and white *recessive*, so the offspring exhibits the color red. However, the offspring still carries both genes, and when two such offspring are crossed, several things can happen in the third generation. The table below, which is called a *Punnet square*, shows the equally likely outcomes.

		Second Parent	
		R	*W*
First Parent	*R*	*RR*	*RW*
	W	*WR*	*WW*

Use the fact that red is dominant over white to find the following. Assume that there are an equal number of red and white genes in the population.

a. P(a flower is red) **b.** P(a flower is white)

63. *Genetics* Mendel found no dominance in snapdragons, with one red gene and one white gene producing pink-flowered offspring. These second-generation pinks, however, still carry one red and one white gene, and when they are crossed, the next generation still yields the Punnet square from Exercise 62. Find each probability.

a. P(red) **b.** P(pink) **c.** P(white)

(Mendel verified these probability ratios experimentally and did the same for many characteristics other than flower color. His work, published in 1866, was not recognized until 1890.)

64. *Genetics* In most animals and plants, it is very unusual for the number of main parts of the organism (such as arms, legs, toes, or flower petals) to vary from generation to generation. Some species, however, have *meristic variability,* in which the number of certain body parts varies from generation to generation. One researcher studied the front feet of certain guinea pigs and produced the following probabilities.[†]

$$P(\text{only four toes, all perfect}) = 0.77$$
$$P(\text{one imperfect toe and four good ones}) = 0.13$$
$$P(\text{exactly five good toes}) = 0.10$$

Find the probability of each event.

a. No more than four good toes

b. Five toes, whether perfect or not

65. *Doctor Visit* The probability that a visit to a primary care physician's (PCP) office results in neither lab work nor referral to a specialist is 35%. Of those coming to a PCP's office, 30% are referred to specialists and 40% require lab work. Determine the probability that a visit to a PCP's office results in both lab work and referral to a specialist.[‡] (Choose one of the following. *Hint:* Use the union rule for probability.)

a. 0.05 **b.** 0.12 **c.** 0.18

d. 0.25 **e.** 0.35

66. *Shoulder Injuries* Among a large group of patients recovering from shoulder injuries, it is found that 22% visit both a physical therapist and a chiropractor, whereas 12% visit neither of these. The probability that a patient visits a chiropractor exceeds by 0.14 the probability that a patient visits a physical therapist. Determine the probability that a randomly chosen member of this group visits a physical

*The probabilities of a person being male or female are from *The World Almanac and Book of Facts*, 1995. The probabilities of a male and female being color-blind are from *Parsons' Diseases of the Eye* (18th ed.) by Stephen J. H. Miller, Churchill Livingstone, 1990, p. 269. This reference gives a range of 3 to 4% for the probability of gross color blindness in men; we used the midpoint of this range.

[†]Wright, J. R., Data from "An Analysis of Variability in Guinea Pigs," *Genetics,* Vol. 19, pp. 506–536.

[‡]Problem 2 from the 2005 Sample Exam P of the *Education and Examination Committee of the Society of Actuaries*. Reprinted by permission of the Society of Actuaries.

therapist.[*] (Choose one of the following. *Hint:* Use the union rule for probability, and let $x = P$(patient visits a physical therapist).)

a. 0.26 **b.** 0.38 **c.** 0.40

d. 0.48 **e.** 0.62

67. *Health Plan* An insurer offers a health plan to the employees of a large company. As part of this plan, the individual employees may choose exactly two of the supplementary coverages A, B, and C, or they may choose no supplementary coverage. The proportions of the company's employees that choose coverages A, B, and C are 1/4, 1/3, and 5/12, respectively. Determine the probability that a randomly chosen employee will choose no supplementary coverage.[†] (Choose one of the following. *Hint:* Draw a Venn diagram with three sets, and let $x = P(A \cap B)$. Use the fact that 4 of the 8 regions in the Venn diagram have a probability of 0.)

a. 0 **b.** 47/144 **c.** 1/2

d. 97/144 **e.** 7/9

SOCIAL SCIENCES

68. *Presidential Candidates* In 2002, *The New York Times* columnist William Safire gave the following odds against various prominent Democrats receiving their party's presidential nomination in 2004.[‡]

Al Gore: 2 to 1
Tom Daschle: 4 to 1
John Kerry: 4 to 1
Chris Dodd: 4 to 1
Joe Lieberman: 5 to 1
Joe Biden: 5 to 1
Pat Leahy: 6 to 1
Russell Feingold: 8 to 1
John Edwards: 9 to 1
Dick Gephardt: 15 to 1

John Allen Paulos observed that there is something wrong with those odds.[§] Translate these odds into probabilities of winning the nomination, and then explain why these are not possible.

69. *Earnings* The following data were gathered for 130 adult U.S. workers: 55 were women; 3 women earned more than $40,000; and 62 men earned $40,000 or less. Find the probability that an individual is

a. a woman earning $40,000 or less;

b. a man earning more than $40,000;

c. a man or is earning more than $40,000;

d. a woman or is earning $40,000 or less.

70. *Expenditures for Music* A survey of 100 people about their music expenditures gave the following information: 38 bought rock music; 20 were teenagers who bought rock music; and 26 were teenagers. Find the probabilities that a person is

a. a teenager who buys nonrock music;

b. someone who buys rock music or is a teenager;

c. not a teenager;

d. not a teenager, but a buyer of rock music.

71. *Refugees* In a refugee camp in southern Mexico, it was found that 90% of the refugees came to escape political oppression, 80% came to escape abject poverty, and 70% came to escape both. What is the probability that a refugee in the camp was not poor nor seeking political asylum?

72. *Community Activities* At the first meeting of a committee to plan a local Lunar New Year celebration, the persons attending are 3 Chinese men, 4 Chinese women, 3 Vietnamese women, 2 Vietnamese men, 4 Korean women, and 2 Korean men. A chairperson is selected at random. Find the probabilities that the chairperson is the following.

a. Chinese

b. Korean or a woman

c. A man or Vietnamese

d. Chinese or Vietnamese

e. Korean and a woman

73. *Elections* If the odds that a given candidate will win an election are 3 to 2, what is the probability that the candidate will lose?

74. *Military* There were 212,156 female military personnel in 2004 in various ranks and military branches, as listed in the table on the next page.[‖]

[*]Problem 8 from the 2005 Sample Exam P of the *Education and Examination Committee of the Society of Actuaries*. Reprinted by permission of the Society of Actuaries.
[†]Problem 15 from the 2005 Sample Exam P of the *Education and Examination Committee of the Society of Actuaries*. Reprinted by permission of the Society of Actuaries.
[‡]Safire, William, "The Henry Poll," *The New York Times*, June 25, 2001, p. A17.
[§]http://abcnews.go.com/sections/scitech/DailyNews/safire010628.html.
[‖]*Selected Manpower Statistics, Fiscal Year 2004*, U.S. Department of Defense, p. 31.

	Army (A)	Air Force (B)	Navy (C)	Marines (D)
Officers (O)	12,309	13,596	8111	1096
Enlisted (E)	60,395	59,436	45,415	9683
Cadets & Midshipmen (M)	639	754	722	0

a. Convert the numbers in the table to probabilities.

b. Find the probability that a randomly selected woman is in the Army.

c. Find the probability that a randomly selected woman is an officer in the Navy or Marine Corps.

d. $P(A \cup B)$

e. $P(E \cup (C \cup D))$

75. *Perceptions of Threat* Research has been carried out to measure the amount of intolerance that citizens of Russia have for left-wing Communists and right-wing Fascists, as indicated in the first table below. Note that the numbers are given as percents and each row sums to 100 (except for rounding).*

a. Find the probability that a randomly chosen citizen of Russia would be somewhat or extremely intolerant of right-wing Fascists.

b. Find the probability that a randomly chosen citizen of Russia would be completely tolerant of left-wing Communists.

c. Compare your answers to parts a and b and provide possible reasons for these numbers.

76. *Perceptions of Threat* Research has been carried out to measure the amount of intolerance that U.S. citizens have for left-wing Communists and right-wing Fascists, as indicated in the second table below. Note that the numbers are given as percents and each row sums to 100 (except for rounding).*

a. Find the probability that a randomly chosen U.S. citizen would have at least some intolerance of right-wing Fascists.

b. Find the probability that a randomly chosen U.S. citizen would have at least some intolerance of left-wing Communists.

c. Compare your answers to parts a and b and provide possible reasons for these numbers.

d. Compare these answers to the answers to Exercise 75.

Russia	None at All	Don't Know	Not Very Much	Somewhat	Extremely
Left-Wing Communists	47.8	6.7	31.0	10.5	4.1
Right-Wing Fascists	3.0	3.2	7.1	27.1	59.5

United States	None at All	Don't Know	Not Very Much	Somewhat	Extremely
Left-Wing Communists	13.0	2.7	33.0	34.2	17.1
Right-Wing Fascists	10.1	3.3	20.7	43.1	22.9

*Gibson, J. L., "Putting Up with Fellow Russians: An Analysis of Political Tolerance in the Fledgling Russian Democracy," *Political Research Quarterly*, Vol. 51, No. 1, Mar. 1998, pp. 37–68.

GENERAL INTEREST

77. *Weather* If the odds that it will rain are 4 to 7, what is the probability of rain? Interpret your answer.

78. *Olympics* In recent winter Olympics, each part of the women's figure skating program has 12 judges, but the scores of only 9 of the judges are randomly selected for the final results.* As we will see in the next chapter, there are 220 possible ways for the 9 judges whose scores are counted to be selected. *The New York Times* examined those 220 possibilities for the short program in the 2006 Olympics, based on the published scores of the judges, and listed what the results would have been for each, as shown below.

a. The winner of the short program was Sasha Cohen. For a random combination of 9 judges, what is the probability of that outcome?

b. The second place finisher in the short program was Irina Slutskaya. For a random combination of 9 judges, what is the probability of that outcome?

c. The third place finisher in the short program was Shizuka Arakawa. For a random combination of 9 judges, what is the probability of that outcome? Do not include outcomes that include a tie.

Outcome	1. Slutskaya 2. Cohen 3. Arakawa	1. Slutskaya 2. Arakawa 3. Cohen	1. Slutskaya 2, 3. Arakawa and Cohen tied	1. Cohen 2. Slutskaya 3. Arakawa	1. Cohen 2. Arakawa 3. Slutskaya
Number of Possible Judging Combinations	92	33	3	67	25

7.5 Conditional Probability; Independent Events

? THINK ABOUT IT

What is the probability that a broker who uses research picks stocks that go up?

The training manager for a large brokerage firm has noticed that some of the firm's stockbrokers use the firm's research advice, while other brokers tend to follow their own feelings of which stocks will go up. To see whether the research department performs better than brokers' feelings, the manager surveyed 100 brokers, with results as shown in the following table.

	Picked Stocks That Went Up (A)	**Didn't Pick Stocks That Went Up (A')**	**Totals**
Used Research (B)	30	15	45
Didn't Use Research (B')	30	25	55
Totals	60	40	100

**The New York Times*, Feb. 23, 2006, p. D3.

Letting A represent the event "picked stocks that went up," and letting B represent the event "used research," we can find the following probabilities.

$$P(A) = \frac{60}{100} = 0.6 \qquad P(A') = \frac{40}{100} = 0.4$$

$$P(B) = \frac{45}{100} = 0.45 \qquad P(B') = \frac{55}{100} = 0.55$$

To answer the question asked at the beginning of this section, suppose we want to find the probability that a broker using research will pick stocks that go up. From the table, of the 45 brokers who use research, 30 picked stocks that went up, with

$$P(\text{broker who uses research picks stocks that go up}) = \frac{30}{45} \approx 0.6667.$$

This is a different number than the probability that a broker picks stocks that go up, 0.6, since we have additional information (the broker uses research) that has *reduced the sample space*. It other words, we found the probability that a broker picks stocks that go up, A, given the additional information that the broker uses research, B. This is called the *conditional probability* of event A, given that event B has occurred, written $P(A|B)$. ($P(A|B)$ may also be read as "the probability of A given B.")

In the example above,

$$P(A|B) = \frac{30}{45},$$

which can be written as

$$P(A|B) = \frac{30/100}{45/100} = \frac{P(A \cap B)}{P(B)},$$

where $P(A \cap B)$ represents, as usual, the probability that both A and B will occur.

To generalize this result, assume that E and F are two events for a particular experiment, and that all events in the sample space S are equally likely. Using the fundamental principle of probability,

$$P(F) = \frac{n(F)}{n(S)} \qquad \text{and} \qquad P(E \cap F) = \frac{n(E \cap F)}{n(S)}.$$

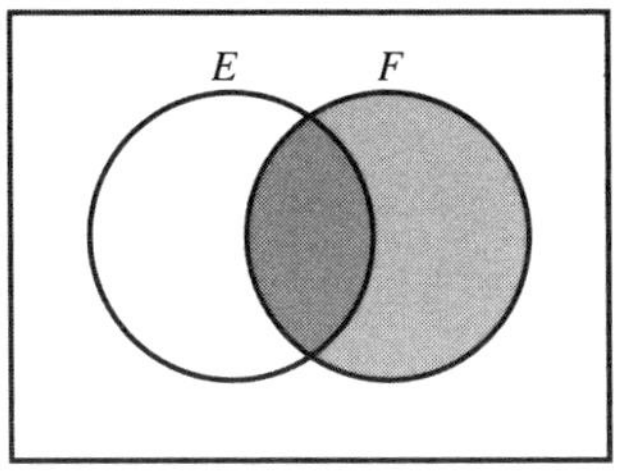

Event F has a total of $n(F)$ elements.

FIGURE 20

We now want $P(E|F)$, the probability that E occurs given that F has occurred. Since we assume F has occurred, reduce the sample space to F: look only at the elements inside F. See Figure 20. Of these $n(F)$ elements, there are $n(E \cap F)$ elements where E also occurs. This makes

$$P(E|F) = \frac{n(E \cap F)}{n(F)}.$$

Divide numerator and denominator by $n(S)$ to get

$$P(E|F) = \frac{n(E \cap F)/n(S)}{n(F)/n(S)} = \frac{P(E \cap F)}{P(F)}.$$

This last result motivates the definition of conditional probability.

CONDITIONAL PROBABILITY

The **conditional probability** of event E given event F, written $P(E|F)$, is

$$P(E|F) = \frac{P(E \cap F)}{P(F)}, \quad \text{where } P(F) \neq 0.$$

This definition tells us that, for equally likely outcomes, conditional probability is found by *reducing the sample space to event F*, and then finding the number of outcomes in F that are also in event E. Thus,

$$P(E|F) = \frac{n(E \cap F)}{n(F)}.$$

Although the definition of conditional probability was motivated by an example with equally likely outcomes, it is valid in all cases. For an intuitive explanation, think of the formula as giving the probability that both E and F occur compared with the entire probability of F.

EXAMPLE 1 **Stocks**

Use the information given in the chart at the beginning of this section to find the following probabilities.

(a) $P(B|A)$

▶**Solution** This represents the probability that the broker used research, given that the broker picked stocks that went up. Reduce the sample space to A. Then find $n(A \cap B)$ and $n(A)$.

$$P(B|A) = \frac{P(B \cap A)}{P(A)} = \frac{n(A \cap B)}{n(A)} = \frac{30}{60} = \frac{1}{2}$$

If a broker picked stocks that went up, then the probability is 1/2 that the broker used research.

(b) $P(A'|B)$

▶**Solution** In words, this is the probability that a broker picks stocks that do not go up, even though he used research.

$$P(A'|B) = \frac{n(A' \cap B)}{n(B)} = \frac{15}{45} = \frac{1}{3}$$

(c) $P(B'|A')$

▶**Solution** Here, we want the probability that a broker who picked stocks that did not go up did not use research.

$$P(B'|A') = \frac{n(B' \cap A')}{n(A')} = \frac{25}{40} = \frac{5}{8}$$

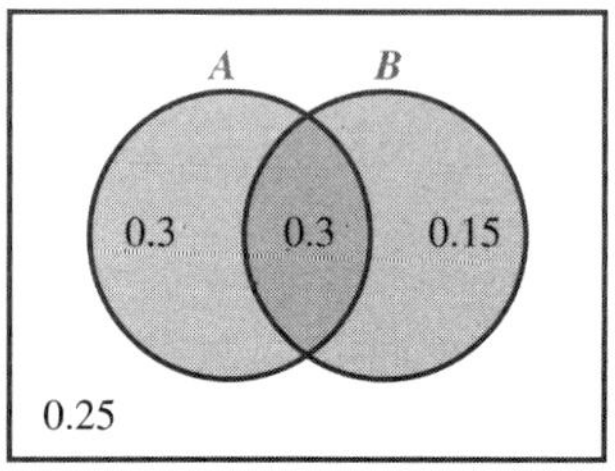

$P(A) = 0.3 + 0.3 = 0.6$

FIGURE 21

Venn diagrams are useful for illustrating problems in conditional probability. A Venn diagram for Example 1, in which the probabilities are used to indicate the number in the set defined by each region, is shown in Figure 21. In the diagram,

$P(B|A)$ is found by reducing the sample space to just set A. Then $P(B|A)$ is the ratio of the number in that part of set B that is also in A to the number in set A, or $0.3/0.6 = 0.5$.

EXAMPLE 2

Conditional Probabilities

Given $P(E) = 0.4$, $P(F) = 0.5$, and $P(E \cup F) = 0.7$, find $P(E|F)$.

▶Solution Find $P(E \cap F)$ first. By the union rule,

$$P(E \cup F) = P(E) + P(F) - P(E \cap F)$$
$$0.7 = 0.4 + 0.5 - P(E \cap F)$$
$$P(E \cap F) = 0.2.$$

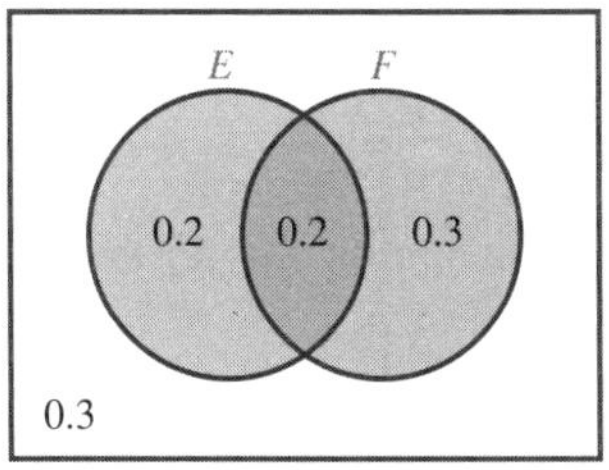

FIGURE 22

$P(E|F)$ is the ratio of the probability of that part of E that is in F to the probability of F, or

$$P(E|F) = \frac{P(E \cap F)}{P(F)} = \frac{0.2}{0.5} = \frac{2}{5}.$$

The Venn diagram in Figure 22 illustrates Example 2.

EXAMPLE 3

Tossing Coins

Two fair coins were tossed, and it is known that at least one was a head. Find the probability that both were heads.

▶Solution At first glance the answer to this question appears to be $1/2$. Using mathematics, however, we will see otherwise. The sample space has four equally likely outcomes, $S = \{hh, ht, th, tt\}$. Define two events:

$$E_1 = \text{at least 1 head} = \{hh, ht, th\},$$

and

$$E_2 = \text{2 heads} = \{hh\}.$$

Since there are four equally likely outcomes, $P(E_1) = 3/4$. Also, $P(E_1 \cap E_2) = 1/4$. We want the probability that both were heads, given that at least one was a head; that is, we want to find $P(E_2|E_1)$. Because of the condition that at least one coin was a head, the reduced sample space is

$$\{hh, ht, th\}.$$

Since only one outcome in this reduced sample space is 2 heads,

$$P(E_2|E_1) = \frac{1}{3}.$$

Alternatively, use the definition given above.

$$P(E_2|E_1) = \frac{P(E_2 \cap E_1)}{P(E_1)} = \frac{1/4}{3/4} = \frac{1}{3}$$

EXAMPLE 4

Playing Cards

Two cards are drawn from a standard deck, one after another without replacement. Find the probability that the second card is red, given that the first card is red.

Solution According to the conditional probability formula,

$$P(\text{second card is red} \mid \text{first card is red}) = \frac{P(\text{second card is red and the first card is red})}{P(\text{first card is red})}.$$

We will soon see how to compute probabilities such as the one in the numerator. But there is a much simpler way to calculate this conditional probability. We only need to observe that with one red card gone, there are 51 cards left, 25 of which are red, so

$$P(\text{second card is red} \mid \text{first card is red}) = \frac{25}{51}.$$

It is important not to confuse $P(A \mid B)$ with $P(B \mid A)$. For example, in a criminal trial, a prosecutor may point out to the jury that the probability of the defendant's DNA profile matching that of a sample taken at the scene of the crime, given that the defendant is innocent, $P(D \mid I)$, is very small. What the jury must decide, however, is the probability that the defendant is innocent, given that the defendant's DNA profile matches the sample, $P(I \mid D)$. Confusing the two is an error sometimes called "the prosecutor's fallacy," and the 1990 conviction of a rape suspect in England was overturned by a panel of judges, who ordered a retrial, because the fallacy made the original trial unfair.*

In the next section, we will see how to compute $P(A \mid B)$ when we know $P(B \mid A)$.

Product Rule

If $P(E) \neq 0$ and $P(F) \neq 0$, then the definition of conditional probability shows that

$$P(E \mid F) = \frac{P(E \cap F)}{P(F)} \quad \text{and} \quad P(F \mid E) = \frac{P(F \cap E)}{P(E)}.$$

Using the fact that $P(E \cap F) = P(F \cap E)$, and solving each of these equations for $P(E \cap F)$, we obtain the following rule.

PRODUCT RULE OF PROBABILITY

If E and F are events, then $P(E \cap F)$ may be found by either of these formulas.

$$\boldsymbol{P(E \cap F) = P(F) \cdot P(E \mid F)} \quad \text{or} \quad \boldsymbol{P(E \cap F) = P(E) \cdot P(F \mid E)}$$

The product rule gives a method for finding the probability that events E and F both occur, as illustrated by the next few examples.

EXAMPLE 5 **Business Majors**

In a class with 2/5 women and 3/5 men, 25% of the women are business majors. Find the probability that a student chosen from the class at random is a female business major.

*Pringle, David, "Who's the DNA Fingerprinting Pointing At?" *New Scientist*, Jan. 29, 1994, pp. 51–52.

▸Solution Let B and W represent the events "business major" and "woman," respectively. We want to find $P(B \cap W)$. By the product rule,

$$P(B \cap W) = P(W) \cdot P(B|W).$$

Using the given information, $P(W) = 2/5 = 0.4$ and $P(B|W) = 0.25$. Thus,

$$P(B \cap W) = 0.4(0.25) = 0.10.$$

The next examples show how a tree diagram is used with the product rule to find the probability of a sequence of events.

EXAMPLE 6

Advertising

A company needs to hire a new director of advertising. It has decided to try to hire either person A or B, who are assistant advertising directors for its major competitor. To decide between A and B, the company does research on the campaigns managed by either A or B (no campaign is managed by both), and finds that A is in charge of twice as many advertising campaigns as B. Also, A's campaigns have satisfactory results 3 out of 4 times, while B's campaigns have satisfactory results only 2 out of 5 times. Suppose one of the competitor's advertising campaigns (managed by A or B) is selected randomly.

We can represent this situation schematically as follows. Let A denote the event "person A does the job" and B the event "person B does the job." Notice that in this situation, A and B are complementary events, that is, $A = B'$, and $A \cup B$ is the sample space. Let S be the event "satisfactory results" and U the event "unsatisfactory results." Then the given information can be summarized in the tree diagram in Figure 23. Since A does twice as many jobs as B, we have $P(A) = 2/3$ and $P(B) = 1/3$, as noted on the first-stage branches of the tree. When A does a job, the probability of satisfactory results is $3/4$, and of unsatisfactory results, $1/4$, as noted on the second-stage branches. Similarly, the probabilities when B does the job are noted on the remaining second-stage branches. The composite branches labeled 1–4 represent the four mutually exclusive possibilities for the running and outcome of the selected campaign.

(a) Find the probability that A is in charge of the selected campaign and that it produces satisfactory results.

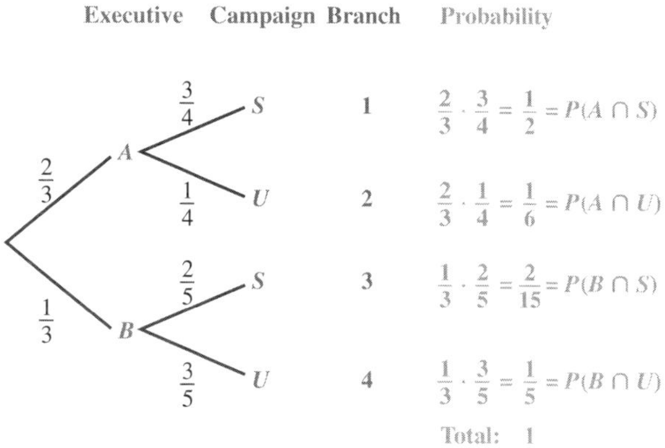

FIGURE 23

▶**Solution** We are asked to find $P(A \cap S)$. We know that when A does the job, the probability of success is 3/4, that is, $P(S|A) = 3/4$. Hence, by the product rule,

$$P(A \cap S) = P(A) \cdot P(S|A) = \frac{2}{3} \cdot \frac{3}{4} = \frac{1}{2}.$$

The event $A \cap S$ is represented by branch 1 of the tree, and as we have just seen, its probability is the product of the probabilities of the pieces that make up that branch.

(b) Find the probability that B runs the campaign and that it produces satisfactory results.

▶**Solution** We must find $P(B \cap S)$. The event is represented by branch 3 of the tree and, as before, its probability is the product of the probabilities of the pieces of that branch:

$$P(B \cap S) = P(B) \cdot P(S|B) = \frac{1}{3} \cdot \frac{2}{5} = \frac{2}{15}.$$

(c) What is the probability that the selected campaign is satisfactory?

▶**Solution** The event S is the union of the mutually exclusive events $A \cap S$ and $B \cap S$, which are represented by branches 1 and 3 of the diagram. By the union rule,

$$P(S) = P(A \cap S) + P(B \cap S) = \frac{1}{2} + \frac{2}{15} = \frac{19}{30}.$$

Thus, the probability of an event that appears on several branches is the sum of the probabilities of each of these branches.

(d) What is the probability that the selected campaign is unsatisfactory?

▶**Solution** $P(U)$ can be read from branches 2 and 4 of the tree.

$$P(U) = \frac{1}{6} + \frac{1}{5} = \frac{11}{30}$$

Alternatively, since U is the complement of S,

$$P(U) = 1 - P(S) = 1 - \frac{19}{30} = \frac{11}{30}.$$

EXAMPLE 7

Environmental Inspections

The Environmental Protection Agency is considering inspecting 6 plants for environmental compliance: 3 in Chicago, 2 in Los Angeles, and 1 in New York. Due to a lack of inspectors, they decide to inspect two plants selected at random, one this month and one next month, with each plant equally likely to be selected, but no plant selected twice. What is the probability that 1 Chicago plant and 1 Los Angeles plant are selected?

▶**Solution** A tree diagram showing the various possible outcomes is given in Figure 24. In this diagram, the events of inspecting a plant in Chicago, Los Angeles, and New York are represented by C, LA, and NY, respectively. For the first inspection, $P(\text{C first}) = 3/6 = 1/2$ because 3 of the 6 plants are in Chicago, and

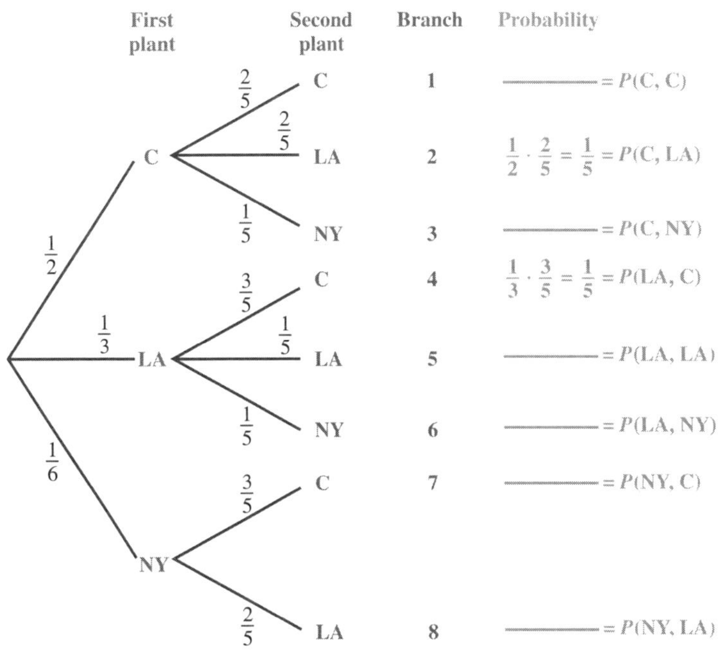

FIGURE 24

all plants are equally likely to be selected. For the second inspection, $P(\text{LA second} \mid \text{C first}) = 2/5$. One plant has been removed, leaving 5, of which 2 are in Los Angeles.

We want to find the probability of selecting exactly 1 Chicago plant and 1 Los Angeles plant. This event can occur in two ways: inspecting Chicago this month and Los Angeles next month (branch 2 of the tree diagram), or inspecting Los Angeles this month and Chicago next month (branch 4). For branch 2,

$$P(\text{C first}) \cdot P(\text{LA second} \mid \text{C first}) = \frac{1}{2} \cdot \frac{2}{5} = \frac{1}{5}.$$

For branch 4, where Los Angeles is inspected first,

$$P(\text{LA first}) \cdot P(\text{C second} \mid \text{LA first}) = \frac{1}{3} \cdot \frac{3}{5} = \frac{1}{5}.$$

Since the two events are mutually exclusive, the final probability is the sum of these two probabilities.

$$\begin{aligned} P(1\ \text{C}, 1\ \text{LA}) &= P(\text{C first}) \cdot P(\text{LA second} \mid \text{C first}) \\ &\quad + P(\text{LA first}) \cdot P(\text{C second} \mid \text{LA first}) \\ &= \frac{2}{5} \end{aligned}$$

The product rule is often used with *stochastic processes*, which are mathematical models that evolve over time in a probabilistic manner. For example, selecting factories at random for inspection is such a process, in which the probabilities change with each successive selection.

EXAMPLE 8 **Playing Cards**

Two cards are drawn from a standard deck, one after another without replacement.

> **FOR REVIEW**
> You may wish to refer to the picture of a deck of cards shown in Figure 17 (Section 7.3) and the description accompanying it.

(a) Find the probability that the first card is a heart and the second card is red.

Solution Start with the tree diagram in Figure 25. On the first draw, since there are 13 hearts among the 52 cards, the probability of drawing a heart is $13/52 = 1/4$. On the second draw, since a (red) heart has been drawn already, there are 25 red cards in the remaining 51 cards. Thus, the probability of drawing a red card on the second draw, given that the first is a heart, is $25/51$. By the product rule of probability,

$$\begin{aligned} &P(\text{heart first and red second}) \\ &\quad = P(\text{heart first}) \cdot P(\text{red second} \mid \text{heart first}) \\ &\quad = \frac{1}{4} \cdot \frac{25}{51} = \frac{25}{204} \approx 0.123. \end{aligned}$$

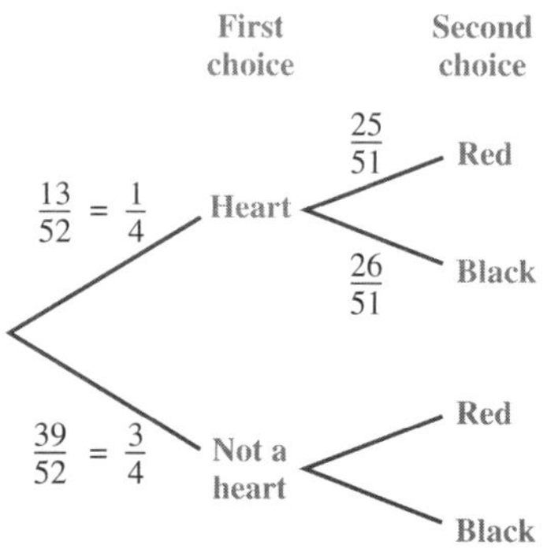

FIGURE 25

(b) Find the probability that the second card is red.

Solution To solve this, we need to fill out the bottom branch of the tree diagram in Figure 25. Unfortunately, if the first card is not a heart, it is not clear how to find the probability that the second card is red, because it depends upon whether the first card is red or black. One way to solve this problem would be to divide the bottom branch into two separate branches: diamond and black card (club or spade). There is a simpler way, however, since we don't care whether or not the first card is a heart, as we did in part (a). Instead, we'll consider whether the first card is red or black, and then do the same for the second card. The result, with the corresponding probabilities, is in Figure 26. The probability that the second card is red is found by multiplying the probabilities along the two branches and adding.

$$\begin{aligned} P(\text{red second}) &= \frac{1}{2} \cdot \frac{25}{51} + \frac{1}{2} \cdot \frac{26}{51} \\ &= \frac{1}{2} \end{aligned}$$

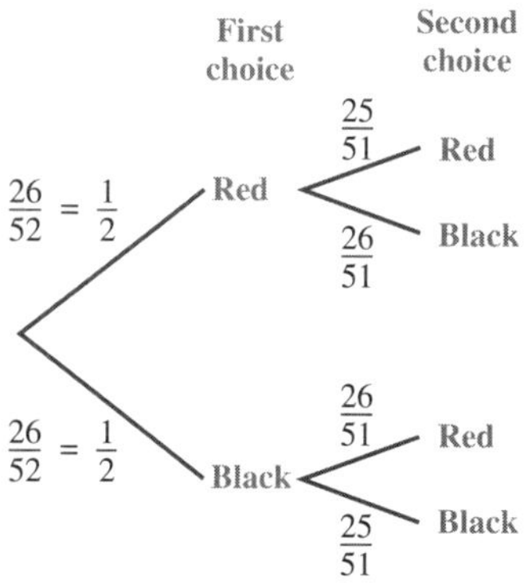

FIGURE 26

The probability is $1/2$, exactly the same as the probability that any card is red. If we know nothing about the first card, there is no reason for the probability of the second card to be anything other than $1/2$.

Independent Events

Suppose, in Example 8(a), that we draw the two cards *with* replacement rather than without replacement (that is, we put the first card back before drawing the second card). If the first card is a heart, then the probability of drawing a red card on the second draw is $26/52$, rather than $25/51$, because there are still 52 cards in the deck, 26 of them red. In this case, $P(\text{red second} \mid \text{heart first})$ is the same as $P(\text{red second})$. The value of the second card is not affected by the value of the first card. We say that the event that the second card is red is *independent* of the event that the first card is a heart since the knowledge of the first card does not influence what happens to the second card. On the other hand, when we draw without replacement, the events that the first card is a heart and that the second card is red are *dependent* events. The fact that the first card is a heart means there is one less red card in the deck, influencing the probability that the second card is red.

As another example, consider tossing a fair coin twice. If the first toss shows heads, the probability that the next toss is heads is still $1/2$. Coin tosses are independent events, since the outcome of one toss does not influence the outcome of the next toss. Similarly, rolls of a fair die are independent events. On the other hand, the events "the milk is old" and "the milk is sour" are dependent events; if the milk is old, there is an increased chance that it is sour. Also, in the example at the beginning of this section, the events A (broker picked stocks that went up) and B (broker used research) are dependent events, because information about the use of research affected the probability of picking stocks that go up. That is, $P(A|B)$ is different from $P(A)$.

If events E and F are independent, then the knowledge that E has occurred gives no (probability) information about the occurrence or nonoccurrence of event F. That is, $P(F)$ is exactly the same as $P(F|E)$, or

$$P(F|E) = P(F).$$

This, in fact, is the formal definition of independent events.

INDEPENDENT EVENTS

Events E and F are **independent events** if

$$P(F|E) = P(F) \quad \text{or} \quad P(E|F) = P(E).$$

If the events are not independent, they are **dependent events**.

When E and F are independent events, then $P(F|E) = P(F)$ and the product rule becomes

$$P(E \cap F) = P(E) \cdot P(F|E) = P(E) \cdot P(F).$$

Conversely, if this equation holds, then it follows that $P(F) = P(F|E)$. Consequently, we have this useful fact:

PRODUCT RULE FOR INDEPENDENT EVENTS

Events E and F are independent events if and only if

$$P(E \cap F) = P(E) \cdot P(F).$$

EXAMPLE 9 **Calculator**

A calculator requires a keystroke assembly and a logic circuit. Assume that 99% of the keystroke assemblies are satisfactory and 97% of the logic circuits are satisfactory. Find the probability that a finished calculator will be satisfactory.

▶**Solution** If the failure of a keystroke assembly and the failure of a logic circuit are independent events, then

$$\begin{aligned} &P(\text{satisfactory calculator}) \\ &\quad = P(\text{satisfactory keystroke assembly}) \cdot P(\text{satisfactory logic circuit}) \\ &\quad = (0.99)(0.97) \approx 0.96. \end{aligned}$$

(The probability of a defective calculator is $1 - 0.96 = 0.04$.)

CAUTION It is common for students to confuse the ideas of *mutually exclusive* events and *independent* events. Events E and F are mutually exclusive if $E \cap F = \emptyset$. For example, if a family has exactly one child, the only possible outcomes are $B = \{\text{boy}\}$ and $G = \{\text{girl}\}$. These two events are mutually exclusive. The events are *not* independent, however, since $P(G|B) = 0$ (if a family with only one child has a boy, the probability it has a girl is then 0). Since $P(G|B) \neq P(G)$, the events are not independent.

Of all the families with exactly two children, the events $G_1 = \{\text{first child is a girl}\}$ and $G_2 = \{\text{second child is a girl}\}$ are independent, since $P(G_2|G_1)$ equals $P(G_2)$. However, G_1 and G_2 are not mutually exclusive, since $G_1 \cap G_2 = \{\text{both children are girls}\} \neq \emptyset$. ■

To show that two events E and F are independent, show that $P(F|E) = P(F)$ or that $P(E|F) = P(E)$ or that $P(E \cap F) = P(E) \cdot P(F)$. Another way is to observe that knowledge of one outcome does not influence the probability of the other outcome, as we did for coin tosses.

NOTE In some cases, it may not be apparent from the physical description of the problem whether two events are independent or not. For example, it is not obvious whether the event that a baseball player gets a hit tomorrow is independent of the event that he got a hit today. In such cases, it is necessary to calculate whether $P(F|E) = P(F)$, or, equivalently, whether $P(E \cap F) = P(E) \cdot P(F)$. ■

EXAMPLE 10 **Snow in Manhattan**

On a typical January day in Manhattan the probability of snow is 0.10, the probability of a traffic jam is 0.80, and the probability of snow or a traffic jam (or both) is 0.82. Are the event "it snows" and the event "a traffic jam occurs" independent?

▶**Solution** Let S represent the event "it snows" and T represent the event "a traffic jam occurs." We must determine whether

$$P(T|S) = P(T) \quad \text{or} \quad P(S|T) = P(S).$$

We know $P(S) = 0.10$, $P(T) = 0.8$, and $P(S \cup T) = 0.82$. We can use the union rule (or a Venn diagram) to find $P(S \cap T) = 0.08$, $P(T|S) = 0.8$, and $P(S|T) = 0.1$. Since

$$P(T|S) = P(T) = 0.8 \quad \text{and} \quad P(S|T) = P(S) = 0.1,$$

the events "it snows" and "a traffic jam occurs" are independent.

Although we showed $P(T|S) = P(T)$ and $P(S|T) = P(S)$ in Example 10, only one of these results is needed to establish independence. It is also important to note that independence of events does not necessarily follow intuition; it is established from the mathematical definition of independence.

7.5 Exercises

If a single fair die is rolled, find the probabilities of the following results.

1. A 2, given that the number rolled was odd

2. A 4, given that the number rolled was even

3. An even number, given that the number rolled was 6

If two fair dice are rolled, find the probabilities of the following results.

4. A sum of 8, given that the sum is greater than 7

5. A sum of 6, given that the roll was a "double" (two identical numbers)

6. A double, given that the sum was 9

If two cards are drawn without replacement from an ordinary deck, find the probabilities of the following results.

7. The second is a heart, given that the first is a heart.

8. The second is black, given that the first is a spade.

9. The second is a face card, given that the first is a jack.

10. The second is an ace, given that the first is not an ace.

11. A jack and a 10 are drawn.

12. An ace and a 4 are drawn.

13. Two black cards are drawn.

14. Two hearts are drawn.

15. In your own words, explain how to find the conditional probability $P(E|F)$.

Decide whether each of the following pairs of events are dependent or independent.

16. A red and a green die are rolled. A is the event that the red die comes up even, and B is the event that the green die comes up even.

17. C is the event that it rains more than 10 days in Chicago next June, and D is the event that it rains more than 15 days.

18. E is the event that a resident of Texas lives in Dallas, and F is the event that a resident of Texas lives in either Dallas or Houston.

19. A coin is flipped. G is the event that today is Tuesday, and H is the event that the coin comes up heads.

In the previous section, we described an experiment in which the numbers 1, 2, 3, 4, *and* 5 *are written on slips of paper, and* 2 *slips are drawn at random one at a time without replacement. Find each probability in Exercises 20 and 21.*

20. The probability that the first number is 3, given the following.

a. The sum is 7.

b. The sum is 8.

21. The probability that the sum is 8, given the following.

a. The first number is 5.

b. The first number is 4.

22. Suppose two dice are rolled. Let A be the event that the sum of the two dice is 7. Find an event B related to numbers on the dice such that A and B are

a. independent;

b. dependent.

23. Your friend asks you to explain how the product rule for independent events differs from the product rule for dependent events. How would you respond?

24. Another friend asks you to explain how to tell whether two events are dependent or independent. How would you reply? (Use your own words.)

25. A student reasons that the probability in Example 3 of both coins being heads is just the probability that the other coin is a head, that is, $1/2$. Explain why this reasoning is wrong.

26. The following problem, submitted by Daniel Hahn of Blairstown, Iowa, appeared in the "Ask Marilyn" column of *Parade* magazine.*

"You discover two booths at a carnival. Each is tended by an honest man with a pair of covered coin shakers. In each shaker is a single coin, and you are allowed to bet upon the chance that both coins in that booth's shakers are heads after the man in the booth shakes them, does an inspection, and can tell you that at least one of the shakers contains a head. The difference is that the man in the first booth always looks inside both of his shakers, whereas the man in the second booth looks inside only one of the shakers. Where will you stand the best chance?"

27. The following question was posed in *Chance News* by Craig Fox and Yoval Rotenstrich.† You are playing a game in which a fair coin is flipped and a fair die is rolled. You win a prize if both the coin comes up heads and a 6 is rolled on the die. Now suppose the coin is tossed and the die is rolled, but you are not allowed to see either result. You are told, however, that either the head or the 6 occurred. You are then offered the chance to cancel the game and play a new game in which a die is rolled (there is no coin), and you win a prize if a 6 is rolled.

a. Is it to your advantage to switch to the new game, or to stick with the original game? Answer this question by calculating your probability of winning in each case.

b. Many people erroneously think that it's better to stick with the original game. Discuss why this answer might seem intuitive, but why it is wrong.

28. Suppose a male defendant in a court trial has a mustache, beard, tattoo, and an earring. Suppose, also, that an eyewitness has identified the perpetrator as someone with these characteristics. If the respective probabilities for the male population in this region are 0.35, 0.30, 0.10, and 0.05, is it fair to multiply these probabilities together to conclude that the probability that a person having these characteristics is 0.000525, or 21 in 40,000, and thus decide that the defendant must be guilty?

29. In a two-child family, if we assume that the probabilities of a male child and a female child are each 0.5, are the events *both children are the same sex* and *at most one male* independent? Are they independent for a three-child family?

30. Let A and B be independent events with $P(A) = \frac{1}{4}$ and $P(B) = \frac{1}{5}$. Find $P(A \cap B)$ and $P(A \cup B)$.

31. If A and B are events such that $P(A) = 0.5$ and $P(A \cup B) = 0.7$, find $P(B)$ when

a. A and B are mutually exclusive;

b. A and B are independent.

32. Laura Johnson, a game show contestant, could win one of two prizes: a shiny new Porsche or a shiny new penny. Laura is given two boxes of marbles. The first box has 50 pink marbles in it and the second box has 50 blue marbles in it. The game show host will pick someone from the audience to be blindfolded and then draw a marble from one of the two boxes. If a pink marble is drawn, she wins the Porsche. Otherwise, Laura wins the penny.‡ Can Laura increase her chances of winning by redistributing some of the marbles from one box to the other? Explain.

*Source: *Parade* magazine, June 12, 1994, p. 18. © 1994 Marilyn vos Savant. Initially published in *Parade* magazine. All rights reserved.

†*Chance News* 10.01, Jan 16, 2001.

‡This problem is based on the "Puzzler of the Week: Prison Marbles" from the week of Sept. 7, 1996, on National Public Radio's *Car Talk*.

Applications

BUSINESS AND ECONOMICS

Banking *The Midtown Bank has found that most customers at the tellers' windows either cash a check or make a deposit. The following table indicates the transactions for one teller for one day.*

	Cash Check	No Check	Totals
Make Deposit	60	20	80
No Deposit	30	10	40
Totals	90	30	120

Letting C represent "cashing a check" and D represent "making a deposit," express each probability in words and find its value.

33. $P(C \mid D)$ **34.** $P(D' \mid C)$ **35.** $P(C' \mid D')$

36. $P(C' \mid D)$ **37.** $P[(C \cap D)']$

38. *Airline Delays* In February 2006, the major U.S. airline with the fewest delays was US Airways, for which 79.2% of their flights arrived on time.* Assume that the event that a given flight arrives on time is independent of the event that another flight arrives on time.

a. Chrissy Jenkins plans to take four separate flights for her publisher next month on US Airways. Assuming that the airline has the same on-time performance as in February 2006, what is the probability that all four flights arrive on time?

b. Discuss how realistic it is to assume that the on-time arrivals of the different flights are independent.

39. *Backup Computers* Corporations where a computer is essential to day-to-day operations, such as banks, often have a second backup computer in case the main computer fails. Suppose there is a 0.003 chance that the main computer will fail in a given time period, and a 0.005 chance that the backup computer will fail while the main computer is being repaired. Assume these failures represent independent events, and find the fraction of the time that the corporation can assume it will have computer service. How realistic is our assumption of independence?

40. *ATM Transactions* Among users of automated teller machines (ATMs), 92% use ATMs to withdraw cash, and 32% use them to check their account balance.† Suppose that 96% use ATMs to either withdraw cash or check their account balance (or both). Given a woman who uses an ATM to check her account balance, what is the probability that she also uses an ATM to get cash?

Quality Control *A bicycle factory runs two assembly lines, A and B. If 95% of line A's products pass inspection, while only 85% of line B's products pass inspection, and 60% of the factory's bikes come off assembly line A (the rest off B), find the probabilities that one of the factory's bikes did not pass inspection and came off the following.*

41. Assembly line A **42.** Assembly line B

43. Find the probability that one of the factory's bikes did not pass inspection.

LIFE SCIENCES

44. *Genetics* Both of a certain pea plant's parents had a gene for red and a gene for white flowers. (See Exercise 62 in Section 7.4.) If the offspring has red flowers, find the probability that it combined a gene for red and a gene for white (rather than 2 for red).

Genetics *Assuming that boy and girl babies are equally likely, fill in the remaining probabilities on the tree diagram on the next page and use the following information to find the probability that a family with three children has all girls, given the following.*

45. The first is a girl.

46. The third is a girl.

47. The second is a girl.

48. At least 2 are girls.

49. At least 1 is a girl.

**Air Travel Consumer Report*, U.S. Department of Transportation, April 2006, http://airconsumer.ost.dot.gov/reports/atcr06.htm.

†*Chicago Tribune*, Dec. 18, 1995, Sec. 4, p. 1.

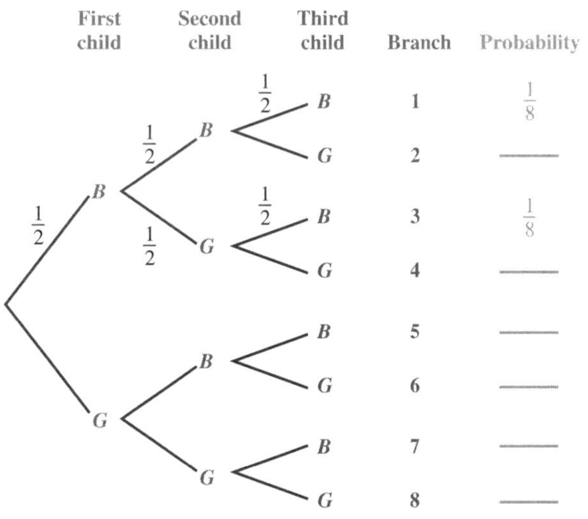

50. *AIDS* The following table gives the estimated numbers of cases of HIV/AIDS diagnosed in 2004 for areas of the United States with confidential name-based HIV infection reporting for adults, based on gender and method of transmission.*

Method of Transmission	Male	Female	Totals
Homosexual contact	18,203	0	18,203
Heterosexual contact	4581	8102	12,683
Injection drug use	3826	2134	5960
Other	1533	174	1707
Totals	28,143	10,410	38,553

a. Find the probability that a male in this group contracted AIDS via homosexual contact.

b. Find the probability that a female in this group contracted AIDS via injection drug use.

51. *Medical Experiment* A medical experiment showed that the probability that a new medicine is effective is 0.75, the probability that a patient will have a certain side effect is 0.4, and the probability that both events occur is 0.3. Decide whether these events are dependent or independent.

Color Blindness The following table shows frequencies for red-green color blindness, where M represents "person is male" and C represents "person is color-blind." Use this table to find the following probabilities. (See Exercise 61, Section 7.4.)

	M	M'	Totals
C	0.035	0.004	0.039
C'	0.452	0.509	0.961
Totals	0.487	0.513	1.000

52. $P(M)$

53. $P(C)$

54. $P(M \cap C)$

55. $P(M \cup C)$

56. $P(M|C)$

57. $P(C|M)$

58. $P(M'|C)$

59. Are the events C and M, described above, dependent? What does this mean?

60. *Color Blindness* A scientist wishes to determine whether there is a relationship between color blindness (C) and deafness (D).

a. Suppose the scientist found the probabilities listed in the table. What should the findings be? (See Exercises 52–59.)

b. Explain what your answer tells us about color blindness and deafness.

	D	D'	Totals
C	0.0008	0.0392	0.0400
C'	0.0192	0.9408	0.9600
Totals	0.0200	0.9800	1.0000

61. *Obesity* In 2002, 66.9% of men and 50.0% of women in the United States were overweight.† Given that 48.3% of adult Americans are men and 51.7% are women,‡ find the probability that a randomly selected adult fits the following description.

a. An overweight man

b. Overweight

c. Are the events that an adult American is a man and that an adult American is overweight independent? Explain.

*Centers for Disease Control and Prevention, HIV/AIDS Surveillance Report, http://www.cdc.gov/hiv/topics/surveillance/resources/reports/2004report/table1.htm.
†*Time Almanac 2006*, p. 553.
‡Ibid, p. 121.

62. ***Hospital Insurance*** An insurance company pays hospital claims. The number of claims that include emergency room or operating room charges is 85% of the total number of claims. The number of claims that do not include emergency room charges is 25% of the total number of claims. The occurrence of emergency room charges is independent of the occurrence of operating room charges on hospital claims. Calculate the probability that a claim submitted to the insurance company includes operating room charges.* (Choose one of the following.)

a. 0.10 **b.** 0.20 **c.** 0.25 **d.** 0.40 **e.** 0.80

63. ***Blood Pressure*** A doctor is studying the relationship between blood pressure and heartbeat abnormalities in her patients. She tests a random sample of her patients and notes their blood pressures (high, low, or normal) and their heartbeats (regular or irregular). She finds that:

(i) 14% have high blood pressure.

(ii) 22% have low blood pressure.

(iii) 15% have an irregular heartbeat.

(iv) Of those with an irregular heartbeat, one-third have high blood pressure.

(v) Of those with normal blood pressure, one-eighth have an irregular heartbeat.

What portion of the patients selected have a regular heartbeat and low blood pressure?[†] (Choose one of the following. *Hint:* Make a table similar to the one for Exercises 52–59.)

a. 2% **b.** 5% **c.** 8% **d.** 9% **e.** 20%

64. ***Breast Cancer*** To explain why the chance of a woman getting breast cancer in the next year goes up each year, while the chance of a woman getting breast cancer in her lifetime goes down, Ruma Falk made the following analogy.[‡] Suppose you are looking for a letter that you may have lost. You have 8 drawers in your desk. There is a probability of 0.1 that the letter is in any one of the 8 drawers, and a probability of 0.2 that the letter is not in any of the drawers.

a. What is the probability that the letter is in drawer 1?

b. Given that the letter is not in drawer 1, what is the probability that the letter is in drawer 2?

c. Given that the letter is not in drawer 1 or 2, what is the probability that the letter is in drawer 3?

d. Given that the letter is not in drawers 1–7, what is the probability that the letter is in drawer 8?

e. Based on your answers to parts a–d, what is happening to the probability that the letter is in the next drawer?

f. What is the probability that the letter is in some drawer?

g. Given that the letter is not in drawer 1, what is the probability that the letter is in some drawer?

h. Given that the letter is not in drawer 1 or 2, what is the probability that the letter is in some drawer?

i. Given that the letter is not in drawers 1–7, what is the probability that the letter is in some drawer?

j. Based on your answers to parts f–i, what is happening to the probability that the letter is in some drawer?

65. ***Drug Screening*** In searching for a new drug with commercial possibilities, drug company researchers use the ratio

$$N_S : N_A : N_P : 1.$$

That is, if the company gives preliminary screening to N_S substances, it may find that N_A of them are worthy of further study, with N_P of these surviving into full-scale development. Finally, 1 of the substances will result in a marketable drug. Typical numbers used by Smith, Kline, and French Laboratories in planning research budgets might be $2000 : 30 : 8 : 1$.[§] Use this ratio for parts a–f.

a. Suppose a compound has been chosen for preliminary screening. Find the probability that the compound will survive and become a marketable drug.

b. Find the probability that the compound will not lead to a marketable drug.

c. Suppose the number of such compounds receiving preliminary screening is a. Set up the probability that none of them produces a marketable drug. (Assume independence throughout these exercises.)

d. Use your results from part c to find the probability that at least one of the drugs will prove marketable.

e. Suppose now that N scientists are employed in the preliminary screening, and that each scientist can screen c compounds per year. Find the probability that no marketable drugs will be discovered in a year.

f. Find the probability that at least one marketable drug will be discovered.

*Problem 37 from May 2003 Course 1 Examination of the *Education and Examination Committee of the Society of Actuaries.* Reprinted by permission of the Society of Actuaries.

[†]Problem 12 from the 2005 Sample Exam P of the *Education and Examination Committee of the Society of Actuaries.* Reprinted by permission of the Society of Actuaries.

[‡]Falk, Ruma, *Chance News*, July 23, 1995.

[§]Pyle, E. B., III, B. Douglas, G. W. Ebright, W. J. Westlake, and A. B. Bender, "Scientific Manpower Allocation to New Drug Screening Programs," *Management Science*, Vol. 19, No. 12, August 1973.

*Hockey The table below lists the number of head and neck injuries for 319 ice hockey players' exposures wearing either a full shield or half shield in the Canadian Inter-University Athletics Union during the 1997–1998 season.**

For a randomly selected injury, find each probability.

66. $P(A)$

67. $P(C|F)$

68. $P(A|H)$

69. $P(B'|H')$

70. Are the events A and H independent events?†

	Half Shield (H)	Full Shield (F)	Totals
Head and Face Injuries (A)	95	34	129
Concussions (B)	41	38	79
Neck Injuries (C)	9	7	16
Other Injuries (D)	202	150	352
Totals	347	229	576

SOCIAL SCIENCES

71. *Working Women* A survey has shown that 52% of the women in a certain community work outside the home. Of these women, 64% are married, while 86% of the women who do not work outside the home are married. Find the probabilities that a woman in that community can be categorized as follows.

a. Married

b. A single woman working outside the home

PHYSICAL SCIENCES

72. *Rain Forecasts* In a letter to the journal *Nature*, Robert A. J. Matthews gives the following table of outcomes of forecast and weather over 1000 1-hour walks, based on the United Kingdom's Meteorological office's 83% accuracy in 24-hour forecasts.‡

	Rain	No Rain	Totals
Forecast of Rain	66	156	222
Forecast of No Rain	14	764	778
Totals	80	920	1000

a. Verify that the probability that the forecast called for rain, given that there was rain, is indeed 83%. Also verify that the probability that the forecast called for no rain, given that there was no rain, is also 83%.

b. Calculate the probability that there was rain, given that the forecast called for rain.

c. Calculate the probability that there was no rain, given that the forecast called for no rain.

d. Observe that your answer to part c is higher than 83%, and that your answer to part b is much lower. Discuss which figure best describes the accuracy of the weather forecast in recommending whether or not you should carry an umbrella.

73. *Earthquakes* There are seven geologic faults (and possibly more) capable of generating a magnitude 6.7 earthquake in the region around San Francisco.§ Their probabilities of rupturing by the year 2032 are 27%, 21%, 11%, 10%, 4%, 3%, and 3%.

a. Calculate the probability that at least one of these faults erupts by the year 2032, assuming that these are independent events.

b. Scientists forecast a 62% chance of an earthquake with magnitude at least 6.7 in the region around San Francisco by the year 2032. Compare this with your answer from part a. Consider the realism of the assumption of independence. Also consider the role of roundoff. For example, the probability of 10% for one of the faults is presumably rounded to the nearest percent, with the actual probability between 9.5% and 10.5%.

*Benson, Brian, Nicholas Nohtaki, M. Sarah Rose, and Willem Meeuwisse, "Head and Neck Injuries Among Ice Hockey Players Wearing Full Face Shields vs. Half Face Shields," *JAMA,* Vol. 282, No. 24, Dec. 22/29, 1999, pp. 2328–2332.

†We are assuming here and in other exercises that the events consist entirely of the numbers given in the table. If the numbers are interpreted as a sample of all people fitting the description of the events, then testing for independence is more complicated, requiring a technique from statistics known as a *contingency table*.

‡Matthews, Robert A. J., *Nature*, Vol. 382, Aug. 29, 1996, p. 3.

§*Science News*, Vol. 169, April 15, 2006, pp. 234–236.

74. ***Reliability*** The probability that a key component of a space rocket will fail is 0.03.

a. How many such components must be used as backups to ensure that the probability of at least one of the components working is 0.999999 or more?

b. Is it reasonable to assume independence here?

GENERAL INTEREST

75. ***Titanic*** The following table lists the number of passengers who were on the Titanic and the number of passengers who survived, according to class of ticket.*

	Children		Women		Men		Totals	
	On	**Survived**	**On**	**Survived**	**On**	**Survived**	**On**	**Survived**
First Class	6	6	144	140	175	57	325	203
Second Class	24	24	165	76	168	14	357	114
Third Class	79	27	93	80	462	75	634	182
Totals	109	57	402	296	805	146	1316	499

Use this information to determine the following (round answers to four decimal places).

a. What is the probability that a randomly selected passenger was second class?

b. What is the overall probability of surviving?

c. What is the probability of a first-class passenger surviving?

d. What is the probability of a child who was also in the third class surviving?

e. Given that the survivor is from first class, what is the probability that she was a woman?

f. Given that a male has survived, what is the probability that he was in third class?

g. Are the events third-class survival and male survival independent events? What does this imply?

76. ***Real Estate*** A real estate agent trying to sell you an attractive beachfront house claims that it will not collapse unless it is subjected simultaneously to extremely high winds and extremely high waves. According to weather service records, there is a 0.001 probability of extremely high winds, and the same for extremely high waves. The real estate agent claims, therefore, that the probability of both occurring is $(0.001)(0.001) = 0.000001$. What is wrong with the agent's reasoning?

77. ***Age and Loans*** Suppose 20% of the population are 65 or over, 26% of those 65 or over have loans, and 53% of those under 65 have loans. Find the probabilities that a person fits into the following categories.

a. 65 or over and has a loan

b. Has a loan

78. ***Women Joggers*** In a certain area, 15% of the population are joggers and 40% of the joggers are women. If 55% of those who do not jog are women, find the probabilities that an individual from that community fits the following descriptions.

a. A woman jogger **b.** Not a jogger **c.** A woman

d. Are the events that a person is a woman and a person is a jogger independent? Explain.

79. ***Diet Soft Drinks*** Two-thirds of the population are on a diet at least occasionally. Of this group, $4/5$ drink diet soft drinks, while $1/2$ of the rest of the (nondieting) population drink diet soft drinks. Find the probabilities that a person fits into the following categories.

a. Drinks diet soft drinks

b. Diets, but does not drink diet soft drinks

80. ***Driver's License Test*** The Motor Vehicle Department has found that the probability of a person passing the test for a driver's license on the first try is 0.75. The probability that an individual who fails on the first test will pass on the second try is 0.80, and the probability that an individual who fails the first and second tests will pass the third time is 0.70. Find the probabilities that an individual will do the following.

a. Fail both the first and second tests

b. Fail three times in a row

c. Require at least two tries

*Takis, Sandra L., "Titanic: A Statistical Exploration," *Mathematics Teacher*, Vol. 92, No. 8, Nov. 1999, pp. 660–664.

81. *Ballooning* A pair of mathematicians in a hot air balloon were told that there are four independent burners, any one of which is sufficient to keep the balloon aloft. If the probability of any one burner failing during a flight is 0.001, what is the probability that the balloon will crash due to all four burners failing?

82. *Speeding Tickets* A smooth-talking young man has a $1/3$ probability of talking a policeman out of giving him a speeding ticket. The probability that he is stopped for speeding during a given weekend is $1/2$. Find the probabilities of the events in parts a and b.

a. He will receive no speeding tickets on a given weekend.

b. He will receive no speeding tickets on 3 consecutive weekends.

c. We have assumed that what happens on the second or third weekend is the same as what happened on the first weekend. Is this realistic? Will driving habits remain the same after getting a ticket?

83. *Luxury Cars* In one area, 4% of the population drive luxury cars. However, 17% of the CPAs drive luxury cars. Are the events "person drives a luxury car" and "person is a CPA" independent?

84. *Studying* A teacher has found that the probability that a student studies for a test is 0.60, the probability that a student gets a good grade on a test is 0.70, and the probability that both occur is 0.52.

a. Are these events independent?

b. Given that a student studies, find the probability that the student gets a good grade.

c. Given that a student gets a good grade, find the probability that the student studied.

85. *Football* A football coach whose team is 14 points behind needs two touchdowns to win. Each touchdown is worth 6 points. After a touchdown, the coach can choose either a 1-point kick, which is almost certain to succeed, or a 2-point conversion, which is roughly half as likely to succeed. After the first touchdown, the coach must decide whether to go for 1 or 2 points. If the 2-point conversion is successful, the almost certain 1-point kick after the second touchdown will win the game. If the 2-point conversion fails, the team can try another 2-point conversion after the second touchdown to tie. Some coaches, however, prefer to go for the almost certain 1-point kick after the first touchdown, hoping that the momentum will help them get a 2-point conversion after the second touchdown and win the game. They fear that an unsuccessful 2-point conversion after the first touchdown will discourage the team, which can then at best tie.*

a. Draw a tree diagram for the 1-point kick after the first touchdown and the 2-point conversion after the second touchdown. Letting the probability of success for the 1-point kick and the 2-point conversion be k and r, respectively, show that

$$P(\text{win}) = kr,$$
$$P(\text{tie}) = r(1 - k), \quad \text{and}$$
$$P(\text{lose}) = 1 - r.$$

b. Consider the case of trying for a 2-point conversion after the first touchdown. If it succeeds, try a 1-point kick after the second touchdown. If the 2-point conversion fails, try another one after the second touchdown. Draw a tree diagram and use it to show that

$$P(\text{win}) = kr,$$
$$P(\text{tie}) = r(2 - k - r), \quad \text{and}$$
$$P(\text{lose}) = (1 - r)^2.$$

c. What can you say about the probability of winning under each strategy?

d. Given that $r < 1$, which strategy has a smaller probability of losing? What does this tell you about the value of the two strategies?

*Schielack, Vincent P., Jr., "The Football Coach's Dilemma: Should We Go for 1 or 2 Points First?" *The Mathematics Teacher*, Vol. 88, No. 9, Dec. 1995, pp. 731–733.

7.6 Bayes' Theorem

? THINK ABOUT IT **What is the probability that a particular defective item was produced by a new machine operator?**

This question and others like it are answered using Bayes' theorem, discussed in this section.

Suppose the probability that a person gets lung cancer, given that the person smokes a pack or more of cigarettes daily, is known. For a research project, it might be necessary to know the probability that a person smokes a pack or more of cigarettes daily, given that the person has lung cancer. More generally, if $P(E|F)$ is known for two events E and F, can $P(F|E)$ be found? The answer is yes, we can find $P(F|E)$ using the formula to be developed in this section. To develop this formula, we can use a tree diagram to find $P(F|E)$. Since $P(E|F)$ is known, the first outcome is either F or F'. Then for each of these outcomes, either E or E' occurs, as shown in Figure 27.

The four cases have the probabilities shown on the right. By the definition of conditional probability and the product rule,

$$P(F|E) = \frac{P(F \cap E)}{P(E)} = \frac{P(F) \cdot P(E|F)}{P(F) \cdot P(E|F) + P(F') \cdot P(E|F')}.$$

Notice that $P(E)$ is the sum of the first and third cases in the tree diagram. This result is a special case of Bayes' theorem, which is generalized later in this section.

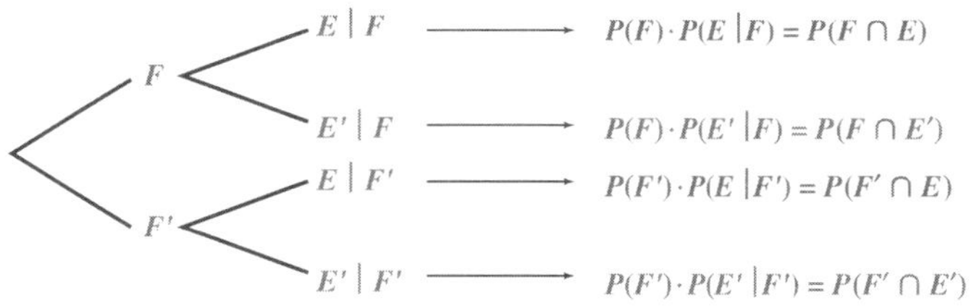

FIGURE 27

BAYES' THEOREM (SPECIAL CASE)

$$P(F|E) = \frac{P(F) \cdot P(E|F)}{P(F) \cdot P(E|F) + P(F') \cdot P(E|F')}$$

EXAMPLE 1 **Worker Errors**

For a fixed length of time, the probability of a worker error on a certain production line is 0.1, the probability that an accident will occur when there is a worker error is 0.3, and the probability that an accident will occur when there is no worker error is 0.2. Find the probability of a worker error if there is an accident.

▶**Solution** Let E represent the event of an accident, and let F represent the event of worker error. From the information given,

$$P(F) = 0.1, \qquad P(E|F) = 0.3, \qquad \text{and} \qquad P(E|F') = 0.2.$$

These probabilities are shown on the tree diagram in Figure 28.

Find $P(F|E)$ by dividing the probability that both E and F occur, given by branch 1, by the probability that E occurs, given by the sum of branches 1 and 3.

$$P(F|E) = \frac{P(F) \cdot P(E|F)}{P(F) \cdot P(E|F) + P(F') \cdot P(E|F')}$$

$$= \frac{(0.1)(0.3)}{(0.1)(0.3) + (0.9)(0.2)} = \frac{0.03}{0.21} = \frac{1}{7} \approx 0.1429$$

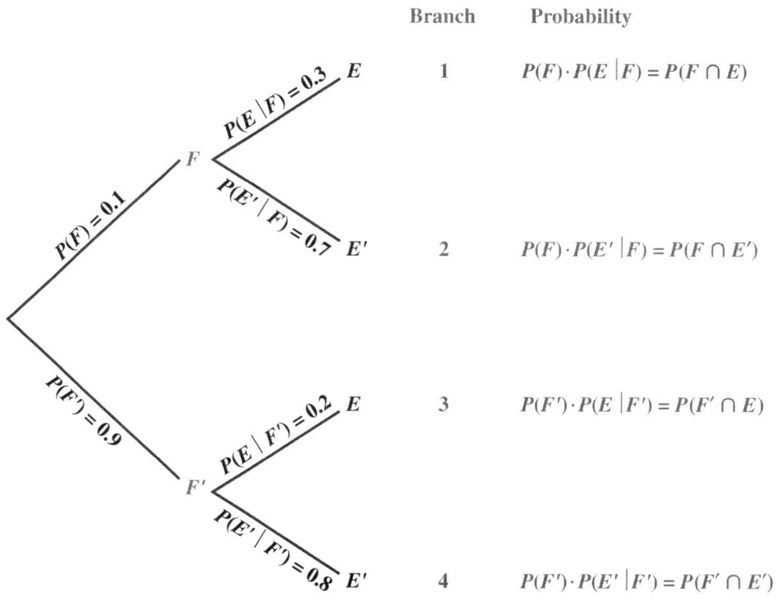

FIGURE 28

The special case of Bayes' theorem can be generalized to more than two events with the tree diagram in Figure 29. This diagram shows the paths that can produce an event E. We assume that the events $F_1, F_2, \ldots, F_n$ are mutually exclusive events (that is, disjoint events) whose union is the sample space, and that E is an event that has occurred. See Figure 30.

The probability $P(F_i|E)$, where $1 \leq i \leq n$, can be found by dividing the probability for the branch containing $P(E|F_i)$ by the sum of the probabilities of all the branches producing event E.

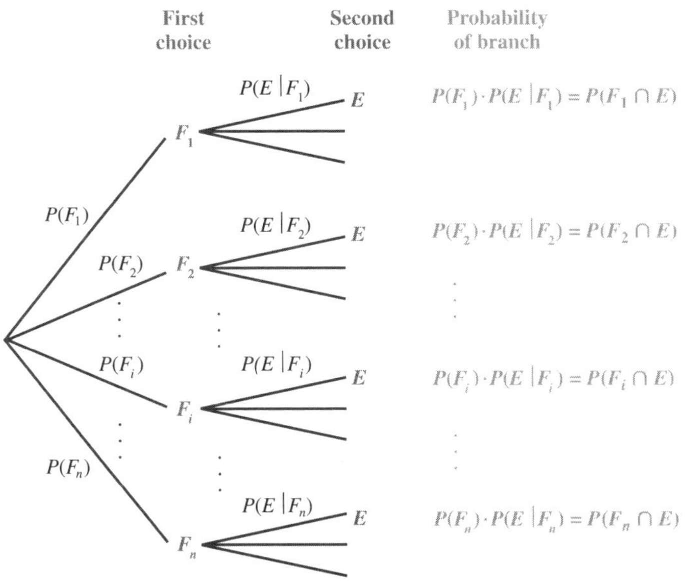

FIGURE 29

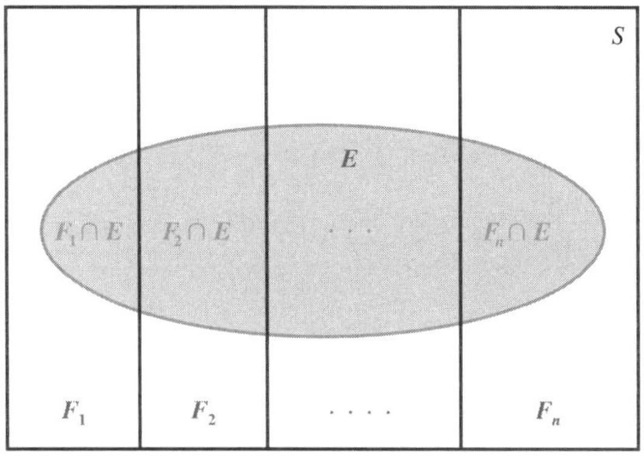

FIGURE 30

BAYES' THEOREM

$$P(F_i|E) = \frac{P(F_i) \cdot P(E|F_i)}{P(F_1) \cdot P(E|F_1) + P(F_2) \cdot P(E|F_2) + \cdots + P(F_n) \cdot P(E|F_n)}$$

This result is known as **Bayes' theorem**, after the Reverend Thomas Bayes (1702–1761), whose paper on probability was published about three years after his death.

The statement of Bayes' theorem can be daunting. Actually, it is easier to remember the formula by thinking of the tree diagram that produced it. Go through the following steps.

USING BAYES' THEOREM

1. Start a tree diagram with branches representing $F_1, F_2, \ldots, F_n$. Label each branch with its corresponding probability.
2. From the end of each of these branches, draw a branch for event E. Label this branch with the probability of getting to it, $P(E|F_i)$.
3. You now have n different paths that result in event E. Next to each path, put its probability—the product of the probabilities that the first branch occurs, $P(F_i)$, and that the second branch occurs, $P(E|F_i)$; that is, the product $P(F_i) \cdot P(E|F_i)$, which equals $P(F_i \cap E)$.
4. $P(F_i|E)$ is found by dividing the probability of the branch for F_i by the sum of the probabilities of all the branches producing event E.

EXAMPLE 2 Machine Operators

Based on past experience, a company knows that an experienced machine operator (one or more years of experience) will produce a defective item 1% of the time. Operators with some experience (up to one year) have a 2.5% defect rate, and new operators have a 6% defect rate. At any one time, the company has 60% experienced operators, 30% with some experience, and 10% new operators. Find the probability that a particular defective item was produced by a new operator.

▶**Solution** Let E represent the event "item is defective," F_1 represent "item was made by an experienced operator," F_2 represent "item was made by an operator with some experience," and F_3 represent "item was made by a new operator." Then

$$P(F_1) = 0.60 \qquad P(E|F_1) = 0.01$$
$$P(F_2) = 0.30 \qquad P(E|F_2) = 0.025$$
$$P(F_3) = 0.10 \qquad P(E|F_3) = 0.06.$$

?

We need to find $P(F_3|E)$, the probability that an item was produced by a new operator, given that it is defective. First, draw a tree diagram using the given information, as in Figure 31. The steps leading to event E are shown in red.

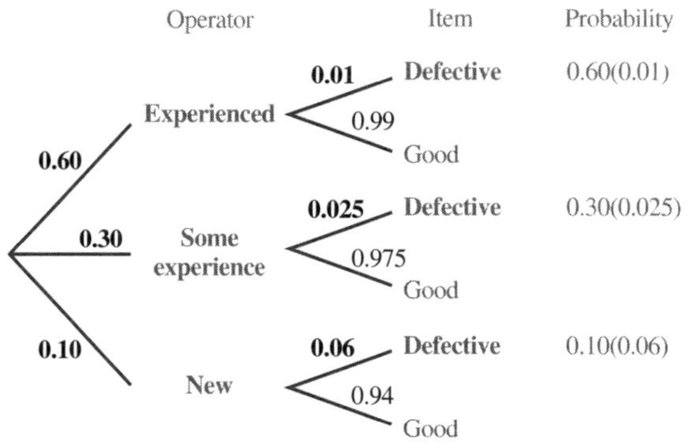

FIGURE 31

Find $P(F_3|E)$ with the bottom branch of the tree in Figure 31: Divide the probability for this branch by the sum of the probabilities of all the branches leading to E, or

$$P(F_3|E) = \frac{0.10(0.06)}{0.60(0.01) + 0.30(0.025) + 0.10(0.06)} = \frac{0.006}{0.0195} = \frac{4}{13} \approx 0.3077.$$

In a similar way, the probability that the defective item was produced by an operator with some experience is

$$P(F_2|E) = \frac{0.30(0.025)}{0.60(0.01) + 0.30(0.025) + 0.10(0.06)} = \frac{0.0075}{0.0195} = \frac{5}{13} \approx 0.3846.$$

Finally, the probability that the defective item was produced by an experienced operator is $P(F_1|E) = 4/13 \approx 0.3077$. Check that $P(F_1|E) + P(F_2|E) + P(F_3|E) = 1$ (that is, the defective item was made by *someone*).

EXAMPLE 3 **Manufacturing**

A manufacturer buys items from six different suppliers. The fraction of the total number of items obtained from each supplier, along with the probability that an item purchased from that supplier is defective, are shown in the following table.

Supplier	Fraction of Total Supplied	Probability of Defect
1	0.05	0.04
2	0.12	0.02
3	0.16	0.07
4	0.23	0.01
5	0.35	0.03
6	0.09	0.05

Find the probability that a defective item came from supplier 5.

▶**Solution** Let F_1 be the event that an item came from supplier 1, with F_2, F_3, F_4, F_5, and F_6 defined in a similar manner. Let E be the event that an item is defective. We want to find $P(F_5|E)$. Use the probabilities in the table above to prepare a tree diagram, or work with the rows of the table to get

$$P(F_5|E) = \frac{(0.35)(0.03)}{(0.05)(0.04) + (0.12)(0.02) + (0.16)(0.07) + (0.23)(0.01) + (0.35)(0.03) + (0.09)(0.05)}$$

$$= \frac{0.0105}{0.0329} \approx 0.319.$$

There is about a 32% chance that a defective item came from supplier 5. Even though supplier 5 has only 3% defectives, his probability of being "guilty" is relatively high, about 32%, because of the large fraction supplied by 5.

CAUTION Notice that the 0.04 in the upper right of the previous table represents the probability of a defective item *given* that the item came from supplier 1. In contrast, the probability of 0.035 in the table for Exercises 52–59 of the previous section represents the probability that a person is color-blind *and* male. The tables in this section represent probability in a different way than those of the previous section. Tables that you encounter outside of this course might represent probability in either way. You can usually tell what is intended by the context, but be careful! ■

▶ 7.6 Exercises

For two events M and N, $P(M) = 0.4$, $P(N|M) = 0.3$, *and* $P(N|M') = 0.4$. *Find the following.*

1. $P(M|N)$

2. $P(M'|N)$

For mutually exclusive events R_1, R_2, and R_3, we have $P(R_1) = 0.15$, $P(R_2) = 0.55$, and $P(R_3) = 0.30$. Also, $P(Q|R_1) = 0.40$, $P(Q|R_2) = 0.20$, and $P(Q|R_3) = 0.70$. Find the following.

3. $P(R_1|Q)$ **4.** $P(R_2|Q)$ **5.** $P(R_3|Q)$ **6.** $P(R_1'|Q)$

Suppose you have three jars with the following contents: 2 black balls and 1 white ball in the first, 1 black ball and 2 white balls in the second, and 1 black ball and 1 white ball in the third. One jar is to be selected, and then 1 ball is to be drawn from the selected jar. If the probabilities of selecting the first, second, or third jar are 1/2, 1/3, and 1/6, respectively, find the probabilities that if a white ball is drawn, it came from the following jars.

7. The second jar **8.** The third jar

Applications

BUSINESS AND ECONOMICS

9. *Employment Test* A manufacturing firm finds that 70% of its new hires turn out to be good workers and 30% become poor workers. All current workers are given a reasoning test. Of the good workers, 85% pass it; 35% of the poor workers pass it. Assume that these figures will hold true in the future. If the company makes the test part of its hiring procedure and only hires people who meet the previous requirements and also pass the test, what percent of the new hires will turn out to be good workers?

Job Qualifications Of all the people applying for a certain job, 75% are qualified and 25% are not. The personnel manager claims that she approves qualified people 85% of the time; she approves an unqualified person 20% of the time. Find each probability.

10. A person is qualified if he or she was approved by the manager.

11. A person is unqualified if he or she was approved by the manager.

Quality Control A building contractor buys 70% of his cement from supplier A, and 30% from supplier B. A total of 90% of the bags from A arrive undamaged, while 95% of the bags from B arrive undamaged. Give the probabilities that a damaged bag is from the following sources.

12. Supplier A **13.** Supplier B

Appliance Reliability Companies A, B, and C produce 15%, 40%, and 45%, respectively, of the major appliances sold in a certain area. In that area, 1% of the company A appliances, $1\frac{1}{2}$% of the company B appliances, and 2% of the company C appliances need service within the first year. Suppose a defective appliance is chosen at random; find the probabilities that it was manufactured by the following companies.

14. Company A **15.** Company B

Television Advertising On a given weekend in the fall, a tire company can buy television advertising time for a college football game, a baseball game, or a professional football game. If the company sponsors the college football game, there is a 70% chance of a high rating, a 50% chance if they sponsor a baseball game, and a 60% chance if they sponsor a professional football game. The probabilities of the company sponsoring these various games are 0.5, 0.2, and 0.3, respectively. Suppose the company does get a high rating; find the probabilities that it sponsored the following.

16. A college football game

17. A professional football game

18. *Auto Insurance* An auto insurance company insures drivers of all ages. An actuary compiled the following statistics on the company's insured drivers:

Age of Driver	Probability of Accident	Portion of Company's Insured Drivers
16–20	0.06	0.08
21–30	0.03	0.15
31–65	0.02	0.49
66–99	0.04	0.28

A randomly selected driver that the company insures has an accident. Calculate the probability that the driver was age 16–20.* (Choose one of the following.)

a. 0.13 **b.** 0.16 **c.** 0.19 **d.** 0.23 **e.** 0.40

*Problem 8 from May 2003 Course 1 Examination of the *Education and Examination Committee of the Society of Actuaries*. Reprinted by permission of the Society of Actuaries.

19. *Life Insurance* An insurance company issues life insurance policies in three separate categories: standard, preferred, and ultra-preferred. Of the company's policyholders, 50% are standard, 40% are preferred, and 10% are ultra-preferred. Each standard policyholder has probability 0.010 of dying in the next year, each preferred policyholder has probability 0.005 of dying in the next year, and each ultra-preferred policyholder has probability 0.001 of dying in the next year. A policyholder dies in the next year. What is the probability that the deceased policyholder was ultra-preferred? * (Choose one of the following.)

a. 0.0001 **b.** 0.0010 **c.** 0.0071 **d.** 0.0141 **e.** 0.2817

20. *Automobile Collisions* An actuary studied the likelihood that different types of drivers would be involved in at least one collision during any one-year period. The results of the study are presented below.

Type of Driver	Percentage of All Drivers	Probability of at Least One Collision
Teen	8%	0.15
Young Adult	16%	0.08
Midlife	45%	0.04
Senior	31%	0.05
Total	100%	

Given that a driver has been involved in at least one collision in the past year, what is the probability that the driver is a young adult driver?† (Choose one of the following.)

a. 0.06 **b.** 0.16 **c.** 0.19 **d.** 0.22 **e.** 0.25

21. *Shipping Errors* The following information pertains to three shipping terminals operated by Krag Corp.‡

Terminal	Percentage of Cargo Handled	Percentage of Error
Land	50	2
Air	40	4
Sea	10	14

Krag's internal auditor randomly selects one set of shipping documents, ascertaining that the set selected contains an error. Which of the following gives the probability that the error occurred in the Land Terminal?

a. 0.02 **b.** 0.10

c. 0.25 **d.** 0.50

22. *Mortgage Defaults* A bank finds that the relationship between mortgage defaults and the size of the down payment is given by the following table.

Down Payment	Number of Mortgages with This Down Payment	Probability of Default
5%	1260	0.06
10%	700	0.04
20%	560	0.02
25%	280	0.01

a. If a default occurs, what is the probability that it is on a mortgage with a 5% down payment?

b. What is the probability that a mortgage that is paid to maturity has a 10% down payment?

LIFE SCIENCES

23. *Colorectal Cancer* Researchers found that only one out of 24 physicians could give the correct answer to the following problem: "The probability of colorectal cancer can be given as 0.3%. If a person has colorectal cancer, the probability that the hemoccult test is positive is 50%. If a person does not have colorectal cancer, the probability that he still tests positive is 3%. What is the probability that a person who tests positive actually has colorectal cancer?"§ What is the correct answer?

24. *Hepatitis Blood Test* The probability that a person with certain symptoms has hepatitis is 0.8. The blood test used to confirm this diagnosis gives positive results for 90% of people with the disease and 5% of those without the disease. What is the probability that an individual who has the symptoms and who reacts positively to the test actually has hepatitis?

*Problem 20 from the 2005 Sample Exam P of the *Education and Examination Committee of the Society of Actuaries*. Reprinted by permission of the Society of Actuaries.

†Problem 23 from the 2005 Sample Exam P of the *Education and Examination Committee of the Society of Actuaries*. Reprinted by permission of the Society of Actuaries.

‡Uniform CPA Examination, Nov. 1989.

§Hoffrage, Ulrich, Samuel Lindsey, Ralph Hertwig, and Gerd Gigerenzer, *Science*, Vol. 290, Dec. 22, 2000, pp. 2261–2262.

25. *Sensitivity and Specificity* The sensitivity of a medical test is defined as the probability that a test will be positive given that a person has a disease, written $P(T^+|D^+)$. The specificity of a test is defined as the probability that a test will be negative given that the person does not have the disease, written $P(T^-|D^-)$. For example, the sensitivity and specificity for breast cancer during a clinical breast examination by a trained expert is approximately 0.54 and 0.94, respectively.*

a. If 2% of U.S. women have breast cancer,† find the probability that a woman who tests positive during a clinical breast examination actually has breast cancer.

b. Given that a woman tests negative during a clinical breast examination, find the probability that she does not have breast cancer.

c. Using the information above, how many false positives would you expect for every 1000 clinical breast examinations?

26. *Test for HIV* A test for the virus that causes AIDS, developed by Octopus Diagnostics Research of Hantsport, Nova Scotia, shows the presence or absence of HIV in a drop of blood in two minutes, compared with five days for other tests.‡ Preliminary results indicate a false positive rate (an indication that the HIV virus is present when it is not) of less than 2%, and a false negative rate (a failure to detect the presence of the HIV virus) of up to 5%. Assume for this exercise that these rates are exactly 2% and 5%. In 2001, there were 950,000 people in North America with the HIV virus, out of a population of 491 million.§ Suppose a resident of North America is chosen at random and given this test. If the result is positive, what is the probability that the person actually has the HIV virus?

27. *Smokers* A health study tracked a group of persons for five years. At the beginning of the study, 20% were classified as heavy smokers, 30% as light smokers, and 50% as nonsmokers. Results of the study showed that light smokers were twice as likely as nonsmokers to die during the five-year study, but only half as likely as heavy smokers. A randomly selected participant from the study died over the five-year period. Calculate the probability that the participant was a heavy smoker.‖ (Choose one of the following. *Hint:* Let $x = P$(a nonsmoker dies).)

a. 0.20 **b.** 0.25 **c.** 0.35 **d.** 0.42 **e.** 0.57

28. *Emergency Room* Upon arrival at a hospital's emergency room, patients are categorized according to their condition as critical, serious, or stable. In the past year:

(i) 10% of the emergency room patients were critical;

(ii) 30% of the emergency room patients were serious;

(iii) the rest of the emergency room patients were stable;

(iv) 40% of the critical patients died;

(v) 10% of the serious patients died; and

(vi) 1% of the stable patients died.

Given that a patient survived, what is the probability that the patient was categorized as serious upon arrival?# (Choose one of the following.)

a. 0.06 **b.** 0.29 **c.** 0.30 **d.** 0.39 **e.** 0.64

29. *Blood Test* A blood test indicates the presence of a particular disease 95% of the time when the disease is actually present. The same test indicates the presence of the disease 0.5% of the time when the disease is not present. One percent of the population actually has the disease. Calculate the probability that a person has the disease, given that the test indicates the presence of the disease.** (Choose one of the following.)

a. 0.324 **b.** 0.657 **c.** 0.945 **d.** 0.950 **e.** 0.995

30. *Circulation* The probability that a randomly chosen male has a circulation problem is 0.25. Males who have a circulation problem are twice as likely to be smokers as those who do not have a circulation problem. What is the conditional probability that a male has a circulation problem, given that he is a smoker? †† (Choose one of the following.)

a. 1/4 **b.** 1/3 **c.** 2/5 **d.** 1/2 **e.** 2/3

*Barton, Mary B., Russell Harris, and Suzanne Fletcher, "Does This Patient Have Breast Cancer, The Screening Clinical Breast Examinations: Should It Be Done? How?" *JAMA,* Vol. 282, No. 13, Oct. 6, 1999, pp. 1270–1280.

†*The World Almanac and Book of Facts 2000*, p. 902.

‡*Maclean's*, Feb. 17, 1997, p. 70.

§*The World Almanac and Book of Facts 2003*, pp. 857, 859.

‖Problem 31 from May 2003 Course 1 Examination of the *Education and Examination Committee of the Society of Actuaries*. Reprinted by permission of the Society of Actuaries.

#Problem 21 from the 2005 Sample Exam P of the *Education and Examination Committee of the Society of Actuaries.* Reprinted by permission of the Society of Actuaries.

**Problem 25 from the 2005 Sample Exam P of the *Education and Examination Committee of the Society of Actuaries.* Reprinted by permission of the Society of Actuaries.

††Problem 26 from the 2005 Sample Exam P of the *Education and Examination Committee of the Society of Actuaries*. Reprinted by permission of the Society of Actuaries.

SOCIAL SCIENCES

31. *Binge Drinking* A 1995 study by the Harvard School of Public Health reported that 86% of male students who live in a fraternity house are binge drinkers. The figure for fraternity members who are not residents of a fraternity house is 71%, while the figure for men who do not belong to a fraternity is 45%.* Suppose that 10% of U.S. male students live in a fraternity house, 15% belong to a fraternity but do not live in a fraternity house, and 75% do not belong to a fraternity.

    a. What is the probability that a randomly selected male student is a binge drinker?

    b. If a randomly selected male student is a binge drinker, what is the probability that he lives in a fraternity house?

32. *Murder* During the murder trial of O. J. Simpson, Alan Dershowitz, an advisor to the defense team, stated on television that only about 0.1% of men who batter their wives actually murder them. Statistician I. J. Good observed that even if, given that a husband is a batterer, the probability he is guilty of murdering his wife is 0.001, what we really want to know is the probability that the husband is guilty, given that the wife was murdered.† Good estimates the probability of a battered wife being murdered, given that her husband is not guilty, as 0.001. The probability that she is murdered if her husband is guilty is 1, of course. Using these numbers and Dershowitz's 0.001 probability of the husband being guilty, find the probability that the husband is guilty, given that the wife was murdered.

Never-Married Adults by Age Group The following tables give the proportion of men and of women 18 and older in each age group in 2003, as well as the proportion in each group who have never been married.‡

	Men	
Age	Proportion of Population	Proportion Never Married
18–24	0.135	0.895
25–34	0.191	0.434
35–44	0.212	0.195
45–64	0.320	0.088
65 or over	0.142	0.074

	Women	
Age	Proportion of Population	Proportion Never Married
18–24	0.123	0.806
25–34	0.179	0.311
35–44	0.203	0.132
45–64	0.316	0.075
65 or over	0.179	0.037

33. Find the probability that a randomly selected man who has never married is between 35 and 44 years old (inclusive).

34. Find the probability that a randomly selected woman who has been married is between 18 and 24 (inclusive).

35. Find the probability that a randomly selected woman who has never been married is between 45 and 64 (inclusive).

Seat Belt Effectiveness A federal study showed that in 1990, 49% of all those involved in a fatal car crash wore seat belts. Of those in a fatal crash who wore seat belts, 44% were injured and 27% were killed. For those not wearing seat belts, the comparable figures were 41% and 50%, respectively.§

36. Find the probability that a randomly selected person who was killed in a car crash was wearing a seat belt.

37. Find the probability that a randomly selected person who was unharmed in a fatal crash was not wearing a seat belt.

GENERAL INTEREST

38. *Automobile Accidents* A study of automobile accidents produced the following data:

Model Year	Proportion of All Vehicles	Probability of Involvement in an Accident
1997	0.16	0.05
1998	0.18	0.02
1999	0.20	0.03
Other	0.46	0.04

*The New York Times, Dec. 6, 1995, p. B16.

†Good, I. J., "When Batterer Turns Murderer," *Nature*, Vol. 375, No. 15, June 15, 1995, p. 541.

‡From data in *The New York Times 2006 Almanac*, p. 289.

§National Highway Traffic Safety Administration, Office of Driver and Pedestrian Research: "Occupant Protection Trends in 19 Cities," Nov. 1989, and "Use of Automatic Safety Belt Systems in 19 Cities, Feb. 1991.

An automobile from one of the model years 1997, 1998, and 1999 was involved in an accident. Determine the probability that the model year of this automobile is 1997.* (Choose one of the following. *Hint:* Be sure to ignore the "Other" branch of the tree.)

a. 0.22 **b.** 0.30 **c.** 0.33 **d.** 0.45 **e.** 0.50

39. *Terrorists* John Allen Paulos has pointed out a problem with massive, untargeted wiretaps.† To illustrate the problem, he supposes that one out of every million Americans has terrorist ties. Furthermore, he supposes that the terrorist profile is 99% accurate, so that if a person has terrorist ties, the profile will pick them up 99% of the time, and if the person does not have terrorist ties, the profile will accidentally pick them up only 1% of the time. Given that the profile has picked up a person, what is the probability that the person actually has terrorist ties? Discuss how your answer affects your opinion on domestic wiretapping.

40. *Three Prisoners* The famous "problem of three prisoners" is as follows.‡

> Three men, A, B, and C, were in jail. A knew that one of them was to be set free and the other two were to be executed. But he didn't know who was the one to be spared. To the jailer who did know, A said, "Since two out of the three will be executed, it is certain that either B or C will be, at least. You will give me no information about my own chances if you give me the name of one man, B or C, who is going to be executed." Accepting this argument after some thinking, the jailer said "B will be executed." Thereupon A felt happier because now either he or C would go free, so his chance had increased from $1/3$ to $1/2$.

a. Assume that initially each of the prisoners is equally likely to be set free. Assume also that if both B and C are to be executed, the jailer is equally likely to name either B or C. Show that A is wrong, and that his probability of being freed, given that the jailer says B will be executed, is still 1/3.

b. Now assume that initially the probabilities of A, B, and C being freed are $1/4$, $1/4$, and $1/2$, respectively. As in part a, assume also that if both B and C are to be executed, the jailer is equally likely to name either B or C. Now show that A's probability of being freed, given that the jailer says B will be executed, actually drops to $1/5$. Discuss the reasonableness of this answer, and why this result might violate someone's intuition.

Chapter 7 Review

Chapter Summary

We began this chapter by introducing sets, which are collections of objects. We introduced the following set operations:

- complement (A' is the set of elements not in A),
- intersection ($A \cap B$ is the set of elements belonging to both set A and set B), and
- union ($A \cup B$ is the set of elements belonging to either set A or set B).

We used tree diagrams and Venn diagrams to define and study concepts in set operations as well as in probability. We introduced the following terms:

- experiment (an activity or occurrence with an observable result),
- trial (a repetition of an experiment),
- outcome (a result of a trial),

*Problem 27 from the 2005 Sample Exam P of the *Education and Examination Committee of the Society of Actuaries.* Reprinted by permission of the Society of Actuaries.

†http://abcnews.go.com/Technology/WhosCounting/story?id=1560771.

‡Shimojo, Shinsuke, and Shin'ichi Ichikawa, "Intuitive Reasoning About Probability: Theoretical and Experimental Analyses of the 'Problem of Three Prisoners,'" *Cognition*, Vol. 32, 1989, pp. 1–24.

- sample space (the set of all possible outcomes for an experiment), and
- event (a subset of a sample space).

We investigated how to compute various probabilities and we explored some of the properties of probability. In particular, we studied the following concepts:

- empirical probability (based on how frequently an event actually occurred),
- conditional probability (in which some other event is assumed to have occurred),
- odds (an alternative way of expressing probability),
- independent events (in which the occurrence of one event does not affect the probability of another), and
- Bayes' theorem (used to calculate certain types of conditional probability).

Throughout the chapter, many applications of probability were introduced and analyzed. In the next two chapters, we will employ these techniques to further our study into the fields of probability and statistics.

SETS SUMMARY

Number of Subsets A set of k distinct elements has 2^k subsets.

Disjoint Sets If sets A and B are disjoint, then

$$A \cap B = \emptyset \qquad \text{and} \qquad n(A \cap B) = 0.$$

Union Rule for Sets For any sets A and B,

$$n(A \cup B) = n(A) + n(B) - n(A \cap B).$$

PROBABILITY SUMMARY

Basic Probability Principle Let S be a sample space of equally likely outcomes, and let event E be a subset of S. Then the probability that event E occurs is

$$P(E) = \frac{n(E)}{n(S)}.$$

Mutually Exclusive Events If E and F are mutually exclusive events,

$$E \cap F = \emptyset \qquad \text{and} \qquad P(E \cap F) = 0.$$

Union Rule For any events E and F from a sample space S,

$$P(E \cup F) = P(E) + P(F) - P(E \cap F).$$

Complement Rule $P(E) = 1 - P(E')$ and $P(E') = 1 - P(E)$

Odds The odds in favor of event E are $\dfrac{P(E)}{P(E')}$, $P(E') \neq 0$.

If the odds favoring event E are m to n, then

$$P(E) = \frac{m}{m+n} \qquad \text{and} \qquad P(E') = \frac{n}{m+n}.$$

Properties of Probability

1. For any event E in sample space S, $0 \leq P(E) \leq 1$.
2. The sum of the probabilities of all possible distinct outcomes is 1.

Conditional Probability The conditional probability of event E, given that event F has occurred, is

$$P(E|F) = \frac{P(E \cap F)}{P(F)}, \qquad \text{where } P(F) \neq 0.$$

For equally likely outcomes, conditional probability is found by reducing the sample space to event F; then

$$P(E|F) = \frac{n(E \cap F)}{n(F)}.$$

Product Rule of Probability If E and F are events, then $P(E \cap F)$ may be found by either of these formulas.

$$P(E \cap F) = P(F) \cdot P(E|F) \quad \text{or} \quad P(E \cap F) = P(E) \cdot P(F|E)$$

Independent Events If E and F are independent events,

$$P(E|F) = P(E), \quad P(F|E) = P(F), \quad \text{and} \quad P(E \cap F) = P(E) \cdot P(F).$$

Bayes' Theorem

$$P(F_i|E) = \frac{P(F_i) \cdot P(E|F_i)}{P(F_1) \cdot P(E|F_1) + P(F_2) \cdot P(E|F_2) + \cdots + P(F_n) \cdot P(E|F_n)}$$

Key Terms

7.1 set
element (member)
empty set
set-builder notation
universal set
subset
tree diagram
Venn diagram
complement
intersection
disjoint sets
union

7.2 union rule for sets

7.3 experiment
trial
outcome
sample space
event
simple event
certain event
impossible event
mutually exclusive events
probability

7.4 union rule for probability
odds
empirical probability
probability distribution

7.5 conditional probability
product rule
independent events
dependent events

7.6 Bayes' theorem

Concept Check

Determine whether each of the following statements is true or false, and explain why.

1. A set is a subset of itself.
2. A set has more subsets than it has elements.
3. The union of two sets always has more elements than either set.
4. The intersection of two sets always has fewer elements than either set.
5. The number of elements in the union of two sets can be found by adding the number of elements in each set.
6. The probability of an event is always at least 0 and no larger than 1.
7. The probability of the union of two events can be found by adding the probability of each event.
8. The probability of drawing the Queen of Hearts from a deck of cards is an example of empirical probability.
9. If two events are mutually exclusive, then they are independent.
10. The probability of two independent events can be found by multiplying the probabilities of each event.
11. The probability of an event E given an event F is the same as the probability of F given E.
12. Bayes' theorem can be useful for calculating conditional probability.

Chapter 7 Review Exercises

Write true or false for each statement.

1. $9 \in \{8, 4, -3, -9, 6\}$

2. $4 \notin \{3, 9, 7\}$

3. $2 \notin \{0, 1, 2, 3, 4\}$

4. $0 \in \{0, 1, 2, 3, 4\}$

5. $\{3, 4, 5\} \subseteq \{2, 3, 4, 5, 6\}$

6. $\{1, 2, 5, 8\} \subseteq \{1, 2, 5, 10, 11\}$

7. $\{3, 6, 9, 10\} \subseteq \{3, 9, 11, 13\}$

8. $\emptyset \subseteq \{1\}$

9. $\{2, 8\} \not\subseteq \{2, 4, 6, 8\}$

10. $0 \subseteq \emptyset$

In Exercises 11–20, let $U = \{\text{a, b, c, d, e, f, g, h}\}$, $K = \{\text{c, d, e, f, h}\}$, *and* $R = \{\text{a, c, d, g}\}$. *Find the following.*

11. The number of subsets of K

12. The number of subsets of R

13. K'

14. R'

15. $K \cap R$

16. $K \cup R$

17. $(K \cap R)'$

18. $(K \cup R)'$

19. $\emptyset'$

20. U'

In Exercises 21–26, let $U = \{\text{all employees of the K. O. Brown Company}\}$;
$A = \{\text{employees in the accounting department}\}$;
$B = \{\text{employees in the sales department}\}$;
$C = \{\text{female employees}\}$;
$D = \{\text{employees with an MBA degree}\}$.

Describe each set in words.

21. $A \cap C$

22. $B \cap D$

23. $A \cup D$

24. $A' \cap D$

25. $B' \cap C'$

26. $(B \cup C)'$

Draw a Venn diagram and shade each set.

27. $A \cup B'$

28. $A' \cap B$

29. $(A \cap B) \cup C$

30. $(A \cup B)' \cap C$

Write the sample space for each experiment.

31. Rolling a die

32. Drawing a card from a deck containing only the 13 spades

33. Measuring the weight of a person to the nearest half pound (the scale will not measure more than 300 lb)

34. Tossing a coin 4 times

A jar contains 5 *balls labeled* 3, 5, 7, 9, *and* 11, *respectively, while a second jar contains* 4 *red and* 2 *green balls. An experiment consists of pulling* 1 *ball from each jar, in turn. In Exercises 35–37, write each set using set notation.*

35. The sample space

36. Event E: the number on the first ball is greater than 5

37. Event F: the second ball is green

38. Are the outcomes in the sample space in Exercise 35 equally likely?

In Exercises 39–45, find the probability of each event when a single card is drawn from an ordinary deck.

39. A heart

40. A red queen

41. A face card

42. Black or a face card

43. Red, given that it is a queen

44. A jack, given that it is a face card

45. A face card, given that it is a king

46. Describe what is meant by disjoint sets.

47. Describe what is meant by mutually exclusive events.

48. How are disjoint sets and mutually exclusive events related?

49. Define independent events.

50. Are independent events always mutually exclusive? Are they ever mutually exclusive?

51. An uproar has raged since September 1990 over the answer to a puzzle* published in *Parade* magazine, a supplement of the Sunday newspaper. In the "Ask Marilyn" column, Marilyn vos Savant answered the following question: "Suppose you're on a game show, and you're given the choice of three doors. Behind one door is a car; behind the others, goats. You pick a door, say number 1, and the host, who knows what's behind the other doors, opens another door, say number 3, which has a goat. He then says to you, 'Do you want to pick door number 2?' Is it to your advantage to take the switch?"

Ms. vos Savant estimates that she has since received some 10,000 letters; most of them, including many from mathematicians and statisticians, disagreed with her answer. Her answer has been debated by both professionals and amateurs, and tested in classes at all levels, from grade school to graduate school. But by performing the experiment repeatedly, it can be shown that vos Savant's answer was correct. Find the probabilities of getting the car if you switch or do not switch, and then answer the question yourself. (*Hint:* Consider the sample space.)

Find the odds in favor of a card drawn from an ordinary deck being the following.

52. A club

53. A black jack

54. A red face card or a queen

Find the probabilities of getting the following sums when two fair dice are rolled.

55. 8

56. 0

57. At least 10

58. No more than 5

59. An odd number greater than 8

60. 12, given that the sum is greater than 10

61. 7, given that at least one die shows a 4

62. At least 9, given that at least one die shows a 5

63. Suppose $P(E) = 0.51$, $P(F) = 0.37$, and $P(E \cap F) = 0.22$. Find the following.

a. $P(E \cup F)$ **b.** $P(E \cap F')$ **c.** $P(E' \cup F)$ **d.** $P(E' \cap F')$

64. Box A contains 5 red balls and 1 black ball; box B contains 2 red balls and 3 black balls. A box is chosen, and a ball is selected from it. The probability of choosing box A is $3/8$. If the selected ball is black, what is the probability that it came from box A?

65. Find the probability that the ball in Exercise 64 came from box B, given that it is red.

66. An urn contains 10 balls: 4 red and 6 blue. A second urn contains 16 red balls and an unknown number of blue balls. A single ball is drawn from each urn. The probability that both balls are the same color is 0.44. Calculate the number of blue balls in the second urn.† (Choose one of the following.)

a. 4 **b.** 20 **c.** 24 **d.** 44 **e.** 64

*Source: *Parade* magazine, Sept. 9, 1990, p. 13. © 1990 Marilyn vos Savant. Initially published in *Parade* magazine. All rights reserved.

†Problem 4 from the 2005 Sample Exam P of the *Education and Examination Committee of the Society of Actuaries*. Reprinted by permission of the Society of Actuaries.

Applications

BUSINESS AND ECONOMICS

Appliance Repairs *Of the appliance repair shops listed in the phone book, 80% are competent and 20% are not. A competent shop can repair an appliance correctly 95% of the time; an incompetent shop can repair an appliance correctly 55% of the time. Suppose an appliance was repaired correctly. Find the probabilities that it was repaired by the following.*

67. A competent shop

68. An incompetent shop

Suppose an appliance was repaired incorrectly. Find the probabilities that it was repaired by the following.

69. A competent shop

70. An incompetent shop

71. Find the probability that an appliance brought to a shop chosen at random is repaired correctly.

72. Are the events that a repair shop is competent and that the repair is done correctly independent? Explain.

73. *Sales* A company sells printers and copiers. Let E be the event "a customer buys a printer," and let F be the event "a customer buys a copier." Write the following using $\cap$, $\cup$, or $'$ as necessary.

a. A customer buys neither machine.

b. A customer buys at least one of the machines.

74. *Defective Items* A sample shipment of five hair dryers is chosen at random. The probability of exactly 0, 1, 2, 3, 4, or 5 hair dryers being defective is given in the following table.

Number Defective	0	1	2	3	4	5
Probability	0.34	0.26	0.18	0.12	0.07	0.03

Find the probabilities that the following numbers of hair dryers are defective.

a. No more than 3

b. At least 3

75. *Defective Items* A manufacturer buys items from four different suppliers. The fraction of the total number of items that is obtained from each supplier, along with the probability that an item purchased from that supplier is defective, is shown in the table below.

Supplier	Fraction of Total Supplied	Probability of Defective
1	0.17	0.01
2	0.39	0.02
3	0.35	0.05
4	0.09	0.03

a. Find the probability that a randomly selected item is defective.

b. Find the probability that a defective item came from supplier 4.

c. Find the probability that a defective item came from supplier 2.

d. Are the events that an item came from supplier 4 and that the item is defective independent? Explain.

76. *Car Buyers* The table shows the results of a survey of buyers of a certain model of car.

Car Type	Satisfied	Not Satisfied	Totals
New	300	100	
Used	450		600
Totals		250	

a. Complete the table.

b. How many buyers were surveyed?

c. How many bought a new car and were satisfied?

d. How many were not satisfied?

e. How many bought used cars?

f. How many of those who were not satisfied had purchased a used car?

g. Rewrite the event stated in part f using the expression "given that."

h. Find the probability of the outcome in parts f and g.

i. Find the probability that a used-car buyer is not satisfied.

j. You should have different answers in parts h and i. Explain why.

k. Are the events that a car is new and that the customer is satisfied independent? Explain.

77. *Auto Insurance* An insurance company examines its pool of auto insurance customers and gathers the following information:

(i) All customers insure at least one car.

(ii) 70% of the customers insure more than one car.

(iii) 20% of the customers insure a sports car.

(iv) Of those customers who insure more than one car, 15% insure a sports car.

Calculate the probability that a randomly selected customer insures exactly one car and that car is not a sports car.* (Choose one of the following. *Hint:* Draw a tree diagram, and let x be the probability that a customer who insures exactly one car insures a sports car.)

a. 0.13 **b.** 0.21 **c.** 0.24 **d.** 0.25 **e.** 0.30

78. *Auto Insurance* An auto insurance company has 10,000 policyholders. Each policyholder is classified as:

(i) young or old;

(ii) male or female; and

(iii) married or single.

Of these policyholders, 3000 are young, 4600 are male, and 7000 are married. The policyholders can also be classified as 1320 young males, 3010 married males, and 1400 young married persons. Finally, 600 of the policyholders are young married males. How many of the company's policyholders are young, female, and single?† (Choose one of the following.)

a. 280 **b.** 423 **c.** 486 **d.** 880 **e.** 896

79. *Auto Insurance* An actuary studying the insurance preferences of automobile owners makes the following conclusions:

(i) An automobile owner is twice as likely to purchase collision coverage as disability coverage.

(ii) The event that an automobile owner purchases collision coverage is independent of the event that he or she purchases disability coverage.

(iii) The probability that an automobile owner purchases both collision and disability coverages is 0.15.

What is the probability that an automobile owner purchases neither collision nor disability coverage?‡ (Choose one of the following.)

a. 0.18 **b.** 0.33 **c.** 0.48 **d.** 0.67 **e.** 0.82

80. *Insurance* An insurance company estimates that 40% of policyholders who have only an auto policy will renew next year and 60% of policyholders who have only a homeowners policy will renew next year. The company estimates that 80% of policyholders who have both an auto and a homeowners policy will renew at least one of these policies next year. Company records show that 65% of policyholders have an auto policy, 50% of policyholders have a homeowners policy, and 15% of policyholders have both an auto and a homeowners policy. Using the company's estimates, calculate the percentage of policyholders that will renew at least one policy next year.§ (Choose one of the following.)

a. 20 **b.** 29 **c.** 41 **d.** 53 **e.** 70

LIFE SCIENCES

81. *Sickle Cell Anemia* The square below shows the four possible (equally likely) combinations when both parents are carriers of the sickle cell anemia trait. Each carrier parent has normal cells (N) and trait cells (T).

a. Complete the table.

b. If the disease occurs only when two trait cells combine, find the probability that a child born to these parents will have sickle cell anemia.

c. The child will carry the trait but not have the disease if a normal cell combines with a trait cell. Find this probability.

d. Find the probability that the child is neither a carrier nor has the disease.

		Second Parent	
		N_2	T_2
First Parent	N_1		N_1T_2
	T_1		

82. *Blood Antigens* In Exercise 44 of Section 7.2, we described the eight types of human blood. The percentage of the population having each type is as follows:‖

O^+: 38%; O^-: 8%; A^+: 32%; A^-: 7%;
B^+: 9%; B^-: 2%; AB^+: 3%; AB^-: 1%.

*Problem 5 from May 2003 Course 1 Examination of the *Education and Examination Committee of the Society of Actuaries.* Reprinted by permission of the Society of Actuaries.

†Problem 5 from the 2005 Sample Exam P of the *Education and Examination Committee of the Society of Actuaries.* Reprinted by permission of the Society of Actuaries.

‡Problem 11 from the 2005 Sample Exam P of the *Education and Examination Committee of the Society of Actuaries.* Reprinted by permission of the Society of Actuaries.

§Problem 18 from May 2003 Course 1 Examination of the *Education and Examination Committee of the Society of Actuaries.* Reprinted by permission of the Society of Actuaries.

‖Young, Victoria, "A Matter of Survival," *The Mathematics Teacher*, Vol. 95, No. 2, Feb. 2002, pp. 100–112.

When a person receives a blood transfusion, it is important that the blood be compatible, which means that it introduces no new antigens into the recipient's blood. The following diagram helps illustrate what blood types are compatible.

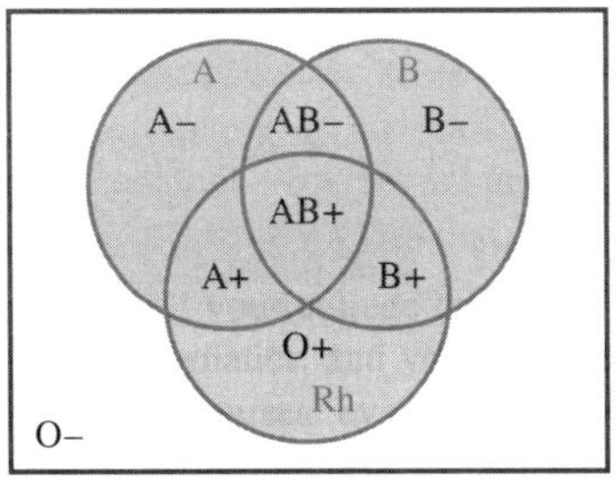

The universal blood type is O^-, since it has none of the additional antigens. The circles labeled A, B, and Rh contain blood types with the A antigen, B antigen, and Rh antigen, respectively. A person with O^- blood can only be transfused with O^- blood, because any other type would introduce a new antigen. Thus the probability that blood from a random donor is compatible is just 8%. A person with AB^+ blood already has all antigens, so the probability that blood from a random donor is compatible is 100%. Find the probability that blood from a random donor is compatible with a person with each blood type.

a. O^+ **b.** A^+ **c.** B^+

d. A^- **e.** B^- **f.** AB^-

83. *Heart Disease* A public health researcher examines the medical records of a group of 937 men who died in 1999 and discovers that 210 of the men died from causes related to heart disease. Moreover, 312 of the 937 men had at least one parent who suffered from heart disease, and, of these 312 men, 102 died from causes related to heart disease. Determine the probability that a man randomly selected from this group died of causes related to heart disease, given that neither of his parents suffered from heart disease.* (Choose one of the following.)

a. 0.115 **b.** 0.173 **c.** 0.224 **d.** 0.327 **e.** 0.514

84. *Risk Factors* An actuary is studying the prevalence of three health risk factors, denoted by A, B, and C, within a population of women. For each of the three factors, the probability is 0.1 that a woman in the population has only this risk factor (and no others). For any two of the three factors, the probability is 0.12 that she has exactly these two risk factors (but not the other). The probability that a woman has all three risk factors, given that she has A and B, is 1/3. What is the probability that a woman has none of the three risk factors, given that she does not have risk factor A?† (Choose one of the following.)

a. 0.280 **b.** 0.311 **c.** 0.467 **d.** 0.484 **e.** 0.700

SOCIAL SCIENCES

85. *Elections* In the 2004 presidential elections, George W. Bush won 50.72% of the popular vote, as opposed to 48.27% for John Kerry and 1.01% for other candidates.‡ According to an election day poll, 50% of those who voted for Bush were male, while 42% of those who voted for Kerry were male.§ For this exercise, assume that 50% of the other voters were male.

a. Find the percentage of the voters who were male. Compare this with the figure of 46% reported in the survey.

b. Find the probability that a randomly selected male voter voted for Bush.

c. Find the probability that a randomly selected female voter voted for Bush.

86. *Television Viewing Habits* A telephone survey of television viewers revealed the following information:

20 watch situation comedies;
19 watch game shows;
27 watch movies;
19 watch movies but not game shows;
15 watch situation comedies but not game shows;
10 watch both situation comedies and movies;
3 watch all three;
7 watch none of these.

a. How many viewers were interviewed?

b. How many viewers watch comedies and movies but not game shows?

c. How many viewers watch only movies?

d. How many viewers do not watch movies?

*Problem 6 from the 2005 Sample Exam P of the *Education and Examination Committee of the Society of Actuaries*. Reprinted by permission of the Society of Actuaries.
†Problem 13 from the 2005 Sample Exam P of the *Education and Examination Committee of the Society of Actuaries*. Reprinted by permission of the Society of Actuaries.
‡*Time Almanac 2006*, p. 45.
§*The New York Times*, Nov. 3, 2004, p. P4.

87. *Randomized Response Method for Getting Honest Answers to Sensitive Questions** Basically, this is a method to guarantee that an individual who answers sensitive question will remain anonymous, thus encouraging a truthful response. This method is, in effect, an application of the formula for finding the probability of an intersection, and operates as follows. Questions A and B are posed, one of which is sensitive and the other not. The probability of receiving a "yes" to the nonsensitive question must be known. For example, one could ask

A: Does your Social Security number end in an odd digit? (Nonsensitive)

B: Have you ever intentionally cheated on your income tax? (Sensitive)

We know that $P(\text{answer yes} \mid \text{answer } A) = 1/2$. We wish to approximate $P(\text{answer yes} \mid \text{answer } B)$. The subject is asked to flip a coin and answer A if the coin comes up heads and otherwise to answer B. In this way, the interviewer does not know which question the subject is answering. Thus, a "yes" answer is not incriminating. There is no way for the interviewer to know whether the subject is saying "Yes, my Social Security number ends in an odd digit" or "Yes, I have intentionally cheated on my income taxes." The percentage of subjects in the group answering "yes" is used to approximate $P(\text{answer yes})$.

a. Use the fact that the event "answer yes" is the union of the event "answer yes and answer A" with the event "answer yes and answer B" to prove that

$$P(\text{answer yes} \mid \text{answer } B) = \frac{P(\text{answer yes}) - P(\text{answer yes} \mid \text{answer } A) \cdot P(\text{answer } A)}{P(\text{answer } B)}.$$

b. If this technique is tried on 100 subjects and 60 answered "yes," what is the approximate probability that a person randomly selected from the group has intentionally cheated on income taxes?

88. *Police Lineup* To illustrate the difficulties with eyewitness identifications from police lineups, John Allen Paulos considers a "lineup" of three pennies, in which we know that two are fair (innocent) and the third (the culprit) has a 75% probability of landing heads.† The probability of picking the culprit by chance is, of course, 1/3. Suppose we observe three heads in a row on one of the pennies. If we then guess that this penny is the culprit, what is the probability that we're right?

89. *SIDS* On July 15, 2005, a panel in England ruled that Roy Meadow, a renowned expert on child abuse and co-founder of London's Royal College of Paediatrics and Child Health, should be erased from the register of physicians in Britain for his faulty statistics at the trial of Sally Clark, who was convicted of murdering her first two babies.‡ Meadow testified at the trial that the probability of a baby dying of sudden death syndrome (SIDS) is $1/8543$. He then calculated that the probability of two babies in a family dying of SIDS is $(1/8543)^2 \approx 1/73{,}000{,}000$. With such a small probability of both babies dying of SIDS, he concluded that the babies were instead murdered. What assumption did Meadow make in doing this calculation? Discuss reasons why this assumption may be invalid. (*Note:* Clark spent three years in prison before her conviction was reversed.)

PHYSICAL SCIENCES

90. *Earthquake* It has been reported that government scientists have predicted that the odds for a major earthquake occurring in the San Francisco Bay area during the next 30 years are 9 to 1.§ What is the probability that a major earthquake will occur during the next 30 years in San Francisco?

GENERAL INTEREST

91. *Making a First Down* A first down is desirable in football—it guarantees four more plays by the team making it, assuming no score or turnover occurs in the plays. After getting a first down, a team can get another by advancing the ball at least 10 yards. During the four plays given by a first down, a team's position will be indicated by a phrase such as "third and 4," which means that the team has already had two of its four plays, and that 4 more yards are needed to get 10 yards necessary for another first down. An article in a management journal‖ offers the following

*Milton, J. S. and J. J. Corbet, *Applied Statistics with Probability*. Copyright © 1979 by Litton Educational Publishing, Inc. Reprinted by permission of Brooks/Cole Publishing Company, Monterey, California.

†John Allen Paulos, "Coins and Confused Eyewitnesses: Calculating the Probability of Picking the Wrong Guy," *Who's Counting*, Feb. 1, 2001. http://more.abcnews.go.com/sections/science/ whoscounting _index/whoscounting_index.html.

‡*Science*, Vol. 309, July 22, 2005, p. 543.

§*The San Francisco Chronicle,* June 8, 1994, p. A1.

‖Carter, Virgil and Robert Machols, "Optimal Strategies on Fourth Down," *Management Science,* Vol. 24, No. 16, Dec. 1978. Copyright © 1978 by The Institute of Management Sciences.

results for 189 games for a particular National Football League season. "Trials" represents the number of times a team tried to make a first down, given that it was currently playing either a third or a fourth down. Here, n represents the number of yards still needed for a first down.

n	Trials	Successes	Probability of Making First Down with n Yards to Go
1	543	388	
2	327	186	
3	356	146	
4	302	97	
5	336	91	

a. Complete the table.

b. Why is the sum of the answers in the table not equal to 1?

92. *States* Of the 50 United States, the following is true:

22 are west of the Mississippi River (western states);*
22 had populations less than 3 million in the 2000 census (small states);
26 begin with the letters A through M (early states);
9 are large late (beginning with the letters N through Z) eastern states;
13 are small western states;
10 are small early states;
5 are small early western states.

a. How many western states had populations more than 3 million in the 2000 census and begin with the letters N through Z?

b. How many states east of the Mississippi had populations more than 3 million in the 2000 census?

93. *Music* Country-western songs often emphasize three basic themes: love, prison, and trucks. A survey of the local country-western radio station produced the following data:

12 songs were about a truckdriver who was in love while in prison;
13 were about a prisoner in love;
28 were about a person in love;
18 were about a truckdriver in love;
33 were about people not in prison;
18 were about prisoners;
15 were about truckdrivers who were in prison;
16 were about truckdrivers who were not in prison.

a. How many songs were surveyed?

Find the number of songs about

b. truckdrivers;

c. prisoners;

d. truckdrivers in prison;

e. people not in prison;

f. people not in love.

94 *Gambling* The following puzzle was featured on the Puzzler part of the radio program *Car Talk* on February 23, 2002.† A con man puts three cards in a bag; one card is green on both sides, one is red on both sides, and the third is green on one side and red on the other. He lets you pick one card out of the bag and put it on a table, so you can see that a red side is face up, but neither of you can see the other side. He offers to bet you even money that the other side is also red. In other words, if you bet $1, you lose if the other side is red but get back $2 if the other side is green. Is this a good bet? What is the probability that the other side is red?

95. *Missiles* In his novel *Debt of Honor*, Tom Clancy writes the following:‡

"There were ten target points—missile silos, the intelligence data said, and it pleased the Colonel [Zacharias] to be eliminating the hateful things, even though the price of that was the lives of other men. There were only three of them [bombers], and his bomber, like the others, carried only eight weapons [smart bombs]. The total number of weapons carried for the mission was only twenty-four, with two designated for each silo, and Zacharias's last four for the last target. Two bombs each. Every bomb had a 95% probability of hitting within four meters of the aim point, pretty good numbers really, except that this sort of mission had precisely no margin for error. Even the paper probability was less than half a percent chance of a double miss, but that number times ten targets meant a 5% chance that [at least] one missile would survive, and that could not be tolerated."

Determine whether the calculations in this quote are correct by the following steps.

a. Given that each bomb had a 95% probability of hitting the missile silo on which it was dropped, and that two bombs were dropped on each silo, what is the probability of a double miss?

*We count here states such as Minnesota, which has more than half of its area to the west of the Mississippi.

†http://www.cartalk.com/content/puzzler/2002.html.

‡Clancy, Tom, *Debt of Honor*, New York: G. P. Putnam's Sons, 1994, pp. 686–687.

b. What is the probability that a specific silo was destroyed (that is, that at least one bomb of the two bombs struck the silo)?

c. What is the probability that all ten silos were destroyed?

d. What is the probability that at least one silo survived? Does this agree with the quote?

e. What assumptions need to be made for the calculations in parts a through d to be valid? Discuss whether these assumptions seem reasonable.

96 *Viewing Habits* A survey of a group's viewing habits over the last year revealed the following information:

(i) 28% watched gymnastics;

(ii) 29% watched baseball;

(iii) 19% watched soccer;

(iv) 14% watched gymnastics and baseball;

(v) 12% watched baseball and soccer;

(vi) 10% watched gymnastics and soccer;

(vii) 8% watched all three sports.

Calculate the percentage of the group that watched none of the three sports during the last year.* (Choose one of the following.)

a. 24 **b.** 36 **c.** 41 **d.** 52 **e.** 60

*Problem 1 from May 2003 Course 1 Examination of the *Education and Examination Committee of the Society of Actuaries*. Reprinted by permission of the Society of Actuaries.

Medical Diagnosis

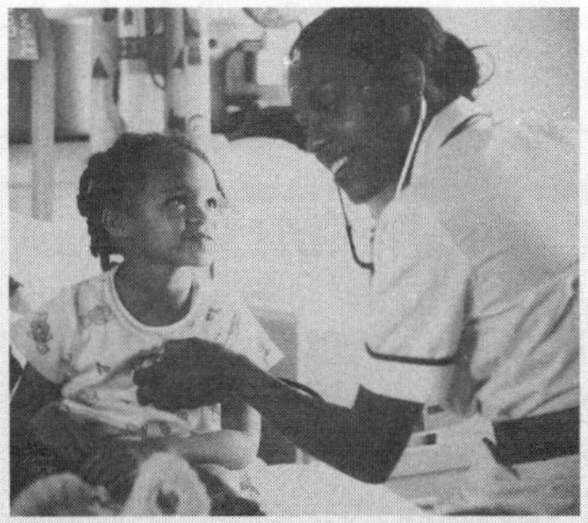

When a patient is examined, information (typically incomplete) is obtained about his or her state of health. Probability theory provides a mathematical model appropriate for this situation, as well as a procedure for quantitatively interpreting such partial information to arrive at a reasonable diagnosis.*

To develop a model, we list the states of health that can be distinguished in such a way that the patient can be in one and only one state at the time of the examination. For each state of health H, we associate a number, $P(H)$, between 0 and 1 such that the sum of all these numbers is 1. This number $P(H)$ represents the probability, before examination, that a patient is in the state of health H, and $P(H)$ may be chosen subjectively from medical experience, using any information available prior to the examination. The probability may be most conveniently established from clinical records; that is, a mean probability is established for patients in general, although the number would vary from patient to patient. Of course, the more information that is brought to bear in establishing $P(H)$, the better the diagnosis.

For example, limiting the discussion to the condition of a patient's heart, suppose there are exactly three states of health, with probabilities as follows.

	State of Health, *H*	***P(H)***
H_1	Patient has a normal heart	0.8
H_2	Patient has minor heart irregularities	0.15
H_3	Patient has a severe heart condition	0.05

Having selected $P(H)$, the information from the examination is processed. First, the results of the examination must be classified. The examination itself consists of observing the state of a number of characteristics of the patient. Let us assume that the examination for a heart condition consists of a stethoscope examination and a cardiogram. The outcome of such an examination, C, might be one of the following:

C_1 = stethoscope shows normal heart and cardiogram shows normal heart;

C_2 = stethoscope shows normal heart and cardiogram shows minor irregularities;

and so on.

It remains to assess for each state of health H the conditional probability $P(C|H)$ of each examination outcome C using only the knowledge that a patient is in a given state of health. (This may be based on the medical knowledge and clinical experience of the doctor.) The conditional probabilities $P(C|H)$ will not vary from patient to patient (although they should be reviewed periodically), so that they may be built into a diagnostic system.

Suppose the result of the examination is C_1. Let us assume the following probabilities:

$$P(C_1|H_1) = 0.9,$$
$$P(C_1|H_2) = 0.4,$$
$$P(C_1|H_3) = 0.1.$$

Now, for a given patient, the appropriate probability associated with each state of health H, after examination, is $P(H|C)$, where C is the outcome of the examination. This can be calculated by using Bayes' theorem. For example, to find $P(H_1|C_1)$—that is, the probability that the patient has a normal heart given that the examination showed a normal stethoscope examination and a normal cardiogram—we use Bayes' theorem as follows:

$$P(H_1|C_1)$$
$$= \frac{P(C_1|H_1)P(H_1)}{P(C_1|H_1)P(H_1) + P(C_1|H_2)P(H_2) + P(C_1|H_3)P(H_3)}$$
$$= \frac{(0.9)(0.8)}{(0.9)(0.8) + (0.4)(0.15) + (0.1)(0.05)} \approx 0.92.$$

Hence, the probability is about 0.92 that the patient has a normal heart on the basis of the examination results. This means that in 8 out of 100 patients, some abnormality will be present and not be detected by the stethoscope or the cardiogram.

*Wright. Roger, "Probabilistic Medical Diagnosis," *Some Mathematical Models in Biology*, rev. ed., Robert M. Thrall, ed., University of Michigan, 1967. Used by permission of Robert M. Thrall.

EXERCISES

1. Find $P(H_2|C_1)$.
2. Assuming the following probabilities, find $P(H_1|C_2)$.

$P(C_2|H_1) = 0.2 \qquad P(C_2|H_2) = 0.8 \qquad P(C_2|H_3) = 0.3$

3. Assuming the probabilities of Exercise 2, find $P(H_3|C_2)$.

DIRECTIONS FOR GROUP PROJECT

Find an article on medical decision making from a medical journal and develop a doctor–patient scenario for that particular decision. Then create a role-playing activity where the doctor and nurse present the various options and the mathematics associated with making such a decision to a patient. Make sure to present the mathematics in a manner that the average patient might understand. (Hint: Many leading medical journals include articles on medical decision making. One particular journal that certainly includes such research is Medical Decision Making.)

8

Counting Principles; Further Probability Topics

If you have 31 ice cream flavors available, how many different three-scoop cones can you make? The answer, which is surprisingly large, involves counting permutations or combinations, the subject of the first two sections in this chapter. The counting formulas we will develop have important applications in probability theory.

In this chapter, we continue our discussion of probability theory. To use the basic definition of probability, $P(E) = n(E)/n(S)$ (where S is the sample space with equally likely outcomes), up to now we have simply listed the outcomes in S and in E. However, when S has many outcomes, listing them all becomes very tedious. In the first two sections of this chapter, we introduce methods for counting the number of outcomes in a set without actually listing them, and then we use this approach in the third section to find probabilities. In the section on binomial probability (repeated independent trials of an experiment with only two possible outcomes), we introduce a formula for finding the probability of a certain number of successes in a number of trials. The final section continues the discussion of probability distributions that we began in Chapter 7.

8.1 The Multiplication Principle; Permutations

? THINK ABOUT IT

In how many ways can seven panelists be seated in a row of seven chairs?

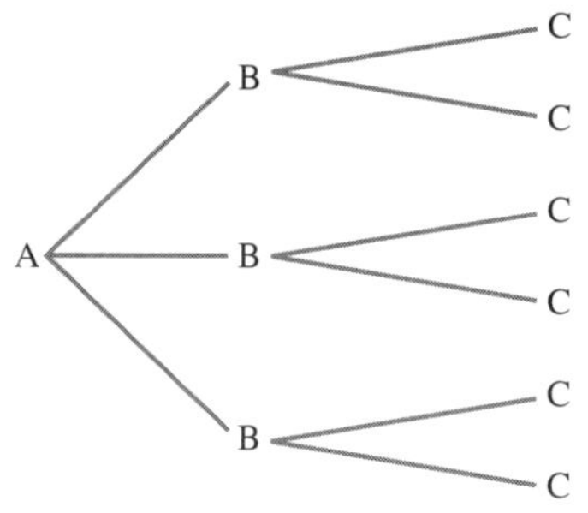

FIGURE 1

Before answering this question, let's begin with a simpler example. If there are 3 roads from town A to town B and 2 roads from town B to town C, in how many ways can a person travel from A to C by way of B? For each of the 3 roads from A there are 2 different routes leading from B to C, or a total of $3 \cdot 2 = 6$ different ways for the trip, as shown in Figure 1. This example illustrates a general principle of counting, called the **multiplication principle**.

MULTIPLICATION PRINCIPLE

Suppose n choices must be made, with

$$m_1 \text{ ways to make choice } 1,$$

and for each of these ways,

$$m_2 \text{ ways to make choice } 2,$$

and so on, with

$$m_n \text{ ways to make choice } n.$$

Then there are

$$m_1 \cdot m_2 \cdot \cdots \cdot m_n$$

different ways to make the entire sequence of choices.

EXAMPLE 1 **Combination Lock**

A certain combination lock can be set to open to any 3-letter sequence.

(a) How many sequences are possible?

▶**Solution** Since there are 26 letters in the alphabet, there are 26 choices for each of the 3 letters. By the multiplication principle, there are $26 \cdot 26 \cdot 26 = 17{,}576$ different sequences.

(b) How many sequences are possible if no letter is repeated?

▶**Solution** There are 26 choices for the first letter. It cannot be used again, so there are 25 choices for the second letter and then 24 choices for the third letter. Consequently, the number of such sequences is $26 \cdot 25 \cdot 24 = 15{,}600$.

EXAMPLE 2

Morse Code

Morse code uses a sequence of dots and dashes to represent letters and words. How many sequences are possible with at most 3 symbols?

▶**Solution** "At most 3" means "1 or 2 or 3" here. Each symbol may be either a dot or a dash. Thus the following number of sequences are possible in each case.

Number of Symbols	Number of Sequences
1	2
2	$2 \cdot 2 = 4$
3	$2 \cdot 2 \cdot 2 = 8$

Altogether, $2 + 4 + 8 = 14$ different sequences are possible.

EXAMPLE 3

I Ching

FIGURE 2

An ancient Chinese philosophical work known as the *I Ching (Book of Changes)* is often used as an oracle from which people can seek and obtain advice. The philosophy describes the duality of the universe in terms of two primary forces: *yin* (passive, dark, receptive) and *yang* (active, light, creative). Figure 2 shows the traditional symbol for yin and yang. The yin energy can also be represented by a broken line (– –) and the yang by a solid line (—). These lines are written on top of one another in groups of three, known as *trigrams.* For example, the trigram ☱ is called *Tui,* the Joyous, and has the image of a lake.

(a) How many trigrams are there altogether?

▶**Solution** Think of choosing between the 2 types of lines for each of the 3 positions in the trigram. There will be 2 choices for each position, so there are $2 \cdot 2 \cdot 2 = 8$ different trigrams.

(b) The trigrams are grouped together, one on top of the other, in pairs known as *hexagrams.* Each hexagram represents one aspect of the *I Ching* philosophy. How many hexagrams are there?

▶**Solution** For each position in the hexagram there are 8 possible trigrams, giving $8 \cdot 8 = 64$ hexagrams.

EXAMPLE 4

Books

A teacher has 5 different books that he wishes to arrange side by side. How many different arrangements are possible?

▶**Solution** Five choices will be made, one for each space that will hold a book. Any of the 5 books could be chosen for the first space. There are 4 choices for the second space, since 1 book has already been placed in the first space; there are 3 choices for the third space, and so on. By the multiplication principle, the number of different possible arrangements is $5 \cdot 4 \cdot 3 \cdot 2 \cdot 1 = 120$.

FOR REVIEW
The natural numbers, also referred to as the positive integers, are the numbers 1, 2, 3, 4, etc.

The use of the multiplication principle often leads to products such as $5 \cdot 4 \cdot 3 \cdot 2 \cdot 1$, the product of all the natural numbers from 5 down to 1. If n is a natural number, the symbol $n!$ (read "*n factorial*") denotes the product of all the natural numbers from n down to 1. If $n = 1$, this formula is understood to give $1! = 1$.

FACTORIAL NOTATION

For any natural number n,

$$n! = n(n-1)(n-2)\cdots(3)(2)(1).$$

Also, by definition,

$$0! = 1.$$

With this symbol, the product $5 \cdot 4 \cdot 3 \cdot 2 \cdot 1$ can be written as $5!$. Also, $3! = 3 \cdot 2 \cdot 1 = 6$. The definition of $n!$ could be used to show that $n[(n-1)]! = n!$ for all natural numbers $n \geq 2$. It is helpful if this result also holds for $n = 1$. This can happen only if $0!$ equals 1, as defined above.

Most calculators have an $n!$ key. A calculator with a 10-digit display and scientific notation capability will usually give the exact value of $n!$ for $n \leq 13$, and approximate values of $n!$ for $14 \leq n \leq 69$. The value of $70!$ is approximately 1.198×10^{100}, which is too large for most calculators. To see how large $70!$ is, suppose a computer counted the numbers from 1 to $70!$ at a rate of 1 billion numbers per second. If the computer started when the universe began, by now it would only be done with a tiny fraction of the total.

On many graphing calculators, the factorial of a number is accessible through a menu. On the TI-83/84 Plus, for example, this menu is found by pressing the `MATH` key, and then selecting `PRB` (for probability).

EXAMPLE 5 **Books**

Suppose the teacher in Example 4 wishes to place only 3 of the 5 books on his desk. How many arrangements of 3 books are possible?

Solution The teacher again has 5 ways to fill the first space, 4 ways to fill the second space, and 3 ways to fill the third. Since he wants to use only 3 books, only 3 spaces can be filled (3 events) instead of 5, for $5 \cdot 4 \cdot 3 = 60$ arrangements.

Permutations The answer 60 in Example 5 is called the number of *permutations* of 5 things taken 3 at a time. A **permutation** of r (where $r \geq 1$) elements from a set of n elements is any specific ordering or arrangement, *without repetition*, of the r elements. Each rearrangement of the r elements is a different permutation. The number of permutations of n things taken r at a time (with $r \leq n$) is written $P(n, r)$. Based on the work in Example 5,

$$P(5,3) = 5 \cdot 4 \cdot 3 = 60.$$

Factorial notation can be used to express this product as follows.

$$5 \cdot 4 \cdot 3 = 5 \cdot 4 \cdot 3 \cdot \frac{2 \cdot 1}{2 \cdot 1} = \frac{5 \cdot 4 \cdot 3 \cdot 2 \cdot 1}{2 \cdot 1} = \frac{5!}{2!} = \frac{5!}{(5-3)!}$$

This example illustrates the general rule of permutations, which can be stated as follows.

PERMUTATIONS

If $P(n, r)$ (where $r \leq n$) is the number of permutations of n elements taken r at a time, then

$$\boldsymbol{P(n,r) = \frac{n!}{(n-r)!}}.$$

CAUTION The letter P here represents *permutations*, not *probability*. In probability notation, the quantity in parentheses describes an *event*. In permutations notation, the quantity in parentheses always comprises *two numbers*. ■

The proof of the permutations rule follows the discussion in Example 5. There are n ways to choose the first of the r elements, $n - 1$ ways to choose the second, and $n - r + 1$ ways to choose the rth element, so that

$$P(n,r) = n(n-1)(n-2)\cdots(n-r+1).$$

Now multiply on the right by $(n-r)!/(n-r)!$.

$$\begin{aligned} P(n,r) &= n(n-1)(n-2)\cdots(n-r+1) \cdot \frac{(n-r)!}{(n-r)!} \\ &= \frac{n(n-1)(n-2)\cdots(n-r+1)(n-r)!}{(n-r)!} \\ &= \frac{n!}{(n-r)!} \end{aligned}$$

NOTE Because we defined 0! equal to 1, the formula for permutations gives the special case

$$P(n,n) = \frac{n!}{(n-n)!} = \frac{n!}{0!} = \frac{n!}{1} = n!.$$

This result also follows from the multiplication principle, because $P(n, n)$ gives the number of permutations of n objects, and there are n choices for the first object, $n - 1$ for the second, and so on, down to just 1 choice for the last object. Example 4 illustrated this idea. ■

The number of permutations of a set with n elements is $n!$.

To find $P(n, r)$, we can use either the permutations formula or direct application of the multiplication principle, as the following example shows.

EXAMPLE 6 **Politics**

In mid 2007, eight candidates sought the Democratic nomination for president. In how many ways could voters rank their first, second, and third choices?

▶**Solution**

METHOD 1
Calculating by Hand

This is the same as finding the number of permutations of 8 elements taken 3 at a time. Since there are 3 choices to be made, the multiplication principle gives $P(8, 3) = 8 \cdot 7 \cdot 6 = 336$. Alternatively, use the permutations formula to get

$$P(8,3) = \frac{8!}{(8-3)!} = \frac{8!}{5!} = \frac{8 \cdot 7 \cdot 6 \cdot \mathbf{5 \cdot 4 \cdot 3 \cdot 2 \cdot 1}}{\mathbf{5 \cdot 4 \cdot 3 \cdot 2 \cdot 1}} = 8 \cdot 7 \cdot 6 = 336.$$

METHOD 2
Graphing Calculator

Graphing calculators have the capacity to compute permutations. For example, on a TI-83/84 Plus, $P(8, 3)$ can be calculated by inputting 8 followed by `nPr` (found in the `MATH-PRB` menu), and a 3 yielding 336, as shown in Figure 3.

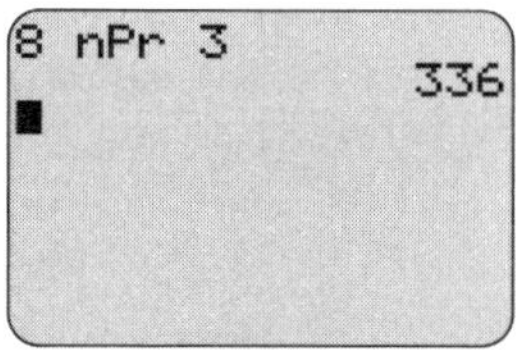

FIGURE 3

METHOD 3
Spreadsheet

Spreadsheets can also compute permutations. For example, in Microsoft Excel, $P(8, 3)$ can be calculated by inputting 8 and 3 in cells, say, A1 and B1, and then typing "`=FACT(A1)/FACT(A1-B1)`" in cell C1 or, for that matter, any other cell.

CAUTION When calculating the number of permutations with the formula, do not try to cancel unlike factorials. For example,

$$\frac{8!}{4!} \neq 2! = 2 \cdot 1 = 2.$$

$$\frac{8!}{4!} = \frac{8 \cdot 7 \cdot 6 \cdot 5 \cdot \mathbf{4 \cdot 3 \cdot 2 \cdot 1}}{\mathbf{4 \cdot 3 \cdot 2 \cdot 1}} = 8 \cdot 7 \cdot 6 \cdot 5 = 1680.$$

Always write out the factors first, then cancel where appropriate. ■

EXAMPLE 7 **Permutations**

Find the following.

(a) The number of permutations of the letters A, B, and C

▶**Solution** By the formula for $P(n, r)$ with both n and r equal to 3,

$$P(3,3) = \frac{3!}{(3-3)!} = \frac{3!}{0!} = \frac{3!}{1} = 3 \cdot 2 \cdot 1 = 6.$$

The 6 permutations (or arrangements) are

ABC, ACB, BAC, BCA, CAB, CBA.

(b) The number of permutations if just 2 of the letters A, B, and C are to be used

▶Solution Find $P(3, 2)$.

$$P(3,2) = \frac{3!}{(3-2)!} = \frac{3!}{1!} = 3! = 6$$

This result is exactly the same answer as in part (a). This is because, in the case of $P(3, 3)$, after the first 2 choices are made, the third is already determined, as shown in the table below.

First Two Letters	AB	AC	BA	BC	CA	CB
Third Letter	C	B	C	A	B	A

EXAMPLE 8

Television

A televised talk show will include 4 women and 3 men as panelists.

(a) In how many ways can the panelists be seated in a row of 7 chairs?

▶Solution Find $P(7, 7)$, the total number of ways to seat 7 panelists in 7 chairs.

?

$$P(7,7) = \frac{7!}{(7-7)!} = \frac{7!}{0!} = \frac{7!}{1} = 7 \cdot 6 \cdot 5 \cdot 4 \cdot 3 \cdot 2 \cdot 1 = 5040$$

There are 5040 ways to seat the 7 panelists.

(b) In how many ways can the panelists be seated if the men and women are to be alternated?

▶Solution Use the multiplication principle. In order to alternate men and women, a woman must be seated in the first chair (since there are 4 women and only 3 men), any of the men next, and so on. Thus there are 4 ways to fill the first seat, 3 ways to fill the second seat, 3 ways to fill the third seat (with any of the 3 remaining women), and so on. This gives

$$4 \cdot 3 \cdot 3 \cdot 2 \cdot 2 \cdot 1 \cdot 1 = 144$$

ways to seat the panelists.

(c) In how many ways can the panelists be seated if the men must sit together, and the women must also sit together?

▶Solution Use the multiplication principle. We first must decide how to arrange the two groups (men and women). There are 2! ways of doing this. Next, there are 4! ways of arranging the women and 3! ways of arranging the men, for a total of

$$2!\,4!\,3! = 2 \cdot 24 \cdot 6 = 288$$

ways.

(d) In how many ways can one woman and one man from the panel be selected?

▶**Solution** There are 4 ways to pick the woman and 3 ways to pick the man, for a total of

$$4 \cdot 3 = 12$$

ways.

NOTE It might help you to add spaces into which you can put the numbers, and below each space denote briefly which choice you're making in each space, such as W_1 for the first woman and M_1 for the first man. For instance, in Example 8(b), we could write the solution in the following manner.

$$\underset{W_1}{\underline{4}} \cdot \underset{M_1}{\underline{3}} \cdot \underset{W_2}{\underline{3}} \cdot \underset{M_2}{\underline{2}} \cdot \underset{W_3}{\underline{2}} \cdot \underset{M_3}{\underline{1}} \cdot \underset{W_4}{\underline{1}} = 144$$

If the n objects in a permutation are not all distinguishable—that is, if there are n_1 of type 1, n_2 of type 2, and so on for r different types, then the number of **distinguishable permutations** is

$$\frac{n!}{n_1!\, n_2! \cdots n_r!}.$$

For example, suppose we want to find the number of permutations of the numbers 1, 1, 4, 4, 4. We cannot distinguish between the two 1's or among the three 4's, so using 5! would give too many distinguishable arrangements. Since the two 1's are indistinguishable and account for 2! of the permutations, we divide 5! by 2!. Similarly, we also divide by 3! to account for the three indistinguishable 4's. This gives

$$\frac{5!}{2!\,3!} = 10$$

permutations.

EXAMPLE 9

Mississippi

In how many ways can the letters in the word *Mississippi* be arranged?

▶**Solution** This word contains 1 m, 4 i's, 4 s's, and 2 p's. To use the formula, let $n = 11$, $n_1 = 1$, $n_2 = 4$, $n_3 = 4$, and $n_4 = 2$ to get

$$\frac{11!}{1!\,4!\,4!\,2!} = 34{,}650$$

arrangements.

NOTE If Example 9 had asked for the number of ways that the letters in a word with 11 *different* letters could be arranged, the answer would be $11! = 39{,}916{,}800$.

EXAMPLE 10

Yogurt

A student buys 3 cherry yogurts, 2 raspberry yogurts, and 2 blueberry yogurts. She puts them in her dormitory refrigerator to eat one a day for the next week. Assuming yogurts of the same flavor are indistinguishable, in how many ways can she select yogurts to eat for the next week?

Solution This problem is again one of distinguishable permutations. The 7 yogurts can be selected in 7! ways, but since the 3 cherry, 2 raspberry, and 2 blueberry yogurts are indistinguishable, the total number of distinguishable orders in which the yogurts can be selected is

$$\frac{7!}{3!\,2!\,2!} = 210.$$

8.1 Exercises

In Exercises 1–12, evaluate the factorial or permutation.

1. $6!$ **2.** $7!$ **3.** $15!$ **4.** $16!$

5. $P(13, 2)$ **6.** $P(12, 3)$ **7.** $P(38, 17)$ **8.** $P(33, 19)$

9. $P(n, 0)$ **10.** $P(n, n)$ **11.** $P(n, 1)$ **12.** $P(n, n - 1)$

13. How many different types of homes are available if a builder offers a choice of 6 basic plans, 3 roof styles, and 2 exterior finishes?

14. A menu offers a choice of 3 salads, 8 main dishes, and 7 desserts. How many different meals consisting of one salad, one main dish, and one dessert are possible?

15. A couple has narrowed down the choice of a name for their new baby to 4 first names and 5 middle names. How many different first- and middle-name arrangements are possible?

16. In a club with 16 members, how many ways can a slate of 3 officers consisting of president, vice-president, and secretary/treasurer be chosen?

17. Define *permutation* in your own words.

18. In Example 7, there are six 3-letter permutations of the letters A, B, and C. How many 3-letter subsets (unordered groups of letters) are there?

19. In Example 7, how many unordered 2-letter subsets of the letters A, B, and C are there?

20. Find the number of distinguishable permutations of the letters in each word.

a. initial **b.** little **c.** decreed

21. A printer has 5 A's, 4 B's, 2 C's, and 2 D's. How many different "words" are possible that use all these letters? (A "word" does not have to have any meaning here.)

22. Wing has different books to arrange on a shelf: 4 blue, 3 green, and 2 red.

a. In how many ways can the books be arranged on a shelf?

b. If books of the same color are to be grouped together, how many arrangements are possible?

c. In how many distinguishable ways can the books be arranged if books of the same color are identical but need not be grouped together?

d. In how many ways can you select 3 books, one of each color, if the order in which the books are selected does not matter?

e. In how many ways can you select 3 books, one of each color, if the order in which the books are selected matters?

23. A child has a set of differently shaped plastic objects. There are 3 pyramids, 4 cubes, and 7 spheres.

a. In how many ways can she arrange the objects in a row if each is a different color?

b. How many arrangements are possible if objects of the same shape must be grouped together and each object is a different color?

c. In how many distinguishable ways can the objects be arranged in a row if objects of the same shape are also the same color, but need not be grouped together?

d. In how many ways can you select 3 objects, one of each shape, if the order in which the objects are selected does not matter and each object is a different color?

e. In how many ways can you select 3 objects, one of each shape, if the order in which the objects are selected matters and each object is a different color?

24. Some students find it puzzling that $0! = 1$, and think that $0!$ should equal 0. If this were true, what would be the value of $P(4, 4)$ using the permutations formula?

25. If you already knew the value of 9!, how could you find the value of 10! quickly?

26. Given that 450! is approximately equal to $1.7333687 \times 10^{1000}$ (to 8 digits of accuracy), find 451! to 7 digits of accuracy.

27. When calculating $n!$, the number of ending zeros in the answer can be determined prior to calculating the actual number by finding the number of times 5 can be factored from $n!$. For example, 7! only has one 5 occurring in its calculation, and so there is only one ending zero in 5040. The number 10! has two 5's (one from the 5 and one from the 10) and so there must be two ending zeros in the answer 3,628,800. Use this idea to determine the number of zeros that occur in the following factorials, and then explain why this works.

a. 13! **b.** 27! **c.** 75!

28. Because of the view screen, calculators only show a fixed number of digits, often 10 digits. Thus, an approximation of a number will be shown by only including the 10 largest place values of the number. Using the ideas from the previous exercise, determine if the following numbers are correct or if they are incorrect by checking if they have the correct number of ending zeros. (*Note:* Just because a number has the correct number of zeros does not imply that it is correct.)

a. 12! = 479,001,610 **b.** 23! = 25,852,016,740,000,000,000,000

c. 15! = 1,307,643,680,000 **d.** 14! = 87,178,291,200

Applications

BUSINESS AND ECONOMICS

29. *Automobile Manufacturing* An automobile manufacturer produces 8 models, each available in 7 different exterior colors, with 4 different upholstery fabrics and 5 interior colors. How many varieties of automobile are available?

LIFE SCIENCES

30. *Drug Sequencing* Twelve drugs have been found to be effective in the treatment of a disease. It is believed that the sequence in which the drugs are administered is important in the effectiveness of the treatment. In how many different sequences can 5 of the 12 drugs be administered?

31. *Insect Classification* A biologist is attempting to classify 52,000 species of insects by assigning 3 initials to each species. Is it possible to classify all the species in this way? If not, how many initials should be used?

32. *Genetics Experiment* In how many ways can 7 of 11 monkeys be arranged in a row for a genetics experiment?

SOCIAL SCIENCES

33. *Social Science Experiment* In an experiment on social interaction, 6 people will sit in 6 seats in a row. In how many ways can this be done?

34. *Election Ballots* In an election with 3 candidates for one office and 6 candidates for another office, how many different ballots may be printed?

GENERAL INTEREST

35. *Course Scheduling* A business school gives courses in typing, shorthand, transcription, business English, technical writing, and accounting. In how many ways can a student arrange a schedule if 3 courses are taken? Assume that the order in which courses are scheduled matters.

36. *Course Scheduling* If your college offers 400 courses, 25 of which are in mathematics, and your counselor arranges your schedule of 4 courses by random selection, how many schedules are possible that do not include a math course? Assume that the order in which courses are scheduled matters.

37. *Baseball Teams* A baseball team has 19 players. How many 9-player batting orders are possible?

38. *Union Elections* A chapter of union Local 715 has 35 members. In how many different ways can the chapter select a president, a vice-president, a treasurer, and a secretary?

39. *Programming Music* A concert to raise money for an economics prize is to consist of 5 works: 2 overtures, 2 sonatas, and a piano concerto.

a. In how many ways can the program be arranged?

b. In how many ways can the program be arranged if an overture must come first?

40. *Programming Music* A zydeco band from Louisiana will play 5 traditional and 3 original Cajun compositions at a concert. In how many ways can they arrange the program if

a. they begin with a traditional piece?

b. an original piece will be played last?

41. *Television Scheduling* The television schedule for a certain evening shows 8 choices from 8 to 9 P.M., 5 choices from 9 to 10 P.M., and 7 choices from 10 to 11 P.M. In how many different ways could a person schedule that evening of television viewing from 8 to 11 P.M.? (Assume each program that is selected is watched for an entire hour.)

42. *Radio Station Call Letters* How many different 4-letter radio station call letters can be made if

a. the first letter must be K or W and no letter may be repeated?

b. repeats are allowed, but the first letter is K or W?

c. the first letter is K or W, there are no repeats, and the last letter is R?

43. *Telephone Numbers* How many 7-digit telephone numbers are possible if the first digit cannot be zero and

a. only odd digits may be used?

b. the telephone number must be a multiple of 10 (that is, it must end in zero)?

c. the telephone number must be a multiple of 100?

d. the first 3 digits are 481?

e. no repetitions are allowed?

Telephone Area Codes Several years ago, the United States began running out of telephone numbers. Telephone companies introduced new area codes as numbers were used up, and eventually almost all area codes were used up.

44. a. Until recently, all area codes had a 0 or 1 as the middle digit, and the first digit could not be 0 or 1. How many area codes are there with this arrangement? How many telephone numbers does the current 7-digit sequence permit per area code? (The 3-digit sequence that follows the area code cannot start with 0 or 1. Assume there are no other restrictions.)

b. The actual number of area codes under the previous system was 152. Explain the discrepancy between this number and your answer to part a.

45. The shortage of area codes was avoided by removing the restriction on the second digit. (This resulted in problems for some older equipment, which used the second digit to determine that a long-distance call was being made.) How many area codes are available under the new system?

46. *License Plates* For many years, the state of California used 3 letters followed by 3 digits on its automobile license plates.

a. How many different license plates are possible with this arrangement?

b. When the state ran out of new numbers, the order was reversed to 3 digits followed by 3 letters. How many new license plate numbers were then possible?

c. Several years ago, the numbers described in b were also used up. The state then issued plates with 1 letter followed by 3 digits and then 3 letters. How many new license plate numbers will this provide?

47. *Social Security Numbers* A social security number has 9 digits. How many social security numbers are there? The U.S. population in 2005 was about 296 million. Is it possible for every U.S. resident to have a unique social security number? (Assume no restrictions.)

48. *Postal Zip Codes* The U.S. Postal Service currently uses 5-digit zip codes in most areas. How many zip codes are possible if there are no restrictions on the digits used? How many would be possible if the first number could not be 0?

49. *Postal Zip Codes* The U.S. Postal Service is encouraging the use of 9-digit zip codes in some areas, adding 4 digits after the usual 5-digit code. How many such zip codes are possible with no restrictions?

50. *Games* The game of Sets* uses a special deck of cards. Each card has either one, two, or three identical shapes, all of the same color and style. There are three possible shapes: squiggle, diamond, and oval. There are three possible colors: green, purple, and red. There are three possible styles: solid, shaded, or outline. The deck consists of all possible combinations of shape, color, style, and number of shapes. How many cards are in the deck?

51. *Games* In the game of Scattergories,† the players take 12 turns. In each turn, a 20-sided die is rolled; each side has a letter. The players must then fill in 12 categories (e.g., vegetable, city, etc.) with a word beginning with the letter rolled. Considering that a game consists of 12 rolls of the 20-sided die, and that rolling the same side more than once is allowed, how many possible games are there?

52. *Games* The game of Twenty Questions consists of asking 20 questions to determine a person, place, or thing that the other person is thinking of. The first question, which is always "Is it an animal, vegetable, or mineral?" has three possible answers. All the other questions must be answered "Yes" or "No." How many possible objects can be distinguished in this game, assuming that all 20 questions are asked? Are 20 questions enough?

53. *Traveling Salesman* In the famous Traveling Salesman Problem, a salesman starts in any one of a set of cities, visits every city in the set once, and returns to the starting city. He would like to complete this circuit with the shortest possible distance.

a. Suppose the salesman has 10 cities to visit. Given that it does not matter what city he starts in, how many different circuits can he take?

b. The salesman decides to check all the different paths in part a to see which is shortest, but realizes that a circuit has the same distance whichever direction it is traveled. How many different circuits must he check?

c. Suppose the salesman has 70 cities to visit. Would it be feasible to have a computer check all the different circuits? Explain your reasoning.

54. *Circular Permutations* Circular permutations arise in applications involving arrangements around a closed loop, as in the previous exercise. Here are two examples.

a. A ferris wheel has 20 seats. How many ways can 20 students arrange themselves on the ferris wheel if each student takes a different seat? We consider two arrangements to be identical if they differ only by rotations of the wheel.

b. A necklace is to be strung with 15 beads, each of a different color. In how many ways can the beads be arranged? We consider two arrangements to be identical if they differ only by rotations of the necklace or by flipping the necklace over. (*Hint:* If every arrangement is counted twice, the correct number of arrangements can be found by dividing by 2.)

8.2 Combinations

? THINK ABOUT IT **In how many ways can a manager select 4 employees for promotion from 12 eligible employees?**

As we shall see, permutations alone cannot be used to answer this question, but combinations will provide the answer.

In the previous section, we saw that there are 60 ways that a teacher can arrange 3 of 5 different books on his desk. That is, there are 60 permutations of 5 books taken 3 at a time. Suppose now that the teacher does not wish to arrange the books on his desk, but rather wishes to choose, without regard to order, any 3 of the 5 books for a book sale to raise money for his school. In how many ways can this be done?

At first glance, we might say 60 again, but this is incorrect. The number 60 counts all possible *arrangements* of 3 books chosen from 5. The following

*Copyright © Marsha J. Falco.
†Copyright © Milton Bradley Company.

6 arrangements, however, would all lead to the same set of 3 books being given to the book sale.

mystery-biography-textbook	biography-textbook-mystery
mystery-textbook-biography	textbook-biography-mystery
biography-mystery-textbook	textbook-mystery-biography

The list shows 6 different *arrangements* of 3 books, but only one *subset* of 3 books. A subset of items listed *without regard to order* is called a **combination**. The number of combinations of 5 things taken 3 at a time is written $\binom{5}{3}$, and read "5 over 3" or "5 choose 3."* Since they are subsets, combinations are *not ordered.*

To evaluate $\binom{5}{3}$, start with the $5 \cdot 4 \cdot 3$ *permutations* of 5 things taken 3 at a time. Since combinations are not ordered, find the number of combinations by dividing the number of permutations by the number of ways each group of 3 can be ordered; that is, divide by 3!.

$$\binom{5}{3} = \frac{5 \cdot 4 \cdot 3}{3!} = \frac{5 \cdot 4 \cdot 3}{3 \cdot 2 \cdot 1} = 10$$

There are 10 ways that the teacher can choose 3 books for the book sale.

Generalizing this discussion gives the following formula for the number of combinations of n elements taken r at a time:

$$\binom{n}{r} = \frac{P(n, r)}{r!}.$$

Another version of this formula is found as follows.

$$\begin{aligned}\binom{n}{r} &= \frac{P(n, r)}{r!} \\ &= \frac{n!}{(n-r)!} \cdot \frac{1}{r!} \\ &= \frac{n!}{(n-r)!\,r!}\end{aligned}$$

The steps above lead to the following result.

COMBINATIONS

If $\binom{n}{r}$ denotes the number of combinations of n elements taken r at a time, where $r \leq n$, then

$$\binom{n}{r} = \frac{n!}{(n-r)!\,r!}.$$

*Other common notations for $\binom{n}{r}$ are ${}_nC_r$, C_r^n, and $C(n, r)$.

EXAMPLE 1 **Committees**

How many committees of 3 people can be formed from a group of 8 people?

▸Solution

METHOD 1
Calculating by Hand

A committee is an unordered group, so use the combinations formula for $\binom{8}{3}$.

$$\binom{8}{3} = \frac{8!}{5!3!} = \frac{8 \cdot 7 \cdot 6 \cdot 5 \cdot 4 \cdot 3 \cdot 2 \cdot 1}{5 \cdot 4 \cdot 3 \cdot 2 \cdot 1 \cdot 3 \cdot 2 \cdot 1} = \frac{8 \cdot 7 \cdot 6}{3 \cdot 2 \cdot 1} = 56$$

METHOD 2
Graphing Calculator

Graphing calculators have the capacity to compute combinations. For example, on a TI-83/84 Plus, $\binom{8}{3}$ can be calculated by inputting 8 followed by `nCr` (found in the `MATH-PRB` menu) and a 3 yielding 56, as shown in Figure 4.

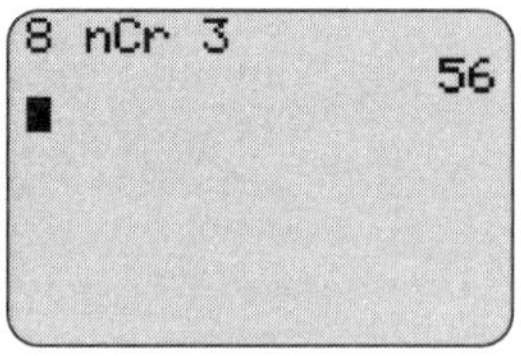

FIGURE 4

METHOD 3
Spreadsheet

Spreadsheets can also compute combinations. For example, in Microsoft Excel, $\binom{8}{3}$ can be calculated by inputting 8 and 3 in cells, say, A1 and B1, and then typing "`=FACT(A1)/(FACT(A1-B1)*FACT(B1))`" in cell C1 or, for that matter, any other cell.

Example 1 shows an alternative way to compute $\binom{n}{r}$. Take r or $n - r$, whichever is smaller. Write the factorial of this number in the denominator. In the numerator, write out a sufficient number of factors of $n!$ so there is one factor in the numerator for each factor in the denominator. For example, to calculate $\binom{8}{3}$ or $\binom{8}{5}$, write

$$\frac{8 \cdot 7 \cdot 6}{3 \cdot 2 \cdot 1} = 56.$$

The factors that are omitted (written in color in Example 1) cancel out of the numerator and denominator, so need not be included.

Notice from the previous discussion that $\binom{8}{3} = \binom{8}{5}$. (See Exercise 25 for a generalization of this idea.) One interpretation of this fact is that the number of ways to form a committee of 3 people chosen from a group of 8 is the same as the number of ways to choose the 5 people who are not on the committee.

Notice that this is *not* true with permutations: $P(8, 3) \neq P(8, 5)$.

EXAMPLE 2 **Lawyers**

Three lawyers are to be selected from a group of 30 to work on a special project.

(a) In how many different ways can the lawyers be selected?

▶**Solution** Here we wish to know the number of 3-element combinations that can be formed from a set of 30 elements. (We want combinations, not permutations, since order within the group of 3 doesn't matter.)

$$\begin{aligned}\binom{30}{3} &= \frac{30!}{27!3!} = \frac{30 \cdot 29 \cdot 28 \cdot 27!}{27! \cdot 3 \cdot 2 \cdot 1} \\ &= \frac{30 \cdot 29 \cdot 28}{3 \cdot 2 \cdot 1} \\ &= 4060\end{aligned}$$

There are 4060 ways to select the project group.

(b) In how many ways can the group of 3 be selected if a certain lawyer must work on the project?

▶**Solution** Since 1 lawyer already has been selected for the project, the problem is reduced to selecting 2 more from the remaining 29 lawyers.

$$\binom{29}{2} = \frac{29!}{27!\,2!} = \frac{29 \cdot 28 \cdot 27!}{27! \cdot 2 \cdot 1} = \frac{29 \cdot 28}{2 \cdot 1} = 29 \cdot 14 = 406$$

In this case, the project group can be selected in 406 ways.

(c) In how many ways can a nonempty group of at most 3 lawyers be selected from these 30 lawyers?

▶**Solution** Here, by "at most 3" we mean "1 or 2 or 3." (The number 0 is excluded because the group is nonempty.) Find the number of ways for each case.

Case	Number of Ways
1	$\binom{30}{1} = \frac{30!}{29!\,1!} = \frac{30 \cdot 29!}{29!\,(1)} = 30$
2	$\binom{30}{2} = \frac{30!}{28!\,2!} = \frac{30 \cdot 29 \cdot 28!}{28! \cdot 2 \cdot 1} = 435$
3	$\binom{30}{3} = \frac{30!}{27!\,3!} = \frac{30 \cdot 29 \cdot 28 \cdot 27!}{27! \cdot 3 \cdot 2 \cdot 1} = 4060$

The total number of ways to select at most 3 lawyers will be the sum

$$30 + 435 + 4060 = 4525.$$

EXAMPLE 3 **Sales**

A salesman has 10 accounts in a certain city.

(a) In how many ways can he select 3 accounts to call on?

▶**Solution** Within a selection of 3 accounts, the arrangement of the calls is not important, so there are

$$\binom{10}{3} = \frac{10!}{7!\,3!} = \frac{10 \cdot 9 \cdot 8}{3 \cdot 2 \cdot 1} = 120$$

ways he can make a selection of 3 accounts.

(b) In how many ways can he select at least 8 of the 10 accounts to use in preparing a report?

▶**Solution** "At least 8" means "8 or more," which is "8 or 9 or 10." First find the number of ways to choose in each case.

Case	Number of Ways
8	$\binom{10}{8} = \frac{10!}{2!\,8!} = \frac{10 \cdot 9}{2 \cdot 1} = 45$
9	$\binom{10}{9} = \frac{10!}{1!\,9!} = \frac{10}{1} = 10$
10	$\binom{10}{10} = \frac{10!}{0!\,10!} = 1$

He can select at least 8 of the 10 accounts in $45 + 10 + 1 = 56$ ways.

FOR REVIEW

Notice in Example 3 that to calculate the number of ways to select 8 or 9 or 10 accounts, we added the three numbers found. The union rule for sets from Chapter 7 says that when *A* and *B* are disjoint sets, the number of elements in *A* or *B* is the number of elements in *A* plus the number in *B*.

CAUTION When we are making a first decision *and* a second decision, we *multiply* to find the total number of ways. When we are making a decision in which the first choice *or* the second choice are valid choices, we *add* to find the total number of ways. ■

The formulas for permutations and combinations given in this section and in the previous section will be very useful in solving probability problems in the next section. Any difficulty in using these formulas usually comes from being unable to differentiate between them. Both permutations and combinations give the number of ways to choose *r* objects from a set of *n* objects. The differences between permutations and combinations are outlined in the following table.

Permutations	Combinations
Different orderings or arrangements of the *r* objects are different permutations.	Each choice or subset of *r* objects gives one combination. Order within the group of *r* objects does not matter.
$P(n, r) = \frac{n!}{(n-r)!}$	$\binom{n}{r} = \frac{n!}{(n-r)!\,r!}$
Clue words: arrangement, schedule, order	Clue words: group, committee, set, sample
Order matters!	**Order does not matter!**

In the next examples, concentrate on recognizing which formula should be applied.

EXAMPLE 4 **Permutations and Combinations**

For each problem, tell whether permutations or combinations should be used to solve the problem.

(a) How many 4-digit code numbers are possible if no digits are repeated?

▶**Solution** Since changing the order of the 4 digits results in a different code, use permutations.

(b) A sample of 3 light bulbs is randomly selected from a batch of 15. How many different samples are possible?

▶**Solution** The order in which the 3 light bulbs are selected is not important. The sample is unchanged if the items are rearranged, so combinations should be used.

(c) In a baseball conference with 8 teams, how many games must be played so that each team plays every other team exactly once?

▶**Solution** Selection of 2 teams for a game is an *unordered* subset of 2 from the set of 8 teams. Use combinations again.

(d) In how many ways can 4 patients be assigned to 6 different hospital rooms so that each patient has a private room?

▶**Solution** The room assignments are an *ordered* selection of 4 rooms from the 6 rooms. Exchanging the rooms of any 2 patients within a selection of 4 rooms gives a different assignment, so permutations should be used.

(e) Solve the problems in parts (a)–(d) above. The answers are given in the footnote.*

EXAMPLE 5 **Promotions**

A manager must select 4 employees for promotion; 12 employees are eligible.

(a) In how many ways can the 4 be chosen?

▶**Solution** Since there is no reason to differentiate among the 4 who are selected, use combinations.

$$\binom{12}{4} = \frac{12!}{8!\,4!} = 495$$

(b) In how many ways can 4 employees be chosen (from 12) to be placed in 4 different jobs?

▶**Solution** In this case, once a group of 4 is selected, they can be assigned in many different ways (or arrangements) to the 4 jobs. Therefore, this problem requires permutations.

$$P(12, 4) = \frac{12!}{8!} = 11{,}880$$

EXAMPLE 6 **Playing Cards**

In how many ways can a full house of aces and eights (3 aces and 2 eights) occur in 5-card poker?

*(**a**) 5040 (**b**) 455 (**c**) 28 (**d**) 360

FOR REVIEW
Examples 6 and 7 involve a standard deck of 52 playing cards, as shown in Figure 17 in Chapter 7. Recall the discussion that accompanies the photograph.

Solution The arrangement of the 3 aces or the 2 eights does not matter, so we use combinations and the multiplication principle. There are $\binom{4}{3}$ ways to get 3 aces from the 4 aces in the deck, and $\binom{4}{2}$ ways to get 2 eights. By the multiplication principle, the number of ways to get 3 aces and 2 eights is

$$\binom{4}{3} \cdot \binom{4}{2} = 4 \cdot 6 = 24.$$

EXAMPLE 7 **Playing Cards**

Five cards are dealt from a standard 52-card deck.

(a) How many such hands have only face cards?

Solution The face cards are the king, queen, and jack of each suit. Since there are 4 suits, there are 12 face cards. The arrangement of the 5 cards is not important, so use combinations to get

$$\binom{12}{5} = \frac{12!}{7!\,5!} = 792.$$

(b) How many such hands have exactly 2 hearts?

Solution There are 13 hearts in the deck, so the 2 hearts will be selected from those 13 cards. The other 3 cards must come from the remaining 39 cards that are not hearts. Use combinations and the multiplication principle to get

$$\binom{13}{2}\binom{39}{3} = 78 \cdot 9139 = 712{,}842.$$

Notice that the two top numbers in the combinations add up to 52, the total number of cards, and the two bottom numbers add up to 5, the number of cards in a hand.

(c) How many such hands have cards of a single suit?

Solution The total number of ways that 5 cards of a particular suit of 13 cards can occur is $\binom{13}{5}$. Since the arrangement of the 5 cards is not important, use combinations. There are four different suits, so the multiplication principle gives

$$4 \cdot \binom{13}{5} = 4 \cdot 1287 = 5148$$

ways to deal 5 cards of the same suit.

As Example 7 shows, often both combinations and the multiplication principle must be used in the same problem.

EXAMPLE 8 **Soup**

To illustrate the differences between permutations and combinations in another way, suppose 2 cans of soup are to be selected from 4 cans on a shelf: noodle (N), bean (B), mushroom (M), and tomato (T). As shown in Figure 5(a) on the next page, there are 12 ways to select 2 cans from the 4 cans if the order matters (if noodle first and bean second is considered different from bean, then noodle, for example). On the other hand, if order is unimportant, then there are 6 ways to choose 2 cans of soup from the 4, as illustrated in Figure 5(b).

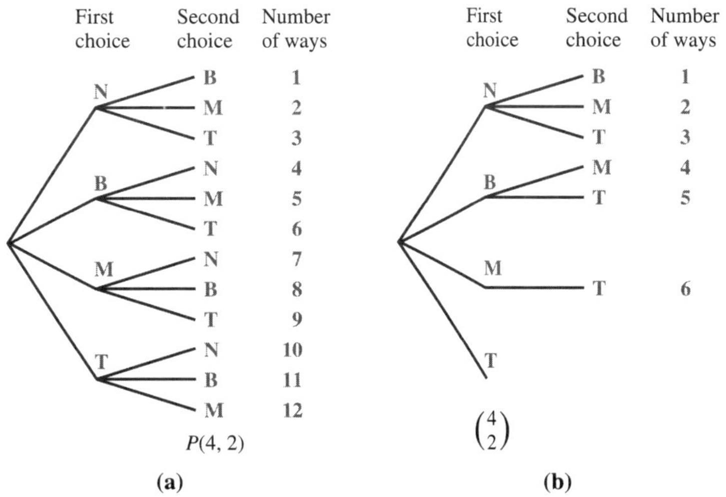

FIGURE 5

CAUTION It should be stressed that not all counting problems lend themselves to either permutations or combinations. Whenever a tree diagram or the multiplication principle can be used directly, it's often best to use it. ■

8.2 Exercises

1. Define combinations in your own words.

Evaluate each combination.

2. $\binom{8}{3}$ **3.** $\binom{12}{5}$ **4.** $\binom{44}{20}$ **5.** $\binom{40}{18}$

6. $\binom{n}{0}$ **7.** $\binom{n}{n}$ **8.** $\binom{n}{1}$ **9.** $\binom{n}{n-1}$

10. In how many ways can a hand of 6 clubs be chosen from an ordinary deck?

11. Five cards are marked with the numbers 1, 2, 3, 4, and 5, then shuffled, and 2 cards are drawn.

a. How many different 2-card combinations are possible?

b. How many 2-card hands contain a number less than 3?

12. An economics club has 31 members.

a. If a committee of 4 is to be selected, in how many ways can the selection be made?

b. In how many ways can a committee of at least 1 and at most 3 be selected?

13. Use a tree diagram for the following.

a. Find the number of ways 2 letters can be chosen from the set {L, M, N} if order is important and repetition is allowed.

b. Reconsider part a if no repeats are allowed.

c. Find the number of combinations of 3 elements taken 2 at a time. Does this answer differ from part a or b?

14. Repeat Exercise 13 using the set {L, M, N, P}.

15. Explain the difference between a permutation and a combination.

16. Padlocks with digit dials are often referred to as "combination locks." According to the mathematical definition of combination, is this an accurate description? Explain.

In Exercises 17–24, decide whether each exercise involves permutations or combinations, and then solve the problem.

17. In a club with 9 male and 11 female members, how many 5-member committees can be chosen that have

a. all men? **b.** all women? **c.** 3 men and 2 women?

18. In Exercise 17, how many committees can be selected that have

a. at least 4 women? **b.** no more than 2 men?

19. In a game of musical chairs, 12 children will sit in 11 chairs arranged in a row (one will be left out). In how many ways can this happen, if we count rearrangements of the children in the chairs as different outcomes?

20. A group of 3 students is to be selected from a group of 14 students to take part in a class in cell biology.

a. In how many ways can this be done?

b. In how many ways can the group who will not take part be chosen?

21. Marbles are being drawn without replacement from a bag containing 16 marbles.

a. How many samples of 2 marbles can be drawn?

b. How many samples of 4 marbles can be drawn?

c. If the bag contains 3 yellow, 4 white, and 9 blue marbles, how many samples of 2 marbles can be drawn in which both marbles are blue?

22. There are 7 rotten apples in a crate of 26 apples.

a. How many samples of 3 apples can be drawn from the crate?

b. How many samples of 3 could be drawn in which all 3 are rotten?

c. How many samples of 3 could be drawn in which there are two good apples and one rotten one?

23. A bag contains 5 black, 1 red, and 3 yellow jelly beans; you take 3 at random. How many samples are possible in which the jelly beans are

a. all black? **b.** all red? **c.** all yellow?

d. 2 black and 1 red? **e.** 2 black and 1 yellow?

f. 2 yellow and 1 black? **g.** 2 red and 1 yellow?

24. In how many ways can 5 out of 9 plants be arranged in a row on a windowsill?

25. Show that $\binom{n}{r} = \binom{n}{n-r}$.

26. The following problem was posed on National Public Radio's *Weekend Edition*: In how many points can 6 circles intersect?*

a. Find the answer for 6 circles.

b. Find the general answer for n circles.

**Weekend Edition*, National Public Radio, Oct. 23, 1994.

27. How many different dominoes can be formed from the numbers 0 ... 6? (*Hint:* A domino may have the same number of dots on both halves of it or it may have a different number of dots on each half.)

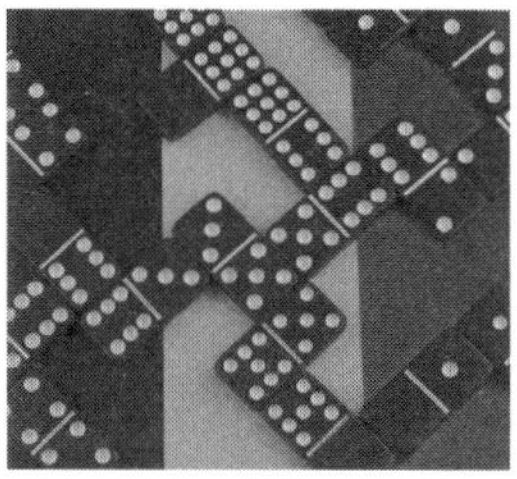

Applications

BUSINESS AND ECONOMICS

28. *Secretarial Assignments* From a pool of 8 secretaries, 3 are selected to be assigned to 3 managers, one per manager. In how many ways can they be selected and assigned?

29. *Sales Schedules* A salesperson has the names of 6 prospects.

  a. In how many ways can she arrange her schedule if she calls on all 6?

  b. In how many ways can she arrange her schedule if she can call on only 4 of the 6?

30. *Worker Grievances* A group of 9 workers decides to send a delegation of 3 to their supervisor to discuss their grievances.

  a. How many delegations are possible?

  b. If it is decided that a particular worker must be in the delegation, how many different delegations are possible?

  c. If there are 4 women and 5 men in the group, how many delegations would include at least 1 woman?

31. *Hamburger Variety* Hamburger Hut sells regular hamburgers as well as a larger burger. Either type can include cheese, relish, lettuce, tomato, mustard, or catsup.

  a. How many different hamburgers can be ordered with exactly three extras?

  b. How many different regular hamburgers can be ordered with exactly three extras?

  c. How many different regular hamburgers can be ordered with at least five extras?

32. *Assembly Line Sampling* Five items are to be randomly selected from the first 50 items on an assembly line to determine the defect rate. How many different samples of 5 items can be chosen?

LIFE SCIENCES

33. *Research Participants* From a group of 16 smokers and 22 nonsmokers, a researcher wants to randomly select 8 smokers and 8 nonsmokers for a study. In how many ways can the study group be selected?

34. *Plant Hardiness* In an experiment on plant hardiness, a researcher gathers 6 wheat plants, 3 barley plants, and 2 rye plants. She wishes to select 4 plants at random.

  a. In how many ways can this be done?

  b. In how many ways can this be done if exactly 2 wheat plants must be included?

SOCIAL SCIENCES

35. *Legislative Committee* A legislative committee consists of 5 Democrats and 4 Republicans. A delegation of 3 is to be selected to visit a small Pacific island republic.

  a. How many different delegations are possible?

  b. How many delegations would have all Democrats?

  c. How many delegations would have 2 Democrats and 1 Republican?

  d. How many delegations would include at least 1 Republican?

36. *Political Committee* From 10 names on a ballot, 4 will be elected to a political party committee. In how many ways can the committee of 4 be formed if each person will have a different responsibility, and different assignments of responsibility are considered different committees?

37. *Judges* When Paul Martinek, publisher of *Lawyers Weekly USA*, was a guest on the television news program *The O'Reilly Factor*, he discussed a decision by a three-judge panel, chosen at random from judges in the Ninth Circuit in California.* The judges had ruled that the mandatory recitation of the Pledge of Allegience is unconstitutional because of the phrase "under God." According to Martinek, "Because there are 45 judges in the Ninth Circuit, there are 3000 different combinations of three-judge panels." Is this true? If not, what is the correct number?

**Mathematics Teacher*, Vol. 96, No. 3, March 2003.

GENERAL INTEREST

38. *Bridge* How many different 13-card bridge hands can be selected from an ordinary deck?

39. *Poker* Five cards are chosen from an ordinary deck to form a hand in poker. In how many ways is it possible to get the following results?

    a. 4 queens
    b. No face card
    c. Exactly 2 face cards
    d. At least 2 face cards
    e. 1 heart, 2 diamonds, and 2 clubs

40. *Poker* In poker, a flush consists of 5 cards with the same suit, such as 5 diamonds.

    a. Find the number of ways of getting a flush consisting of cards with values from 5 to 10 by listing all the possibilities.

    b. Find the number of ways of getting a flush consisting of cards with values from 5 to 10 by using combinations.

41. *Baseball* If a baseball coach has 5 good hitters and 4 poor hitters on the bench and chooses 3 players at random, in how many ways can he choose at least 2 good hitters?

42. *Softball* The coach of the Morton Valley Softball Team has 6 good hitters and 8 poor hitters. He chooses 3 hitters at random.

    a. In how many ways can he choose 2 good hitters and 1 poor hitter?

    b. In how many ways can he choose 3 good hitters?

    c. In how many ways can he choose at least 2 good hitters?

43. *Flower Selection* Five orchids from a collection of 20 are to be selected for a flower show.

    a. In how many ways can this be done?

    b. In how many ways can the 5 be selected if 2 special plants must be included?

44. *Ice Cream Flavors* Baskin-Robbins advertises that it has 31 flavors of ice cream.

    a. How many different double-scoop cones can be made? Assume that the order of the scoops matters.

    b. How many different triple-scoop cones can be made?

    c. How many different double-scoop cones can be made if order doesn't matter?

45. *Lottery* A state lottery game requires that you pick 6 different numbers from 1 to 99. If you pick all 6 winning numbers, you win the jackpot.

    a. How many ways are there to choose 6 numbers if order is not important?

    b. How many ways are there to choose 6 numbers if order matters?

46. *Lottery* In Exercise 45, if you pick 5 of the 6 numbers correctly, you win \$250,000. In how many ways can you pick exactly 5 of the 6 winning numbers without regard to order?

47. *Pizza Varieties* A television commercial for Little Caesars pizza announced that with the purchase of two pizzas, one could receive free any combination of up to five toppings on each pizza. The commercial shows a young child waiting in line at Little Caesars who calculates that there are 1,048,576 possibilities for the toppings on the two pizzas.*

    a. Verify the child's calculation. Use the fact that Little Caesars has 11 toppings to choose from. Assume that the order of the two pizzas matters; that is, if the first pizza has combination 1 and the second pizza has combination 2, that is different from combination 2 on the first pizza and combination 1 on the second.

    b. In a letter to *The Mathematics Teacher*, Joseph F. Heiser argued that the two combinations described in part a should be counted as the same, so the child has actually overcounted. Give the number of possibilities if the order of the two pizzas doesn't matter.

48. *Pizza* In an ad for Pizza Hut, Jessica Simpson explains to the Muppets that there are more than 6 million possibilities for their 4forAll Pizza. Griffin Weber and Glenn Weber wrote an article explaining that the number of possibilities is far more than 6 million, as described below.†

    a. Each pizza can have up to 3 toppings, out of 17 possible choices, or can be one of four specialty pizzas. Calculate the number of different pizzas possible.

    b. Out of the total possible pizzas calculated in the first part of this exercise, a 4forAll Pizza consists of four pizzas in a box. Keeping in mind that the four pizzas could all be different, or there could be two or three different pizzas in the box, or all four pizzas could be the same, calculate the total number of 4forAll Pizzas possible.

    c. The article considers another way of counting the number in part b. Suppose that only 8 pizzas were available, and they were listed in a row with lines separating each type, as in the following diagram:

    $$A|B|C|D|E|F|G|H.$$

    A person orders 4 pizzas by placing 4 X's in the desired places on the diagram, after which the letters can be

*Heiser, Joseph F., "Pascal and Gauss Meet Little Caesars," *Mathematics Teacher*, Vol. 87, Sept. 1994, p. 389.

†Weber, Griffin and Glenn Weber, "Pizza Combinatorics Revisited," *The College Mathematics Journal*, Vol. 37, No. 1, Jan. 2006, pp. 43–44.

ignored. For example, an order for 2 of A, 1 of C, and 1 of G would look like the following diagram.

$$XX||X||||X|$$

The number of ways this can be done is then the number of ways of arranging 11 objects, 4 of which are X and the other 7 of which are vertical lines, or

$$\binom{11}{4} = 330.$$

Use similar reasoning to verify the answer to part b.

49. *Cereal* The Post Corporation has introduced the cereal, *Create a Crunch*™, in which the consumers can combine ingredients to create their own unique cereal. Each box contains 8 packets of food goods. There are four types of cereal: Frosted Alpha Bits®, Cocoa Pebbles®, Fruity Pebbles®, and Honey Comb®. Also included in the box are four "Add-Ins": granola, blue rice cereal, marshmallows, and sprinkles.

a. What is the total number of breakfasts that can be made if a breakfast is defined as any one or more cereals or add-ins?

b. If Melissa Fischer chooses to mix one type of cereal with one add-in, how many different breakfasts can she make?

c. If Jane Fleming chooses to mix two types of cereal with three add-ins, how many different breakfasts can she make?

d. If Kevin Flores chooses to mix at least one type of cereal with at least one type of add-in, how many breakfasts can he make?

e. If Nicole Girrbach's favorite cereal is Fruity Pebbles®, how many different cereals can she make if each of her mixtures must include this cereal?

50. *Football* Writer Gregg Easterbrook, discussing ESPN's unsuccessful attempt to predict the winners for the six National Football League (NFL) divisions and the six wild-card slots, claimed that there were 180 different ways to make this forecast.* Reader Milton Eisner wrote in to tell him that the actual number is much larger.[†] To make the calculation, note that the NFL consists of two conferences, each of which consists of three divisions. Five of the divisions have five teams, while the other has six. There is one winner from each of the six divisions, plus three wild-card slots from each of the two conferences. How many ways can the six division winners and six wild-card slots be chosen?

51. *Music* In the opera *Amahl and the Night Visitors*, the shepherds sing a chorus involving 18 different names, a challenge for singers trying to remember the names in the correct order. (Two of the three authors of this textbook have sung this chorus in public.)

a. In how many ways can the names be arranged?

b. Not all the arrangements of names in part a could be sung, because 10 of the names have 3 syllables, 4 have 2 syllables, and 4 have 4 syllables. Of the 6 lines in the chorus, 4 lines consist of a 3-syllable name repeated, followed by a 2-syllable and then a 4-syllable name (e.g., Emily, Emily, Michael, Bartholomew), and 2 lines consist of a 3-syllable name repeated, followed by two more 3-syllable names (e.g., Josephine, Josephine, Angela, Jeremy). No names are repeated except where we've indicated. (If you think this is confusing, you should try memorizing the chorus.) How many arrangements of the names could fit this pattern?

52. *Committees* Suppose that out of 19 members of a club, two committees are to be formed. A nominating committee is to consist of 7 members, and a public relations committee is to consist of 5 members. No one can be on both committees.

a. Calculate the number of ways that the two committees can be formed, assuming that the nominating committee is formed first.

b. Calculate the number of ways that the two committees can be formed, assuming that the public relations committee is formed first. Verify that this answer is the same as that of part a.

c. Suppose the 7 members of the nominating committee wear red T-shirts, the 5 members of the public relations committee wear yellow T-shirts, and the remaining members of the club wear white T-shirts. A photographer lines up the members of the club to take a picture, but the picture is so blurry that people wearing the same color T-shirt are indistinguishable. In how many distinguishable ways can the club members line up? Explain why this answer is the same as the answers to parts a and b.

53. *Committee* A small department of 5 people decides to form a hiring committee. The only restriction on the size of the committee is that it must have at least 2 members.

a. Calculate the number of different committees possible by adding up the number of committees of different sizes.

b. Calculate the number of different committees possible by taking the total number of subsets of the 5 members and subtracting the number of committees that are invalid because they have too few members.

*http://slate.msn.com/id/95622/.
[†]http://slate.msn.com/id/96439/.

54. *Olympics* In recent Winter Olympics, there were 12 judges for each part of the women's figure skating program, but the scores of only 9 of the judges were randomly selected for the final results.*

a. In how many ways can the 9 judges whose scores are counted be selected?

b. Women's figure skating consists of a short program and a long program, with different judges for each part. How many different sets of judges' scores are possible for the entire event?

8.3 Probability Applications of Counting Principles

? THINK ABOUT IT

If 3 engines are tested from a shipping container packed with 12 diesel engines, 2 of which are defective, what is the probability that at least 1 of the defective engines will be found (in which case the container will not be shipped)?

This problem theoretically could be solved with a tree diagram, but it would require a tree with a large number of branches. Many of the probability problems involving *dependent* events that were solved earlier by using tree diagrams can also be solved by using permutations or combinations. Permutations and combinations are especially helpful when the numbers involved are large.

To compare the method of using permutations or combinations with the method of tree diagrams used in Section 7.5, the first example repeats Example 7 from that section.

EXAMPLE 1 **Environmental Inspections**

The Environmental Protection Agency is considering inspecting 6 plants for environmental compliance: 3 in Chicago, 2 in Los Angeles, and 1 in New York. Due to a lack of inspectors, they decide to inspect 2 plants selected at random, 1 this month and 1 next month, with each plant equally likely to be selected, but no plant is selected twice. What is the probability that 1 Chicago plant and 1 Los Angeles plant are selected?

▶**Solution** Because the plants are selected one at a time, with one labeled as the first plant and the other as the second, we use permutations. There are two ways to select a Chicago plant and a Los Angeles plant. The first way is to select the Chicago plant followed by the Los Angeles plant, and the second is to select the Los Angeles plant followed by the Chicago plant. The Chicago plant can be selected from the 3 Chicago plants in $\binom{3}{1}$ ways, and the Los Angeles plant can be selected from the 2 Los Angeles plants in $\binom{2}{1}$ ways. By the multiplication principle and the union rule for sets, both results can occur in

$$\binom{3}{1}\binom{2}{1} + \binom{2}{1}\binom{3}{1} \text{ ways,}$$

**The New York Times*, Feb. 23, 2006, p. D3.

> **FOR REVIEW**
> The use of combinations to solve probability problems depends on the basic probability principle introduced earlier and repeated here:
> Let S be a sample space with equally likely outcomes, and let event E be a subset of S. Then the probability that event E occurs, written $P(E)$, is
> $$P(E) = \frac{n(E)}{n(S)},$$
> where $n(E)$ and $n(S)$ represent the number of elements in sets E and S.

giving the numerator of the probability fraction, $P(E) = n(E)/n(S)$. For the denominator, there are 6 ways to select the first plant and 5 ways to select the second, for a total of $6 \cdot 5$ ways. The required probability is

$$P(1\text{ C and }1\text{ LA}) = \frac{\binom{3}{1}\binom{2}{1} + \binom{2}{1}\binom{3}{1}}{6 \cdot 5}$$
$$= \frac{3 \cdot 2 + 2 \cdot 3}{30} = \frac{12}{30} = \frac{2}{5}.$$

This agrees with the answer found earlier.

This example can be solved more simply by observing that the probability that 1 Chicago plant and 1 Los Angeles plant are selected should not depend upon the order in which the plants are selected, so we may use combinations. The numerator is simply the number of ways of selecting 1 Chicago plant out of 3 Chicago plants and 1 Los Angeles plant out of 2 Los Angeles plants. The denominator is just the number of ways of selecting 2 plants out of 6. Then

$$P(1\text{ C and }1\text{ LA}) = \frac{\binom{3}{1}\binom{2}{1}}{\binom{6}{2}} = \frac{6}{15} = \frac{2}{5}.$$

This helps explain why combinations tend to be used more often than permutations in probability. Even if order matters in the original problem, it is sometimes possible to ignore order and use combinations. Be careful to do this only when the final result does not depend on the order of events. Order often does matter. (If you don't believe this, try getting dressed tomorrow morning and then taking your shower.)

Example 1 could also be solved using the tree diagram shown in Figure 6. Two of the branches correspond to drawing 1 Chicago plant and 1 Los Angeles

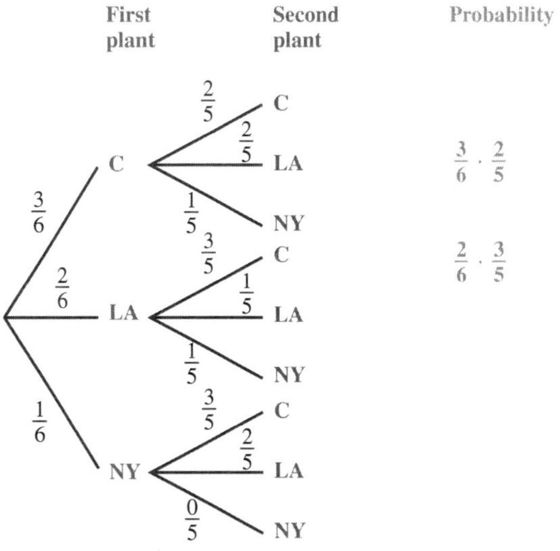

FIGURE 6

plant. The probability for each branch is calculated by multiplying the probabilities along the branch, as we did in the previous chapter. The resulting probabilities for the two branches are then added, giving the result

$$P(1 \text{ C and } 1 \text{ LA}) = \frac{3}{6} \cdot \frac{2}{5} + \frac{2}{6} \cdot \frac{3}{5} = \frac{2}{5}.$$

CAUTION The problems in the first two sections of this chapter asked how many ways a certain operation can be done. The problems in this section ask what is the probability that a certain event occurs; the solution involves answering questions about how many ways the event and the operation can be done.

- If a problem asks how many ways something can be done, the answer must be a nonnegative integer.
- If a problem asks for a probability, the answer must be a number between 0 and 1. ■

EXAMPLE 2 **Nursing**

From a group of 22 nurses, 4 are to be selected to present a list of grievances to management.

(a) In how many ways can this be done?

▶**Solution** Four nurses from a group of 22 can be selected in $\binom{22}{4}$ ways. (Use combinations, since the group of 4 is an unordered set.)

$$\binom{22}{4} = \frac{22!}{18!\,4!} = \frac{(22)(21)(20)(19)}{(4)(3)(2)(1)} = 7315$$

There are 7315 ways to choose 4 people from 22.

(b) One of the nurses is Julie Davis. Find the probability that Davis will be among the 4 selected.

▶**Solution** The probability that Davis will be selected is given by $n(E)/n(S)$, where E is the event that the chosen group includes Davis, and S is the sample space for the experiment of choosing a group of 4. There is only $\binom{1}{1} = 1$ way to choose Davis. The number of ways that the other 3 nurses can be chosen from the remaining 21 nurses is

$$\binom{21}{3} = \frac{21!}{18!\,3!} = 1330.$$

The probability that Davis will be one of the 4 chosen is

$$P(\text{Davis is chosen}) = \frac{n(E)}{n(S)} = \frac{\binom{1}{1}\binom{21}{3}}{\binom{22}{4}} = \frac{1330}{7315} \approx 0.1818.$$

Notice that the two numbers in red in the numerator, 1 and 21, add up to the number in red in the denominator, 22. This indicates that the 22 nurses have been split into two groups, one of size 1 (Davis) and the other of size 21 (the other nurses). Similarly, the green numbers indicate that the 4 nurses chosen consist of two groups of size 1 (Davis) and size 3 (the other nurses chosen).

(c) Find the probability that Davis will not be selected.

▶**Solution** The probability that Davis will not be chosen is $1 - 0.1818 = 0.8182$.

EXAMPLE 3 **Diesel Engines**

When shipping diesel engines abroad, it is common to pack 12 engines in one container that is then loaded on a rail car and sent to a port. Suppose that a company has received complaints from its customers that many of the engines arrive in nonworking condition. To help solve this problem, the company decides to make a spot check of containers after loading. The company will test 3 engines from a container at random; if any of the 3 are nonworking, the container will not be shipped until each engine in it is checked. Suppose a given container has 2 nonworking engines. Find the probability that the container will not be shipped.

?

▶**Solution** The container will not be shipped if the sample of 3 engines contains 1 or 2 defective engines. If $P(1 \text{ defective})$ represents the probability of exactly 1 defective engine in the sample, then

$$P(\text{not shipping}) = P(1 \text{ defective}) + P(2 \text{ defective}).$$

There are $\binom{12}{3}$ ways to choose the 3 engines for testing:

$$\binom{12}{3} = \frac{12!}{9!\,3!} = 220.$$

There are $\binom{2}{1}$ ways of choosing 1 defective engine from the 2 in the container, and for each of these ways, there are $\binom{10}{2}$ ways of choosing 2 good engines from among the 10 in the container. By the multiplication principle, there are

$$\binom{2}{1}\binom{10}{2} = \frac{2!}{1!\,1!} \cdot \frac{10!}{8!\,2!} = 2 \cdot 45 = 90$$

ways of choosing a sample of 3 engines containing 1 defective engine with

$$P(1 \text{ defective}) = \frac{90}{220} = \frac{9}{22}.$$

There are $\binom{2}{2}$ ways of choosing 2 defective engines from the 2 defective engines in the container, and $\binom{10}{1}$ ways of choosing 1 good engine from among the 10 good engines, for

$$\binom{2}{2}\binom{10}{1} = 1 \cdot 10 = 10$$

ways of choosing a sample of 3 engines containing 2 defective engines. Finally,

$$P(2\text{ defective}) = \frac{10}{220} = \frac{1}{22}$$

and

$$\begin{aligned} P(\text{not shipping}) &= P(1\text{ defective}) + P(2\text{ defective}) \\ &= \frac{\binom{2}{1}\binom{10}{2}}{\binom{12}{3}} + \frac{\binom{2}{2}\binom{10}{1}}{\binom{12}{3}} \\ &= \frac{9}{22} + \frac{1}{22} = \frac{10}{22} \approx 0.4545. \end{aligned}$$

Notice that the probability is $1 - 0.4545 = 0.5455$ that the container will be shipped, even though it has 2 defective engines. The management must decide whether this probability is acceptable; if not, it may be necessary to test more than 3 engines from a container.

FOR REVIEW

Recall that if E and E' are complements, then $P(E') = 1 - P(E)$. In Example 3, the event "0 defective in the sample" is the complement of the event "1 or 2 defective in the sample," since there are only 0 or 1 or 2 defective engines possible in the sample of 3 engines.

Observe that in Example 3, the complement of finding 1 or 2 defective engines is finding 0 defective engines. Then instead of finding the sum $P(1\text{ defective}) + P(2\text{ defective})$, the result in Example 3 could be found as $1 - P(0\text{ defective})$.

$$\begin{aligned} P(\text{not shipping}) &= 1 - P(0\text{ defective in sample}) \\ &= 1 - \frac{\binom{2}{0}\binom{10}{3}}{\binom{12}{3}} \\ &= 1 - \frac{1(120)}{220} \\ &= 1 - \frac{120}{220} = \frac{100}{220} \approx 0.4545 \end{aligned}$$

EXAMPLE 4 **Poker**

In a common form of the card game *poker*, a hand of 5 cards is dealt to each player from a deck of 52 cards. There are a total of

$$\binom{52}{5} = \frac{52!}{47!\,5!} = 2{,}598{,}960$$

such hands possible. Find the probability of getting each of the following hands.

(a) A hand containing only hearts, called a *heart flush*

Solution There are 13 hearts in a deck, with

$$\binom{13}{5} = \frac{13!}{8!\,5!} = \frac{(13)(12)(11)(10)(9)}{(5)(4)(3)(2)(1)} = 1287$$

different hands containing only hearts. The probability of a heart flush is

$$P(\text{heart flush}) = \frac{\binom{13}{5}\binom{39}{0}}{\binom{52}{5}} = \frac{1287}{2{,}598{,}960} \approx 0.0004952.$$

You don't really need the $\binom{39}{0}$, since this just equals 1, but it might help to remind you that you are choosing none of the 39 cards that remain after the hearts are removed.

(b) A flush of any suit (5 cards of the same suit)

▶**Solution** There are 4 suits in a deck, so

$$P(\text{flush}) = 4 \cdot P(\text{heart flush}) = 4 \cdot 0.0004952 \approx 0.001981.$$

(c) A full house of aces and eights (3 aces and 2 eights)

▶**Solution** There are $\binom{4}{3}$ ways to choose 3 aces from among the 4 in the deck, and $\binom{4}{2}$ ways to choose 2 eights.

$$P(3 \text{ aces, } 2 \text{ eights}) = \frac{\binom{4}{3} \cdot \binom{4}{2} \cdot \binom{44}{0}}{\binom{52}{5}} = \frac{4 \cdot 6 \cdot 1}{2{,}598{,}960} \approx 0.000009234$$

(d) Any full house (3 cards of one value, 2 of another)

▶**Solution**

METHOD 1
Standard Procedure

The 13 values in a deck give 13 choices for the first value. As in part (c), there are $\binom{4}{3}$ ways to choose the 3 cards from among the 4 cards that have that value. This leaves 12 choices for the second value (order *is* important here, since a full house of 3 aces and 2 eights is not the same as a full house of 3 eights and 2 aces). From the 4 cards that have the second value, there are $\binom{4}{2}$ ways to choose 2. The probability of any full house is then

$$P(\text{full house}) = \frac{13 \cdot \binom{4}{3} \cdot 12 \cdot \binom{4}{2}}{2{,}598{,}960} \approx 0.001441.$$

METHOD 2
Alternative Procedure

As an alternative way of counting the numerator, first count the number of different values in the hand.* Since there are 13 values from which to choose, and we need 2 different values (one for the set of 3 cards and one for the set of 2), there are $\binom{13}{2}$ ways to choose the values. Next, of the two values chosen, select the value for which there are 3 cards, which can be done $\binom{2}{1}$ ways. This automatically determines that the other value is the one for which there are 2 cards. Next, choose the suits for each value. For the value with 3 cards, there are $\binom{4}{3}$ values of the suits, and for the value with 2 cards, there are $\binom{4}{2}$ values. Putting this all together,

*We learned this approach from Professor Peter Grassi of Hofstra University.

$$P(\text{full house}) = \frac{\binom{13}{2}\cdot\binom{2}{1}\cdot\binom{4}{3}\cdot\binom{4}{2}}{2{,}598{,}960} \approx 0.001441.$$

EXAMPLE 5 **Music**

A music teacher has 3 violin pupils, Fred, Carl, and Helen. For a recital, the teacher selects a first violinist and a second violinist. The third pupil will play with the others, but not solo. If the teacher selects randomly, what is the probability that Helen is first violinist, Carl is second violinist, and Fred does not solo?

▶**Solution** Use *permutations* to find the number of arrangements in the sample space.

$$P(3, 3) = 3! = 6$$

(Think of this as filling the positions of first violin, second violin, and no solo.) The 6 arrangements are equally likely, since the teacher will select randomly. Thus, the required probability is $1/6$.

EXAMPLE 6 **Birthdays**

Suppose a group of n people is in a room. Find the probability that at least 2 of the people have the same birthday.

▶**Solution** "Same birthday" refers to the month and the day, not necessarily the same year. Also, ignore leap years, and assume that each day in the year is equally likely as a birthday. To see how to proceed, we first look at the case in which $n = 5$ and find the probability that *no 2 people* from among 5 people have the same birthday. There are 365 different birthdays possible for the first of the 5 people, 364 for the second (so that the people have different birthdays), 363 for the third, and so on. The number of ways the 5 people can have different birthdays is thus the number of permutations of 365 days taken 5 at a time or

$$P(365, 5) = 365 \cdot 364 \cdot 363 \cdot 362 \cdot 361.$$

The number of ways that 5 people can have the same birthday or different birthdays is

$$365 \cdot 365 \cdot 365 \cdot 365 \cdot 365 = (365)^5.$$

Finally, the *probability* that none of the 5 people have the same birthday is

$$\frac{P(365, 5)}{365^5} = \frac{365 \cdot 364 \cdot 363 \cdot 362 \cdot 361}{365 \cdot 365 \cdot 365 \cdot 365 \cdot 365} \approx 0.9729.$$

The probability that at least 2 of the 5 people *do* have the same birthday is $1 - 0.9729 = 0.0271$.

Now this result can be extended to more than 5 people. Generalizing, the probability that no 2 people among n people have the same birthday is

$$\frac{P(365, n)}{365^n}.$$

The probability that at least 2 of the n people *do* have the same birthday is

$$1 - \frac{P(365, n)}{365^n}.$$

The following table shows this probability for various values of n.

Number of People, n	Probability That Two Have the Same Birthday
5	0.0271
10	0.1169
15	0.2529
20	0.4114
22	0.4757
23	0.5073
25	0.5687
30	0.7063
35	0.8144
40	0.8912
50	0.9704
366	1

The probability that 2 people among 23 have the same birthday is 0.5073, a little more than half. Many people are surprised at this result; it seems that a larger number of people should be required.

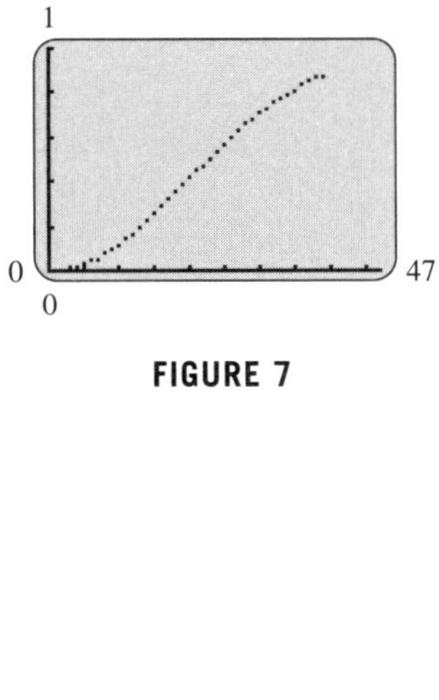

FIGURE 7

Using a graphing calculator, we can graph the probability formula in the previous example as a function of n, but care must be taken that the graphing calculator evaluates the function at integer points. Figure 7 was produced on a TI-83/84 Plus by letting `Y1 = 1 − (365 nPr X)/365^X` on $0 \le x \le 47$. (This domain ensures integer values for x.) Notice that the graph does not extend past $x = 39$. This is because $P(365, n)$ and 365^n are too large for the calculator when $n \ge 40$.

An alternative way of doing the calculations that does not run into such large numbers is based on the concept of conditional probability. The probability that the first person's birthday does not match any so far is $365/365$. The probability that the second person's birthday does not match the first's is $364/365$. The probability that the third person's birthday does not match the first's or the second's is $363/365$. By the product rule of probability, the probability that none of the first 3 people have matching birthdays is

$$\frac{365}{365} \cdot \frac{364}{365} \cdot \frac{363}{365}.$$

Similarly, the probability that no two people in a group of 40 have the same birthday is

$$\frac{365}{365} \cdot \frac{364}{365} \cdot \frac{363}{365} \cdot \cdots \cdot \frac{326}{365}.$$

This probability can be calculated (and then subtracted from 1 to get the probability we seek) without overflowing the calculator by multiplying each fraction times the next, rather then trying to compute the entire numerator and the entire denominator. The calculations are somewhat tedious to do by hand, but can be programmed on a graphing calculator or computer.

As we saw in Examples 1 and 4(d), probability can sometimes be calculated in more than one way. We now look at one more example of this.

EXAMPLE 7 **Fruit**

Ray and Nate are arranging a row of fruit at random on a table. They have 5 apples, 6 oranges, and 7 lemons. What is the probability that all fruit of the same kind are together?

▶Solution

METHOD 1
Distinguishable Permutations

Ray can't tell individual pieces of fruit of the same kind apart. All apples look the same to him, as do all oranges and all lemons. So in the denominator of the probability, he calculates the number of distinguishable ways to arrange the 18 pieces of fruit, given that all apples are indistinguishable, as are all oranges and all lemons.

$$\frac{18!}{5!\,6!\,7!} = 14{,}702{,}688$$

As for the numerator, the only choice is how to arrange the 3 kinds of fruit, for which there are $3! = 6$ ways. Thus

$$P(\text{all fruit of the same kind are together}) = \frac{6}{14{,}702{,}688} = 0.4081 \times 10^{-7}.$$

METHOD 2
Permutations

Nate has better eyesight than Ray and can tell the individual pieces of fruit apart. So in the denominator of the probability, he calculates the number of ways to arrange the 18 pieces of fruit, which is

$$18! = 6.4024 \times 10^{15}.$$

For the numerator, he first must choose how to arrange the 3 kinds of fruit, for which there are $3!$ ways. Then there are $5!$ ways to arrange the apples, $6!$ ways to arrange the oranges, and $7!$ ways to arrange the lemons, for a total number of possibilities of

$$3!\,5!\,6!\,7! = 2{,}612{,}736{,}000.$$

Therefore,

$$P(\text{all fruit of the same kind are together}) = \frac{2{,}612{,}736{,}000}{6.4024 \times 10^{15}} = 0.4081 \times 10^{-7}.$$

The results for Method 1 and Method 2 are the same. The probability does not depend on whether a person can distinguish individual pieces of the same kind of fruit.

8.3 Exercises

A basket contains 7 red apples and 4 yellow apples. A sample of 3 apples is drawn. Find the probabilities that the sample contains the following.

1. All red apples

2. All yellow apples

3. 2 yellow and 1 red apple

4. More red than yellow apples

Two cards are drawn at random from an ordinary deck of 52 cards.

5. How many 2-card hands are possible?

Find the probability that the 2-card hand described above contains the following.

6. 2 aces

7. At least 1 ace

8. All spades

9. 2 cards of the same suit

10. Only face cards

11. No face cards

12. No card higher than 8 (count ace as 1)

Twenty-six slips of paper are each marked with a different letter of the alphabet and placed in a basket. A slip is pulled out, its letter recorded (in the order in which the slip was drawn), and the slip is replaced. This is done 5 times. Find the probabilities that the following "words" are formed.

13. Chuck

14. A word that starts with "p"

15. A word with no repetition of letters

16. A word that contains no "*x*," "*y*," or "*z*"

17. Discuss the relative merits of using tree diagrams versus combinations to solve probability problems. When would each approach be most appropriate?

18. Several examples in this section used the rule $P(E') = 1 - P(E)$. Explain the advantage (especially in Example 6) of using this rule.

For Exercises 19–22, refer to Example 6 in this section.

19. A total of 42 men have served as president through 2007.* Set up the probability that, if 42 men were selected at random, at least 2 have the same birthday.†

20. Set up the probability that at least 2 of the 100 U.S. senators have the same birthday.

21. What is the probability that at least 2 of the 435 members of the House of Representatives have the same birthday?

22. Argue that the probability that in a group of n people *exactly one pair* have the same birthday is

$$\binom{n}{2} \cdot \frac{P(365, n-1)}{365^n}.$$

*Although Bush is the 43rd President, the 22nd and 24th Presidents were the same man: Grover Cleveland.

†In fact, James Polk and Warren Harding were both born on November 2.

23. An elevator has 4 passengers and stops at 7 floors. It is equally likely that a person will get off at any one of the 7 floors. Find the probability that at least 2 passengers leave at the same floor. (*Hint:* Compare this with the birthday problem.)

24. On National Public Radio, the *Weekend Edition* program on Sunday, September 7, 1991, posed the following probability problem: Given a certain number of balls, of which some are blue, pick 5 at random. The probability that all 5 are blue is 1/2. Determine the original number of balls and decide how many were blue.

25. A reader wrote to the "Ask Marilyn" column* in *Parade* magazine, "You have six envelopes to pick from. Two-thirds (that is, four) are empty. One-third (that is, two) contain a \$100 bill. You're told to choose 2 envelopes at random. Which is more likely: (1) that you'll get at least one \$100 bill, or (2) that you'll get no \$100 bill at all?" Find the two probabilities.

26. After studying all night for a final exam, a bleary-eyed student randomly grabs 2 socks from a drawer containing 9 black, 6 brown, and 2 blue socks, all mixed together. What is the probability that she grabs a matched pair?

27. Three crows, 4 blue jays, and 5 starlings sit in a random order on a section of telephone wire. Find the probability that birds of a feather flock together, that is, that all birds of the same type are sitting together.

28. If the letters l, i, t, t, l, and e are chosen at random, what is the probability that they spell the word "little"?

29. If the letters M, i, s, s, i, s, s, i, p, p, and i are chosen at random, what is the probability that they spell the word "Mississippi"?

Applications

BUSINESS AND ECONOMICS

Quality Control A shipment of 11 *printers contains* 2 *that are defective. Find the probability that a sample of the following sizes, drawn from the* 11, *will not contain a defective printer.*

30. 1 **31.** 2 **32.** 3 **33.** 4

Refer to Example 3. *The managers feel that the probability of* 0.5455 *that a container will be shipped even though it contains* 2 *defective engines is too high. They decide to increase the sample size chosen. Find the probabilities that a container will be shipped even though it contains* 2 *defective engines, if the sample size is increased to the following.*

34. 4 **35.** 5

SOCIAL SCIENCES

36. *Election Ballots* Five names are put on a ballot in a randomly selected order. What is the probability that they are not in alphabetical order?

37. *Native American Council* At the first meeting of a committee to plan a Northern California pow-wow, there were 3 women and 3 men from the Miwok tribe, 2 men and 3 women from the Hoopa tribe, and 4 women and 5 men from the Pomo tribe. If the ceremony subcouncil consists of 5 people, and is randomly selected, find the probabilities that the subcouncil contains the following:

a. 3 men and 2 women;

b. exactly 3 Miwoks and 2 Pomos;

c. 2 Miwoks, 2 Hoopas, and a Pomo;

d. 2 Miwoks, 2 Hoopas, and 2 Pomos;

e. more women than men;

f. exactly 3 Hoopas;

g. at least 2 Pomos.

38. *Education* A school in Bangkok requires that students take an entrance examination. After the examination, there is a drawing in which 5 students are randomly selected from each group of 40 for automatic acceptance into the school, regardless of their performance on the examination.

*Source: *Parade* magazine, Apr. 30, 1995, p. 8.

The drawing consists of placing 35 red and 5 green pieces of paper into a box. Each student picks a piece of paper from the box and then does not return the piece of paper to the box. The 5 lucky students who pick the green pieces are automatically accepted into the school.*

a. What is the probability that the first person wins automatic acceptance?

b. What is the probability that the last person wins automatic acceptance?

c. If the students are chosen by the order of their seating, does this give the student who goes first a better chance of winning than the second, third, . . . person? (*Hint:* Imagine that the 40 pieces of paper have been mixed up and laid in a row so that the first student picks the first piece of paper, the second student picks the second piece of paper, and so on.)

GENERAL INTEREST

Poker Find the probabilities of the following hands at poker. Assume aces are either high or low.

39. Royal flush (5 highest cards of a single suit)

40. Straight flush (5 in a row in a single suit, but not a royal flush)

41. Four of a kind (4 cards of the same value)

42. Straight (5 cards in a row, not all of the same suit), with ace either high or low

43. Three of a kind (3 cards of one value, with the other cards of two different values)

44. Two pairs (2 cards of one value, 2 of another value, and 1 of a third value)

45. One pair (2 cards of one value, with the other cards of three different values)

Bridge A bridge hand is made up of 13 *cards from a deck of* 52. *Find the probabilities that a hand chosen at random contains the following.*

46. Only hearts

47. At least 3 aces

48. Exactly 2 aces and exactly 2 kings

49. 6 of one suit, 4 of another, and 3 of another

50. *Writers* At a conference of African American writers in Detroit, special-edition books were selected to be given away in contests. There were 9 books written by Langston Hughes, 5 books by James Baldwin, and 7 books by Toni Morrison. The judge of one contest selected 6 books at random for prizes. Find the probabilities that the selection consisted of the following.

a. 3 Hughes and 3 Morrison books

b. Exactly 4 Baldwin books

c. 2 Hughes, 3 Baldwin, and 1 Morrison book

d. At least 4 Hughes books

e. Exactly 4 books written by males (Morrison is female)

f. No more than 2 books written by Baldwin

51. *Lottery* In the previous section, we found the number of ways to pick 6 different numbers from 1 to 99 in a state lottery. Assuming order is unimportant, what is the probability of picking all 6 numbers correctly to win the big prize?

52. *Lottery* In Exercise 51, what is the probability of picking exactly 5 of the 6 numbers correctly?

53. *Lottery* An article in *The New York Times* discussing the odds of winning the lottery stated, "And who cares if a game-theory professor once calculated the odds of winning as equal to a poker player's chance of drawing four royal flushes in a row, all in spades—then getting up from the card table and meeting four strangers, all with the same birthday?"† Calculate this probability. Does this probability seem comparable to the odds of winning the lottery? (Ignore February 29 as a birthday, and assume that all four strangers have the same birthday as each other, not necessarily the same as the poker player.)

54. *Lottery* A reader wrote to the "Ask Marilyn" column‡ in *Parade* magazine, "A dozen glazed doughnuts are riding on the answer to this question: Are the odds of winning in a lotto drawing higher when picking 6 numbers out of 49 or when picking 5 numbers out of 52?" Calculate each probability to answer the question.

55. *Lottery* On October 22, 2005, the Powerball Lottery had a record jackpot of $340 million. To enter the lottery, 5 numbers are picked between 1 and 55, plus a bonus number between 1 and 42. All 6 numbers must be correct to win the jackpot.

a. What is the probability of winning the jackpot with a single ticket?

b. In an article for the *Minneapolis Star Tribune*, mathematician Douglas Arnold was quoted as saying, "If you were to select a group of Powerball numbers every minute for 138 years, you would have about a 50 percent

*Letter to the editor, *Mathematics Teacher*, Vol. 92, No. 8, Nov. 1999.

†Gould, Lois, "Ticket to Trouble," *The New York Times Magazine,* Apr. 23, 1995, p. 39.

‡Source: *Parade* magazine, Dec. 10, 2000, p. 11. © 2000 Marilyn vos Savant. Initially published in *Parade* magazine. All rights reserved.

chance of picking the winning Powerball ticket."[*] Calculate the actual probability, using an estimate of 365.25 for the number of days in the year. (Arnold later told *Chance News* that this was an "off-the-top-of-my-head calculation" made when a reporter called.[†])

56. *Canadian Lottery* In June 2004, Canada introduced a change in its lottery that violated the usual convention that the smaller the probability of an event, the bigger the prize.[‡] In this lottery, participants have to guess six numbers from 1 to 49. Six numbers between 1 and 49 are then drawn at random, plus a seventh "bonus number."

a. A fifth prize of \$10 goes to those who correctly guess exactly three of the six numbers, but do not guess the bonus number. Find the probability of winning fifth prize.

b. A sixth prize of \$5 goes to those who correctly guess exactly two of the six numbers plus the bonus number. Find the probability of winning sixth prize, and compare this with the probability of winning fifth prize.

57. *Barbie* A controversy arose in 1992 over the Teen Talk Barbie doll, each of which was programmed with four sayings randomly picked from a set of 270 sayings. The controversy was over the saying, "Math class is tough," which some felt gave a negative message toward girls doing well in math. In an interview with *Science*, a spokeswoman for Mattel, the makers of Barbie, said that "There's a less than 1% chance you're going to get a doll that says math class is tough."[§] Is this figure correct? If not, give the correct figure.

58. *Football* During the 1988 college football season, the Big Eight Conference ended the season in a "perfect progression," as shown in the following table.[||]

Won	Lost	Team
7	0	Nebraska (NU)
6	1	Oklahoma (OU)
5	2	Oklahoma State (OSU)
4	3	Colorado (CU)
3	4	Iowa State (ISU)
2	5	Missouri (MU)
1	6	Kansas (KU)
0	7	Kansas State (KSU)

Someone wondered what the probability of such an outcome might be.

a. How many games do the 8 teams play?

b. Assuming no ties, how many different outcomes are there for all the games together?

c. In how many ways could the 8 teams end in a perfect progression?

d. Assuming that each team had an equally likely probability of winning each game, find the probability of a perfect progression with 8 teams.

e. Find a general expression for the probability of a perfect progression in an n-team league with the same assumptions.

59. *Bingo* Bingo has become popular in the United States, and it is an efficient way for many organizations to raise money. The bingo card has 5 rows and 5 columns of numbers from 1 to 75, with the center given as a free cell. Balls showing one of the 75 numbers are picked at random from a container. If the drawn number appears on a player's card, then the player covers the number. In general, the winner is the person who first has a card with an entire row, column, or diagonal covered.[#]

a. Find the probability that a person will win bingo after just four numbers are called.

b. An L occurs when the first column and the bottom row are both covered. Find the probability that an L will occur in the fewest number of calls.

c. An X-out occurs when both diagonals are covered. Find the probability that an X-out occurs in the fewest number of calls.

d. If bingo cards are constructed so that column one has 5 of the numbers from 1 to 15, column two has 5 of the numbers from 16 to 30, column three has 4 of the numbers from 31 to 45, column four has 5 of the numbers from 46 to 60, and column five has 5 of the numbers from 61 to 75, how many different bingo cards could be constructed? (*Hint:* Order matters!)

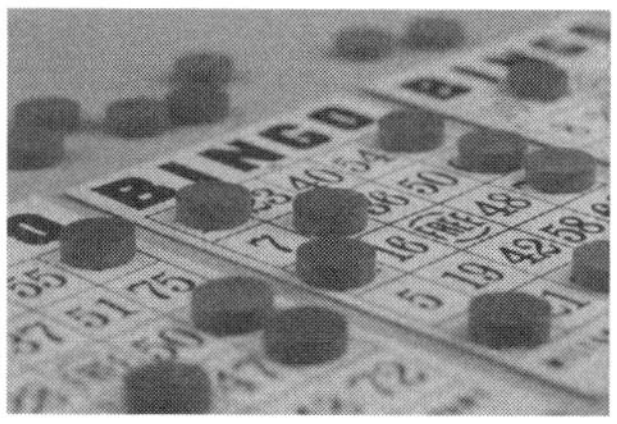

*Furst, Randy, "Advice from 2 number crunchers: Don't spend that \$340 million just yet," *Minneapolis Star Tribune*, Oct. 18, 2005.

†*Chance News 8*, Oct. 15–30, 2005.

‡Helman, Danny, "Reversal of Fortunes," *Chance*, Vol. 18, No. 3, Summer 2005, pp. 20–22.

§*Science*, Vol. 258, Oct. 16, 1992, p. 398.

||Madsen, Richard, "On the Probability of a Perfect Progression," *The American Statistician*, Aug. 1991, Vol. 45, No. 3, p. 214.

#Bay, Jennifer M., Robert E. Reys, Ken Simms, and P. Mark Taylor, "Bingo Games: Turning Student Intuitions into Investigations in Probability and Number Sense," *Mathematics Teacher*, Vol. 93, No. 3, Mar. 2000, pp. 200–206.

8.4 Binomial Probability

? THINK ABOUT IT

What is the probability that 3 out of 6 randomly selected college students attend more than one institution during their college career?

This question involves an experiment that is repeated 6 times. Many probability problems are concerned with experiments in which an event is repeated many times. Other examples include finding the probability of getting 7 heads in 8 tosses of a coin, of hitting a target 6 times out of 6, and of finding 1 defective item in a sample of 15 items. Probability problems of this kind are called **Bernoulli trials** problems, or **Bernoulli processes**, named after the Swiss mathematician Jakob Bernoulli (1654–1705), who is well known for his work in probability theory. In each case, some outcome is designated a success, and any other outcome is considered a failure. This labeling is arbitrary, and does not necessarily have anything to do with real success or failure. Thus, if the probability of a success in a single trial is p, the probability of failure will be $1 - p$. A Bernoulli trials problem, or **binomial experiment**, must satisfy the following conditions.

BINOMIAL EXPERIMENT

1. The same experiment is repeated a fixed number of times.
2. There are only two possible outcomes, success and failure.
3. The repeated trials are independent, so that the probability of success remains the same for each trial.

EXAMPLE 1

Sleep

The chance that an American falls asleep with the TV on at least three nights a week is $1/4$.* Suppose a researcher selects 5 Americans at random and is interested in the probability that all 5 are "TV sleepers."

FOR REVIEW
Recall that if A and B are independent events,

$$P(A \text{ and } B) = P(A)P(B).$$

Solution Here the experiment, selecting a person, is repeated 5 times. If selecting a TV sleeper is labeled a success, then getting a "non-TV sleeper" is labeled a failure. The 5 trials are almost independent. There is a very slight dependence; if, for example, the first person selected is a TV sleeper, then there is one less TV sleeper to choose from when we select the next person (assuming we never select the same person twice). When selecting a small sample out of a large population, however, the probability changes negligibly, so researchers consider such trials to be independent. Thus, the probability that all 5 in our sample are sleepers is

$$\frac{1}{4}\cdot\frac{1}{4}\cdot\frac{1}{4}\cdot\frac{1}{4}\cdot\frac{1}{4} = \left(\frac{1}{4}\right)^5 \approx 0.0009766.$$

Now suppose the problem in Example 1 is changed to that of finding the probability that exactly 4 of the 5 people in the sample are TV sleepers. This

Harper's Magazine, Mar. 1996, p. 13.

outcome can occur in more than one way, as shown below, where s represents a success (a TV sleeper) and f represents a failure (a non-TV sleeper).

outcome 1:	s	s	s	s	f
outcome 2:	s	s	s	f	s
outcome 3:	s	s	f	s	s
outcome 4:	s	f	s	s	s
outcome 5:	f	s	s	s	s

Keep in mind that since the probability of success is $1/4$, the probability of failure is $1 - 1/4 = 3/4$. The probability, then, of each of these 5 outcomes is

$$\left(\frac{1}{4}\right)^4\left(\frac{3}{4}\right).$$

Since the 5 outcomes represent mutually exclusive events, add the 5 identical probabilities, which is equivalent to multiplying the above probability by 5. The result is

$$P(4 \text{ of the 5 people are TV sleepers}) = 5\left(\frac{1}{4}\right)^4\left(\frac{3}{4}\right) = \frac{15}{4^5} \approx 0.01465.$$

In the same way, we can compute the probability of selecting 3 TV sleepers in our sample of 5. The probability of any one way of achieving 3 successes and 2 failures will be

$$\left(\frac{1}{4}\right)^3\left(\frac{3}{4}\right)^2.$$

Rather than list all the ways of achieving 3 successes out of 5 trials, we will count this number using combinations. The number of ways to select 3 elements out of a set of 5 is $\binom{5}{3} = 5!/(2!\,3!) = 10$, giving

$$P(3 \text{ of the 5 people are TV sleepers}) = 10\left(\frac{1}{4}\right)^3\left(\frac{3}{4}\right)^2 = \frac{90}{4^5} \approx 0.08789.$$

A similar argument works in the general case.

BINOMIAL PROBABILITY

If p is the probability of success in a single trial of a binomial experiment, the probability of x successes and $n - x$ failures in n independent repeated trials of the experiment, known as **binomial probability**, is

$$\binom{n}{x} \cdot p^x \cdot (1 - p)^{n-x}.$$

EXAMPLE 2 **College Students**

A recent survey found that 59% of college students attend more than one institution during their college career.* Suppose a sample of 6 students is chosen.

*"Education Life," *The New York Times*, April 23, 2006, p. 24.

Assuming that each student's college attendance pattern is independent of the others, find the probability of each of the following.

?

(a) Exactly 3 of the 6 students attend more than one institution.

▶**Solution** Think of the 6 students chosen as 6 independent trials. A success occurs if the student attends more than one institution. Then this is a binomial experiment with $n = 6$ and $p = P(\text{attend more than one institution}) = 0.59$. To find the probability that exactly 3 students attend more than one institution, let $x = 3$ and use the formula given in the box on the previous page.

$$\begin{aligned} P(\text{exactly } 3) &= \binom{6}{3}(0.59)^3(1 - 0.59)^{6-3} \\ &= 20(0.59)^3(0.41)^3 \\ &= 20(0.2054)(0.06892) \\ &\approx 0.2831 \end{aligned}$$

(b) None of the 6 students attend more than one institution.

▶**Solution** Let $x = 0$.

$$\begin{aligned} P(\text{exactly } 0) &= \binom{6}{0}(0.59)^0(1 - 0.59)^6 \\ &= 1(1)(0.41)^6 \approx 0.00475 \end{aligned}$$

EXAMPLE 3

Coin Toss

Find the probability of getting exactly 7 heads in 8 tosses of a fair coin.

▶**Solution** The probability of success (getting a head in a single toss) is $1/2$. The probability of a failure (getting a tail) is $1 - 1/2 = 1/2$. Thus,

$$P(7 \text{ heads in } 8 \text{ tosses}) = \binom{8}{7}\left(\frac{1}{2}\right)^7\left(\frac{1}{2}\right)^1 = 8\left(\frac{1}{2}\right)^8 = 0.03125.$$

EXAMPLE 4

Defective Items

Assuming that selection of items for a sample can be treated as independent trials, and that the probability that any 1 item is defective is 0.01, find the following.

(a) The probability of 1 defective item in a random sample of 15 items from a production line

▶**Solution** Here, a "success" is a defective item. Since selecting each item for the sample is assumed to be an independent trial, the binomial probability formula applies. The probability of success (a defective item) is 0.01, while the probability of failure (an acceptable item) is 0.99. This makes

$$\begin{aligned} P(1 \text{ defective in } 15 \text{ items}) &= \binom{15}{1}(0.01)^1(0.99)^{14} \\ &= 15(0.01)(0.99)^{14} \\ &\approx 0.1303. \end{aligned}$$

(b) The probability of at most 1 defective item in a random sample of 15 items from a production line

▶**Solution** "At most 1" means 0 defective items or 1 defective item. Since 0 defective items is equivalent to 15 acceptable items,

$$P(0 \text{ defective}) = (0.99)^{15} \approx 0.8601.$$

Use the union rule, noting that 0 defective and 1 defective are mutually exclusive events, to get

$$\begin{aligned} P(\text{at most 1 defective}) &= P(0 \text{ defective}) + P(1 \text{ defective}) \\ &\approx 0.8601 + 0.1303 \\ &= 0.9904. \end{aligned}$$

EXAMPLE 5 **Supermarket Scanners**

A survey by *Money* magazine found that supermarket scanners are overcharging customers at 30% of stores.*

(a) If you shop at 3 supermarkets that use scanners, what is the probability that you will be overcharged in at least one store?

▶**Solution** We can treat this as a binomial experiment, letting $n = 3$ and $p = 0.3$. At least 1 of 3 means 1 or 2 or 3. It will be simpler here to find the probability of being overcharged in none of the 3 stores, that is, $P(0 \text{ overcharges})$, and then find $1 - P(0 \text{ overcharges})$.

$$\begin{aligned} P(0 \text{ overcharges}) &= \binom{3}{0}(0.3)^0(0.7)^3 \\ &= 1(1)(0.343) = 0.343 \\ P(\text{at least one}) &= 1 - P(0 \text{ overcharges}) \\ &= 1 - 0.343 = 0.657 \end{aligned}$$

(b) If you shop at 3 supermarkets that use scanners, what is the probability that you will be overcharged in at most one store?

▶**Solution** "At most one" means 0 or 1, so

$$\begin{aligned} P(0 \text{ or } 1) &= P(0) + P(1) \\ &= \binom{3}{0}(0.3)^0(0.7)^3 + \binom{3}{1}(0.3)^1(0.7)^2 \\ &= 1(1)(0.343) + 3(0.3)(0.49) = 0.784. \end{aligned}$$

The triangular array of numbers shown on the next page is called **Pascal's triangle** in honor of the French mathematician Blaise Pascal (1623–1662), who was one of the first to use it extensively. The triangle was known long before Pascal's time and appears in Chinese and Islamic manuscripts from the eleventh century.

*O'Connell, Vanessa, "Don't Get Cheated by Supermarket Scanners," *Money*, Apr. 1993, pp. 132–138.

PASCAL'S TRIANGLE

$$
\begin{array}{ccccccccccc}
 & & & & & 1 & & & & & \\
 & & & & 1 & & 1 & & & & \\
 & & & 1 & & 2 & & 1 & & & \\
 & & 1 & & 3 & & 3 & & 1 & & \\
 & 1 & & 4 & & 6 & & 4 & & 1 & \\
1 & & 5 & & 10 & & 10 & & 5 & & 1 \\
\vdots & & \vdots & & \vdots & & \vdots & & \vdots & & \vdots
\end{array}
$$

The array provides a quick way to find binomial probabilities. The nth row of the triangle, where $n = 0, 1, 2, 3, \ldots$, gives the coefficients $\binom{n}{r}$ for $r = 0, 1, 2, 3, \ldots, n$. For example, for $n = 4$, $1 = \binom{4}{0}$, $4 = \binom{4}{1}$, $6 = \binom{4}{2}$, and so on. Each number in the triangle is the sum of the two numbers directly above it. For example, in the row for $n = 4$, 1 is the sum of 1, the only number above it, 4 is the sum of 1 and 3, 6 is the sum of 3 and 3, and so on. Adding in this way gives the sixth row:

$$1 \quad 6 \quad 15 \quad 20 \quad 15 \quad 6 \quad 1.$$

Notice that Pascal's triangle tells us, for example, that $\binom{4}{1} + \binom{4}{2} = \binom{5}{2}$ (that is, $4 + 6 = 10$). Using the combinations formula, it can be shown that, in general, $\binom{n}{r} + \binom{n}{r+1} = \binom{n+1}{r+1}$. This is left as an exercise.

EXAMPLE 6

Pascal's Triangle

Use Pascal's triangle to find the probability in Example 5 that if you shop at 6 supermarkets, at least 3 will overcharge you.

▶**Solution** The probability of success is 0.3. Since at least 3 means 3, 4, 5, or 6,

$$
\begin{aligned}
P(\text{at least } 3) &= P(3) + P(4) + P(5) + P(6) \\
&= \binom{6}{3}(0.3)^3(0.7)^3 + \binom{6}{4}(0.3)^4(0.7)^2 \\
&\quad + \binom{6}{5}(0.3)^5(0.7)^1 + \binom{6}{6}(0.3)^6(0.7)^0.
\end{aligned}
$$

Use the sixth row of Pascal's triangle for the combinations to get

$$
\begin{aligned}
P(\text{at least } 3) &= \mathbf{20}(0.3)^3(0.7)^3 + \mathbf{15}(0.3)^4(0.7)^2 \\
&\quad + \mathbf{6}(0.3)^5(0.7)^1 + \mathbf{1}(0.3)^6(0.7)^0 \\
&= 0.1852 + 0.0595 + 0.0102 + 0.0007 \\
&= 0.2556.
\end{aligned}
$$

EXAMPLE 7 **Independent Jury**

If each member of a 9-person jury acts independently of each other and makes the correct determination of guilt or innocence with probability 0.65, find the probability that the majority of jurors will reach a correct verdict.*

▶Solution

METHOD 1 Calculating by Hand

Since the jurors in this particular situation act independently, we can treat this as a binomial experiment. Thus, the probability that the majority of the jurors will reach the correct verdict is given by

$$\begin{aligned} P(\text{at least } 5) &= \binom{9}{5}(0.65)^5(0.35)^4 + \binom{9}{6}(0.65)^6(0.35)^3 \\ &\quad + \binom{9}{7}(0.65)^7(0.35)^2 + \binom{9}{8}(0.65)^8(0.35)^1 + \binom{9}{9}(0.65)^9 \\ &= 0.2194 + 0.2716 + 0.2162 + 0.1004 + 0.0207 \\ &= 0.8283. \end{aligned}$$

METHOD 2 Graphing Calculator

Some graphing calculators provide binomial probabilities. On a TI-83/84 Plus, for example, the command `binompdf(9, .65, 5)`, found in the `DISTR` menu, gives 0.21939, which is the probability that $x = 5$. Alternatively, the command `binomcdf(9, .65, 4)` gives 0.17172 as the probability that 4 or fewer jurors will make the correct decision. Subtract 0.17172 from 1 to get 0.82828 as the probability that the majority of the jurors will make the correct decision. This value rounds to 0.8283, which is in agreement with Method 1. Often, Method 2 is more accurate than Method 1 due to the accumulation of rounding errors when doing successive calculations by hand.

METHOD 3 Spreadsheet

Some spreadsheets also provide binomial probabilities. In Microsoft Excel, for example, the command "`=BINOMDIST(5, 9, .65, 0)`" gives 0.21939, which is the probability that $x = 5$. Alternatively, the command "`=BINOMDIST(4, 9, .65, 1)`" gives 0.17172 as the probability that 4 or fewer jurors will make the correct decision. Subtract 0.17172 from 1 to get 0.82828 as the probability that the majority of the jurors will make the correct decision. This value agrees with the value found in Methods 1 and 2.

▶ 8.4 Exercises

Suppose that a family has 5 children. Also, suppose that the probability of having a girl is 1/2. *Find the probabilities that the family has the following children.*

1. Exactly 2 girls and 3 boys
2. Exactly 3 girls and 2 boys
3. No girls

*Grofman, Bernard, "A Preliminary Model of Jury Decision Making as a Function of Jury Size, Effective Jury Decision Rule, and Mean Juror Judgmental Competence," *Frontiers in Economics*, 1979, pp. 98–110.

4. No boys

5. At least 4 girls

6. At least 3 boys

7. No more than 3 boys

8. No more than 4 girls

A die is rolled 12 *times. Find the probabilities of rolling the following.*

9. Exactly 12 ones

10. Exactly 6 ones

11. Exactly 1 one

12. Exactly 2 ones

13. No more than 3 ones

14. No more than 1 one

A coin is tossed 6 times. Find the probabilities of getting the following.

15. All heads

16. Exactly 3 heads

17. No more than 3 heads

18. At least 3 heads

19. How do you identify a probability problem that involves a binomial experiment?

20. How is Pascal's triangle used to find probabilities?

21. Using the definition of combination in Section 8.2, prove that

$$\binom{n}{r} + \binom{n}{r+1} = \binom{n+1}{r+1}.$$

(This is the formula underlying Pascal's triangle.)

In Exercises 22 and 23, argue that the use of binomial probabilities is not applicable and thus the probabilities that are computed are not correct.

22. In England, a woman was found guilty of smothering her two infant children. Much of the Crown's case against the lady was based on the testimony from a pediatrician who indicated that the chances of two crib deaths occurring in both siblings was only about 1 in 73 million. This number was calculated by assuming that the probability of a single crib death is 1 in 8543 and the probability of two crib deaths is 1 in 8543^2 (i.e., binomial).*

23. A contemporary radio station in Boston has a contest in which a caller is asked his or her date of birth. If the caller's date of birth, including the day, month, and year of birth, matches a predetermined date, the caller wins $1 million. Assuming that there were 36,525 days in the twentieth century and the contest was run 51 times on consecutive days, the probability that the grand prize will be won is

$$1 - \left(1 - \frac{1}{36{,}525}\right)^{51} \approx 0.0014.^{\dagger}$$

**Science*, Vol. 309, July 22, 2005, p. 543. Also see Chapter 7 Review Exercise 89.

†Snell, J. Laurie, "40-Million-Dollar Thursday," *Chance News 9.04*, Mar. 7–April 5, 2000.

Applications

BUSINESS AND ECONOMICS

Management *The survey discussed in Example 5 also found that customers overpay for 1 out of every 10 items, on average. Suppose a customer purchases 15 items. Find the following probabilities.*

24. A customer overpays on 3 items.

25. A customer does not overpay for any item.

26. A customer overpays on at least one item.

27. A customer overpays on at least 2 items.

28. A customer overpays on at most 2 items.

Credit Cards *A survey of consumer finance found that 25.6% of credit-card-holding families hardly ever pay off the balance.* Suppose a random sample of 20 credit-card-holding families is taken. Find the probabilities of each of the following results.*

29. Exactly 6 families hardly ever pay off the balance.

30. Exactly 9 families hardly ever pay off the balance.

31. At least 4 families hardly ever pay off the balance.

32. At most 5 families hardly ever pay off the balance.

Personnel Screening *A company gives prospective workers a 6-question, multiple-choice test. Each question has 5 possible answers, so that there is a 1/5 or 20% chance of answering a question correctly just by guessing. Find the probabilities of getting the following results by chance.*

33. Exactly 2 correct answers

34. No correct answers

35. At least 4 correct answers

36. No more than 3 correct answers

Quality Control *A factory tests a random sample of 20 transistors for defects. The probability that a particular transistor will be defective has been established by past experience as 0.05.*

37. What is the probability that there are no defective transistors in the sample?

38. What is the probability that the number of defective transistors in the sample is at most 2?

39. *Quality Control* The probability that a certain machine turns out a defective item is 0.05. Find the probabilities that in a run of 75 items, the following results are obtained.

a. Exactly 5 defective items

b. No defective items

c. At least 1 defective item

40. *Survey Results* A company is taking a survey to find out whether people like its product. Its last survey indicated that 70% of the population like the product. Based on that, in a sample of 58 people, find the probabilities of the following.

a. All 58 like the product.

b. From 28 to 30 (inclusive) like the product.

41. *Pecans* Pecan producers blow air through the pecans so that the lighter ones are blown out. The lighter-weight pecans are generally bad and the heavier ones tend to be better. These "blow outs" and "good nuts" are often sold to tourists along the highway. Suppose 60% of the "blow outs" are good, and 80% of the "good nuts" are good.†

a. What is the probability that if you crack and check 20 "good nuts" you will find 8 bad ones?

b. What is the probability that if you crack and check 20 "blow outs" you will find 8 bad ones?

c. If we assume that 70% of the roadside stands sell "good nuts," and that out of 20 nuts we find 8 that are bad, what is the probability that the nuts are "blow outs"?

42. *Hurricane Insurance* A company prices its hurricane insurance using the following assumptions:

(i) In any calendar year, there can be at most one hurricane.

(ii) In any calendar year, the probability of a hurricane is 0.05.

(iii) The number of hurricanes in any calendar year is independent of the number of hurricanes in any other calendar year.

**Statistical Abstract of the United States*, 2006, Table 1176.
†Submitted by Professor Irvin R. Hentzel, Iowa State University.

Using the company's assumptions, calculate the probability that there are fewer than 3 hurricanes in a 20-year period.[*] (Choose one of the following.)

a. 0.06 **b.** 0.19 **c.** 0.38 **d.** 0.62 **e.** 0.92

LIFE SCIENCES

Breast Cancer A recent study found that 85% of breast-cancer cases are detectable by mammogram.[†] *Suppose a random sample of 15 women with breast cancer are given mammograms. Find the probability of each of the following results, assuming that detection in the cases is independent.*

43. All of the cases are detectable.

44. None of the cases are detectable.

45. Not all cases are detectable.

46. More than half of the cases are detectable.

Births of Twins The probability that a birth will result in twins is 0.012. Assuming independence (perhaps not a valid assumption), what are the probabilities that out of 100 births in a hospital, there will be the following numbers of sets of twins?

47. Exactly 2 sets of twins

48. At most 2 sets of twins

49. *Effects of Radiation* The probability of a mutation of a given gene under a dose of 1 roentgen of radiation is approximately 2.5×10^{-7}. What is the probability that in 10,000 genes, at least 1 mutation occurs?

50. *Flu Inoculations* A flu vaccine has a probability of 80% of preventing a person who is inoculated from getting the flu. A county health office inoculates 83 people. Find the probabilities of the following.

a. Exactly 10 of the people inoculated get the flu.

b. No more than 4 of the people inoculated get the flu.

c. None of the people inoculated get the flu.

51. *Color Blindness* The probability that a male will be color-blind is 0.042. Find the probabilities that in a group of 53 men, the following will be true.

a. Exactly 5 are color-blind.

b. No more than 5 are color-blind.

c. At least 1 is color-blind.

52. *Pharmacology* In placebo-controlled trials of Pravachol®, a drug that is prescribed to lower cholesterol, 7.3% of the patients who were taking the drug experienced nausea/vomiting, whereas 7.1% of the patients who were taking the placebo experienced nausea/vomiting.[‡]

a. If 100 patients who are taking Pravachol® are selected, what is the probability that 10 or more will experience nausea/vomiting?

b. If a second group of 100 patients receives a placebo, what is the probability that 10 or more will experience nausea/vomiting?

c. Since 7.3% is larger than 7.1%, do you believe that the Pravachol® causes more people to experience nausea/vomiting than a placebo? Explain.

53. *Genetic Fingerprinting* The use of DNA has become an integral part of many court cases. When DNA is extracted from cells and body fluids, genetic information is represented by bands of information, which look similar to a bar code at a grocery store. It is generally accepted that in unrelated people, the probability of a particular band matching is 1 in 4.[§]

a. If 5 bands are compared in unrelated people, what is the probability that all 5 of the bands match? (Express your answer in terms of "1 chance in ?".)

b. If 20 bands are compared in unrelated people, what is the probability that all 20 of the bands match? (Express your answer in terms of "1 chance in ?".)

c. If 20 bands are compared in unrelated people, what is the probability that 16 or more bands match? (Express your answer in terms of "1 chance in ?".)

d. If you were deciding paternity and there were 16 matches out of 20 bands compared, would you believe that the person being tested was the father? Explain.

54. *Salmonella* According to *The Salt Lake Tribune*, the Coffee Garden in Salt Lake City ran into trouble because of their four-egg quiche:[||]

"A Salt Lake County Health Department inspector paid a visit recently and pointed out that research by the Food and Drug Administration indicates that one in four eggs

[*]Problem 39 from the 2005 Sample Exam P of the *Education and Examination Committee of the Society of Actuaries*. Reprinted by permission of the Society of Actuaries.

[†]*Harper's Index*, April 2006.

[‡]Advertisement in *Time*, July 17, 2000, for Pravachol®, developed and marketed by Bristol-Myers Squibb Company.

[§]"Genetic Fingerprinting Worksheet," Centre for Innovation in Mathematics Teaching, http://www.ex.ac.uk/cimt/resource/fgrprnts.htm.

[||]Rolly, Paul and JoAnn Jacobsen-Wells, "Bureaucrat's Math Makes Dizzy Dozen," *The Salt Lake Tribune*, Oct. 11, 2002.

carries *Salmonella* bacterium, so restaurants should never use more than three eggs when preparing quiche.

The manager on duty wondered aloud if simply throwing out three eggs from each dozen and using the remaining nine in four-egg quiches would serve the same purpose.

The inspector wasn't sure, but she said she would research it."

a. Assuming that one in four eggs carries *Salmonella*, and that the event that any one egg is infected is independent of whether any other egg is infected, find the probability that at least one of the eggs in a four-egg quiche carries *Salmonella*.

b. Repeat part a for a three-egg quiche.

c. Discuss whether the assumption of independence is justified.

d. Discuss whether the inspector's reasoning makes sense.

55. *Herbal Remedies* According to Dr. Peter A.G.M. De Smet of the Netherlands, "If an herb caused an adverse reaction in 1 in 1,000 users, a traditional healer would have to treat 4,800 patients with that herb (i.e., one new patient every single working day for more than 18 years) to have a 95 percent chance of observing the reaction in more than one user."* Verify this calculation by finding the probability of observing more than one reaction in 4800 patients, given that 1 in 1000 has a reaction.

56. *Vaccines* A hospital receives $1/5$ of its flu vaccine shipments from Company X and the remainder of its shipments from other companies. Each shipment contains a very large number of vaccine vials. For Company X's shipments, 10% of the vials are ineffective. For every other company, 2% of the vials are ineffective. The hospital tests 30 randomly selected vials from a shipment and finds that one vial is ineffective. What is the probability that this shipment came from Company X?[†] (Choose one of the following. *Hint:* Find the probability that one out of 30 vials is ineffective, given that the shipment came from Company X, and also that the shipment came from other companies. Then use Bayes' theorem.)

a. 0.10 **b.** 0.14 **c.** 0.37 **d.** 0.63 **e.** 0.86

57. *Health Study* A study is being conducted in which the health of two independent groups of ten policyholders is being monitored over a one-year period of time. Individual participants in the study drop out before the end of the study with probability 0.2 (independently of the other participants). What is the probability that at least 9 participants complete the study in one of the two groups, but not in both groups?[‡] (Choose one of the following.)

a. 0.096 **b.** 0.192 **c.** 0.235 **d.** 0.376 **e.** 0.469

SOCIAL SCIENCES

58. *Women Working* A recent study found that 33% of women would prefer to work part-time rather than full-time if money were not a concern.[§] Find the probability that if 10 women are selected at random, at least 3 of them would prefer to work part-time.

Volunteering A recent survey found that 83% of first-year college students were involved in volunteer work at least occasionally.[‖] *Suppose a random sample of 12 college students is taken. Find the probabilities of each of the following results.*

59. Exactly 7 students volunteered at least occasionally.

60. Exactly 9 students volunteered at least occasionally.

61. At least 9 students volunteered at least occasionally.

62. At most 9 students volunteered at least occasionally.

63. *Community College Population* According to the state of California, 33% of all state community college students belong to ethnic minorities. Find the probabilities of the following results in a random sample of 10 California community college students.

a. Exactly 2 belong to an ethnic minority.

b. Three or fewer belong to an ethnic minority.

c. Exactly 5 do not belong to an ethnic minority.

d. Six or more do not belong to an ethnic minority.

64. *Cheating* According to a poll conducted by *U.S. News and World Report*, 84% of college students believe they need to cheat to get ahead in the world today.[#]

*De Smet, Peter A.G.M., "Drug Therapies: Herbal Remedies," *The New England Journal of Medicine*, Vol. 347, Dec. 19, 2002, pp. 2046–2056.

[†]Problem 28 from the 2005 Sample Exam P of the *Education and Examination Committee of the Society of Actuaries*. Reprinted by permission of the Society of Actuaries.

[‡]Problem 41 from the 2005 Sample Exam P of the *Education and Examination Committee of the Society of Actuaries*. Reprinted by permission of the Society of Actuaries.

[§]Ferraro, Cathleen, "Feelings of the Working Women," *The Sacramento Bee*, May 11, 1995, pp. A1, A22.

[‖]"Education Life," *The New York Times*, April 23, 2006, p. 7.

[#]Kleiner, Carolyn and Mary Lord, "The Cheating Game," *U.S. News and World Report*, Nov. 22, 1999, pp. 55–66.

a. Do the results of this poll indicate that 84% of all college students cheat? Explain.

b. If this result is accurate and 100 college students are asked if they believe that cheating is necessary to get ahead in the world, what is the probability that 90 or more of the students will answer affirmatively to the question?

65. *Education* In the "Numbers" section of a recent *Time* magazine, it was reported that 15.2% of low-birth-weight babies graduate from high school by age 19. On the other hand, it was reported that 57.5% of the normal-birth-weight siblings graduated from high school.*

a. If 40 low-birth-weight babies were tracked through high school, what is the probability that fewer than 15 will graduate from high school by age 19?

b. What are some of the factors that may contribute to the wide difference in high school success between these siblings? Do you believe that low birth weight is the primary cause of the difference? What other information do you need to better answer these questions?

66. *War Dead* A newspaper article questioned whether soldiers and marines from some states bear greater risks in Afghanistan and Iraq than those from others.† Out of 644,066 troops deployed as of the time of the article, 1174 had been killed, for a probability of being killed of $p = 1174/644{,}066$. Assume the deaths are independent.

a. Vermont had 9 deaths out of 1613 troops deployed. Find the probability of at least this many deaths.

b. Massachusetts had 28 deaths out of 7146 troops deployed. Find the probability of at least this many deaths.

c. Florida had 54 deaths out of 62,572 troops deployed. Find the probability of at most this many deaths.

d. Discuss why the assumption of independence may be questionable.

67. *Sports* In many sports championships, such as the World Series in baseball and the Stanley Cup final series in hockey, the winner is the first team to win four games. For this exercise, assume that each game is independent of the others, with a constant probability p that one specified team (say, the National League team) wins.

a. Find the probability that the series lasts for four, five, six, and seven games when $p = 0.5$. (*Hint:* Suppose the National League wins the series, so they must win the last game. Consider how the previous games might come out. Then consider the probability that the American League wins.)

b. Morrison and Schmittlein have found that the Stanley Cup finals can be described by letting $p = 0.73$ be the probability that the better team wins each game.‡ Find the probability that the series lasts for four, five, six, and seven games.

c. Some have argued that the assumption of independence does not apply.§ Discuss this issue.

8.5 Probability Distributions; Expected Value

? THINK ABOUT IT

What is the expected payback for someone who buys one ticket in a raffle?

In this section we shall see that the *expected value* of a probability distribution is a type of average. Probability distributions were introduced briefly in the chapter on Sets and Probability. Now we take a more complete look at probability distributions. A probability distribution depends on the idea of a *random variable*, so we begin with that.

*"Numbers," *Time,* July 17, 2000, p. 21.

†Tillman, Jodie, "The Price They Paid: By Several Measures, Vermont Bears Heavy War Burden," *Valley News*, Jan. 30, 2005. For further statistical analysis, see http://www.dartmouth.edu/~chance/ForWiki/GregComments.pdf.

‡Morrison, Donald G. and David C. Schmittlein, "It Takes a Hot Goalie to Raise the Stanley Cup," *Chance*, Vol. 11, No. 1, 1998, pp. 3–7.

§For example, see Groeneveld, Richard A. and Glen Meeden, "Seven Game Series in Sports," *Mathematics Magazine*, Vol. 48, No. 4, Sept. 1975, pp. 187–192.

Random Variables Suppose that the shipping manager at a company receives a package of one dozen computer monitors, of which, unknown to him, three are broken. He checks four of the monitors at random to see how many are broken in his sample of 4. The answer, which we will label x, is one of the numbers 0, 1, 2, or 3. Since the value of x is random, x is called a random variable.

> **RANDOM VARIABLE**
>
> A **random variable** is a function that assigns a real number to each outcome of an experiment.

Probability Distribution In the example with the shipping manager, we can calculate the probability that 0, 1, 2, or 3 monitors in his sample of 4 are broken using the methods of Section 8.3. There are 3 broken monitors and 9 unbroken monitors, so the number of ways of choosing 0 broken monitors (which implies 4 unbroken monitors) is $\binom{3}{0}\binom{9}{4}$. The number of ways of choosing a sample of 4 monitors is $\binom{12}{4}$. Therefore, the probability of choosing 0 broken monitors is

$$P(0) = \frac{\binom{3}{0}\binom{9}{4}}{\binom{12}{4}} = \frac{1\left(\frac{9 \cdot 8 \cdot 7 \cdot 6}{4 \cdot 3 \cdot 2 \cdot 1}\right)}{\left(\frac{12 \cdot 11 \cdot 10 \cdot 9}{4 \cdot 3 \cdot 2 \cdot 1}\right)} = \frac{126}{495} = \frac{14}{55}.$$

Similarly, the probability of choosing 1 broken monitor is

$$P(1) = \frac{\binom{3}{1}\binom{9}{3}}{\binom{12}{4}} = \frac{3 \cdot 84}{495} = \frac{252}{495} = \frac{28}{55}.$$

The probability of choosing 2 broken monitors is

$$P(2) = \frac{\binom{3}{2}\binom{9}{2}}{\binom{12}{4}} = \frac{3 \cdot 36}{495} = \frac{108}{495} = \frac{12}{55}.$$

The probability of choosing 3 broken monitors is

$$P(3) = \frac{\binom{3}{3}\binom{9}{1}}{\binom{12}{4}} = \frac{1 \cdot 9}{495} = \frac{9}{495} = \frac{1}{55}.$$

The results can be put in a table.

x	0	1	2	3
$P(x)$	$14/55$	$28/55$	$12/55$	$1/55$

Such a table that lists the possible values of a random variable, together with the corresponding probabilities, is called a **probability distribution**. The sum of the probabilities in a probability distribution must always equal 1. (The sum in some distributions may vary slightly from 1 because of rounding.)

Instead of writing the probability distribution as a table, we could write the same information as a set of ordered pairs:

$$\{(0, 14/55), (1, 28/55), (2, 12/55), (3, 1/55)\}.$$

There is just one probability for each value of the random variable. Thus, a probability distribution defines a function, called a **probability distribution function**, or simply a **probability function**. We shall use the terms "probability distribution" and "probability function" interchangeably.

The information in a probability distribution is often displayed graphically as a special kind of bar graph called a **histogram**. The bars of a histogram all have the same width, usually 1. (The widths might be different from 1 when the values of the random variable are not consecutive integers.) The heights of the bars are determined by the probabilities. A histogram for the data in the table above is given in Figure 8. A histogram shows important characteristics of a distribution that may not be readily apparent in tabular form, such as the relative sizes of the probabilities and any symmetry in the distribution.

The area of the bar above $x = 0$ in Figure 8 is the product of 1 and $14/55$, or $1 \cdot 14/55 = 14/55$. Since each bar has a width of 1, its area is equal to the probability that corresponds to that value of x. The probability that a particular value will occur is thus given by the area of the appropriate bar of the graph. For example, the probability that one or more monitors is broken is the sum of the areas for $x = 1$, $x = 2$, and $x = 3$. This area, shown in pink in Figure 9, corresponds to $41/55$ of the total area, since

$$P(x \geq 1) = P(x = 1) + P(x = 2) + P(x = 3)$$
$$= 28/55 + 12/55 + 1/55 = 41/55.$$

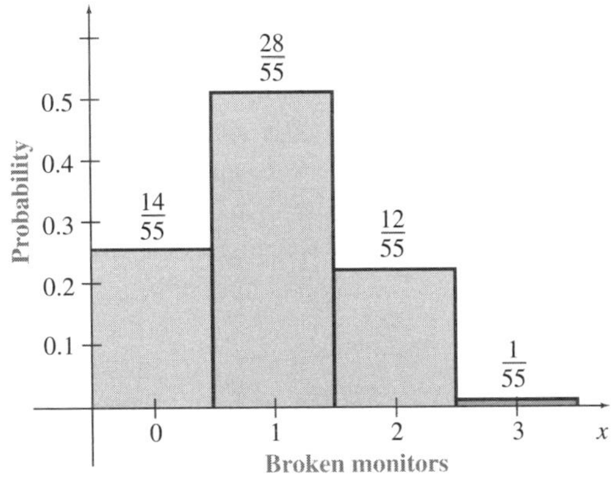

FIGURE 8

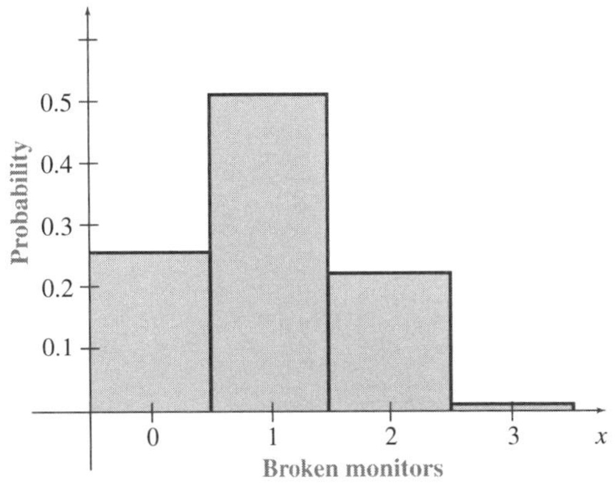

FIGURE 9

EXAMPLE 1 Probability Distributions

(a) Give the probability distribution for the number of heads showing when two coins are tossed.

▶**Solution** Let x represent the random variable "number of heads." Then x can take on the values 0, 1, or 2. Now find the probability of each outcome. To find the probability of 0, 1, or 2 heads, we can either use binomial probability, or notice that there are 4 outcomes in the sample space: $\{hh, ht, th, tt\}$. The results are shown in the table with Figure 10.

x	0	1	2
$P(x)$	1/4	1/2	1/4

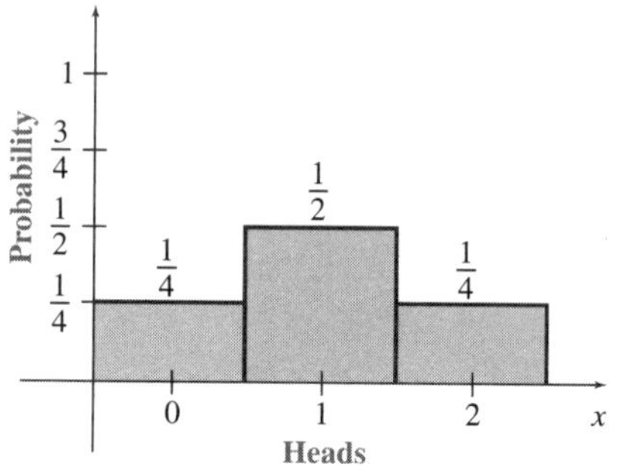

FIGURE 10

(b) Draw a histogram for the distribution in the table. Find the probability that at least one coin comes up heads.

▶**Solution** The histogram is shown in Figure 10. The portion in pink represents

$$P(x \geq 1) = P(x = 1) + P(x = 2)$$
$$= \frac{3}{4}.$$

Expected Value In working with probability distributions, it is useful to have a concept of the typical or average value that the random variable takes on. In Example 1, for instance, it seems reasonable that, on the average, one head shows when two coins are tossed. This does not tell what will happen the next time we toss two coins; we may get two heads, or we may get none. If we tossed two coins many times, however, we would expect that, in the long run, we would average about one head for each toss of two coins.

A way to solve such problems in general is to imagine flipping two coins 4 times. Based on the probability distribution in Example 1, we would expect that 1 of the 4 times we would get 0 heads, 2 of the 4 times we would get 1 head, and 1 of the 4 times we would get 2 heads. The total number of heads we would get, then, is

$$0 \cdot 1 + 1 \cdot 2 + 2 \cdot 1 = 4.$$

The expected numbers of heads per toss is found by dividing the total number of heads by the total number of tosses, or

$$\frac{0 \cdot 1 + 1 \cdot 2 + 2 \cdot 1}{4} = 0 \cdot \frac{1}{4} + 1 \cdot \frac{1}{2} + 2 \cdot \frac{1}{4} = 1.$$

Notice that the expected number of heads turns out to be the sum of the three values of the random variable x multiplied by their corresponding probabilities. We can use this idea to define the *expected value* of a random variable as follows.

> **EXPECTED VALUE**
>
> Suppose the random variable x can take on the n values $x_1, x_2, x_3, \ldots, x_n$. Also, suppose the probabilities that these values occur are, respectively, $p_1, p_2, p_3, \ldots, p_n$. Then the **expected value** of the random variable is
>
> $$E(x) = x_1p_1 + x_2p_2 + x_3p_3 + \cdots + x_np_n$$

EXAMPLE 2 **Computer Monitors**

In the example with the computer monitors, find the expected number of broken monitors that the shipping manager finds.

▶**Solution** Using the values in the first table in this section and the definition of expected value, we find that

$$E(x) = 0 \cdot \frac{14}{55} + 1 \cdot \frac{28}{55} + 2 \cdot \frac{12}{55} + 3 \cdot \frac{1}{55} = 1.$$

On the average, the shipping manager will find 1 broken monitor in the sample of 4. On reflection, this seems natural; 3 of the 12 monitors, or 1/4 of the total, are broken. We should expect, then, that 1/4 of the sample of 4 monitors are broken.

Physically, the expected value of a probability distribution represents a balance point. If we think of the histogram in Figure 8 as a series of weights with magnitudes represented by the heights of the bars, then the system would balance if supported at the point corresponding to the expected value.

EXAMPLE 3 **Symphony Orchestra**

Suppose a local symphony decides to raise money by raffling a microwave oven worth \$400, a dinner for two worth \$80, and 2 books worth \$20 each. A total of 2000 tickets are sold at \$1 each. Find the expected payback for a person who buys one ticket in the raffle.

?

▶**Solution**

METHOD 1
Direct Calculation

Here the random variable represents the possible amounts of payback, where payback = amount won − cost of ticket. The payback of the person winning the oven is \$400 (amount won) − \$1 (cost of ticket) = \$399. The payback for each losing ticket is \$0 − \$1 = −\$1.

The paybacks of the various prizes, as well as their respective probabilities, are shown in the table below. The probability of winning \$19 is 2/2000 because there are 2 prizes worth \$20. We have not reduced the fractions in order to keep all the denominators equal. Because there are 4 winning tickets, there are 1996 losing tickets, so the probability of winning −\$1 is 1996/2000.

x	\$399	\$79	\$19	−\$1
$P(x)$	1/2000	1/2000	2/2000	1996/2000

The expected payback for a person buying one ticket is

$$399\left(\frac{1}{2000}\right) + 79\left(\frac{1}{2000}\right) + 19\left(\frac{2}{2000}\right) + (-1)\left(\frac{1996}{2000}\right) = -\frac{1480}{2000}$$
$$= -0.74.$$

On the average, a person buying one ticket in the raffle will lose \$0.74, or 74¢.

It is not possible to lose 74¢ in this raffle: either you lose \$1, or you win a prize worth \$400, \$80, or \$20, minus the \$1 you pay to play. But if you bought tickets in many such raffles over a long period of time, you would lose 74¢ per ticket on the average. It is important to note that the expected value of a random variable may be a number that can never occur in any one trial of the experiment.

METHOD 2
Alternate Procedure

An alternative way to compute expected value in this and other examples is to calculate the expected amount won and then subtract the cost of the ticket afterward. The amount won is either \$400 (with probability 1/2000), \$80 (with probability 1/2000), \$20 (with probability 2/2000), or \$0 (with probability 1996/2000). The expected payback for a person buying one ticket is then

$$400\left(\frac{1}{2000}\right) + 80\left(\frac{1}{2000}\right) + 20\left(\frac{2}{2000}\right) + 0\left(\frac{1996}{2000}\right) - 1 = -\frac{1480}{2000}$$
$$= -0.74.$$

EXAMPLE 4

Friendly Wager

Each day Donna and Mary toss a coin to see who buys coffee (\$1.20 a cup). One tosses and the other calls the outcome. If the person who calls the outcome is correct, the other buys the coffee; otherwise the caller pays. Find Donna's expected winnings.

▶**Solution** Assume that an honest coin is used, that Mary tosses the coin, and that Donna calls the outcome. The possible results and corresponding probabilities are shown below.

	Possible Results			
Result of Toss	Heads	Heads	Tails	Tails
Call	Heads	Tails	Heads	Tails
Caller Wins?	Yes	No	No	Yes
Probability	1/4	1/4	1/4	1/4

Donna wins a \$1.20 cup of coffee whenever the results and calls match, and she loses a \$1.20 cup when there is no match. Her expected winnings are

$$(1.20)\left(\frac{1}{4}\right) + (-1.20)\left(\frac{1}{4}\right) + (-1.20)\left(\frac{1}{4}\right) + (1.20)\left(\frac{1}{4}\right) = 0.$$

On the average, over the long run, Donna neither wins nor loses.

A game with an expected value of 0 (such as the one in Example 4) is called a **fair game**. Casinos do not offer fair games. If they did, they would win (on the average) \$0, and have a hard time paying the help! Casino games have expected winnings for the house that vary from 1.5 cents per dollar to 60 cents per dollar. Exercises 47–52 at the end of the section ask you to find the expected winnings for certain games of chance.

The idea of expected value can be very useful in decision making, as shown by the next example.

EXAMPLE 5

Life Insurance

At age 50, you receive a letter from Mutual of Mauritania Insurance Company. According to the letter, you must tell the company immediately which of the following two options you will choose: take \$20,000 at age 60 (if you are alive, \$0 otherwise) or \$30,000 at age 70 (again, if you are alive, \$0 otherwise). Based *only* on the idea of expected value, which should you choose?

▶**Solution** Life insurance companies have constructed elaborate tables showing the probability of a person living a given number of years into the future. From a recent such table, the probability of living from age 50 to 60 is 0.88, while the probability of living from age 50 to 70 is 0.64. The expected values of the two options are given below.

$$\text{First option: } (20{,}000)(0.88) + (0)(0.12) = 17{,}600$$
$$\text{Second option: } (30{,}000)(0.64) + (0)(0.36) = 19{,}200$$

Based strictly on expected values, choose the second option.

EXAMPLE 6

Bachelor's Degrees

According to the National Center for Education Statistics, 78.5% of those earning bachelor's degrees in education in the United States in 2003–2004 were female.* Suppose 5 holders of bachelor's degrees in education from 2003 to 2004 are picked at random.

(a) Find the probability distribution for the number that are female.

▶**Solution** We first note that each of the 5 people in the sample is either female (with probability 0.785) or male (with probability 0.215). As in the previous section, we may assume that the probability for each member of the sample is independent of that of any other. Such a situation is described by binomial probability with $n = 5$ and $p = 0.785$, for which we use the binomial probability formula

$$\binom{n}{x} \cdot p^x \cdot (1 - p)^{n-x},$$

where x is the number of females in the sample. For example,

$$P(x = 0) = \binom{5}{0}(0.785)^0(0.215)^5 \approx 0.0005.$$

Similarly, we could calculate the probability that x is any value from 0 to 5, resulting in the probability distribution on the next page (with all probabilities rounded to four places).

*http://nces.ed.gov/programs/digest/d05/tables/dt05_280.asp.

x	0	1	2	3	4	5
$P(x)$	0.0005	0.0084	0.0612	0.2236	0.4082	0.2981

(b) Find the expected number of females in the sample of 5 people.

▶**Solution** Using the formula for expected value, we have

$$E(x) = 0(0.0005) + 1(0.0084) + 2(0.0612) + 3(0.2236) + 4(0.4082) + 5(0.2981) = 3.925.$$

On the average, 3.925 of the people in the sample of 5 will be female.

There is another way to get the answer in part (b) of the previous example. Because 78.5% of those earning bachelor's degrees in education in the United States in 2003–2004 are female, it is reasonable to expect 78.5% of our sample to be female. Thus, 78.5% of 5 is $5(0.785) = 3.925$. Notice that what we have done is to multiply n by p. It can be shown that this method always gives the expected value for binomial probability.

EXPECTED VALUE FOR BINOMIAL PROBABILITY

For binomial probability, $E(x) = np$. In other words, the expected number of successes is the number of trials times the probability of success in each trial.

EXAMPLE 7 **Female Children**

Suppose a family has 3 children.

(a) Find the probability distribution for the number of girls.

▶**Solution** Assuming girls and boys are equally likely, the probability distribution is binomial with $n = 3$ and $p = 1/2$. Letting x be the number of girls in the formula for binomial probability, we find, for example,

$$P(x = 0) = \binom{3}{0}\left(\frac{1}{2}\right)^0\left(\frac{1}{2}\right)^3 = \frac{1}{8}.$$

The other values are found similarly, and the results are shown in the following table.

x	0	1	2	3
$P(x)$	1/8	3/8	3/8	1/8

We can verify this by noticing that in the sample space S of all 3-child families, there are eight equally likely outcomes: $S = \{ggg, ggb, gbg, gbb, bgg, bgb, bbg, bbb\}$. One of the outcomes has 0 girls, three have 1 girl, three have 2 girls, and one has 3 girls.

(b) Find the expected number of girls in a 3-child family using the distribution from part (a).

▶**Solution** Using the formula for expected value, we have

$$\text{Expected number of girls} = 0\left(\frac{1}{8}\right) + 1\left(\frac{3}{8}\right) + 2\left(\frac{3}{8}\right) + 3\left(\frac{1}{8}\right)$$

$$= \frac{12}{8} = 1.5.$$

On average, a 3-child family will have 1.5 girls. This result agrees with our intuition that, on the average, half the children born will be girls.

(c) Find the expected number of girls in a 3-child family using the formula for expected value for binomial probability.

▶**Solution** Using the formula $E(x) = np$ with $n = 3$ and $p = 1/2$, we have

$$\text{Expected number of girls} = 3\left(\frac{1}{2}\right) = 1.5.$$

This agrees with our answer from part (b), as it must.

8.5 Exercises

For each experiment described below, let x determine a random variable, and use your knowledge of probability to prepare a probability distribution.

1. Four coins are tossed, and the number of heads is noted.
2. Two dice are rolled, and the total number of points is recorded.
3. Three cards are drawn from a deck. The number of aces is counted.
4. Two balls are drawn from a bag in which there are 4 white balls and 2 black balls. The number of black balls is counted.

Draw a histogram for the following, and shade the region that gives the indicated probability.

5. Exercise 1; $P(x \leq 2)$
6. Exercise 2; $P(x \geq 11)$
7. Exercise 3; $P(\text{at least one ace})$
8. Exercise 4; $P(\text{at least one black ball})$

Find the expected value for each random variable.

9.

x	2	3	4	5
$P(x)$	0.1	0.4	0.3	0.2

10.

y	4	6	8	10
$P(y)$	0.4	0.4	0.05	0.15

11.

z	9	12	15	18	21
$P(z)$	0.14	0.22	0.38	0.19	0.07

12.

x	30	32	36	38	44
$P(x)$	0.31	0.29	0.26	0.09	0.05

Find the expected value for the random variable x having the probability function shown in each graph.

13.

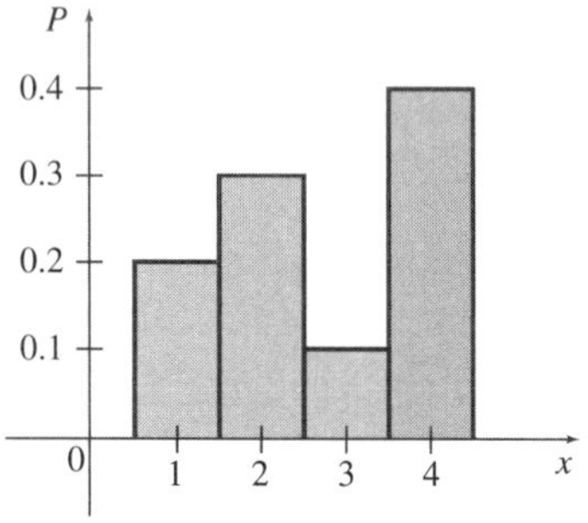

14.

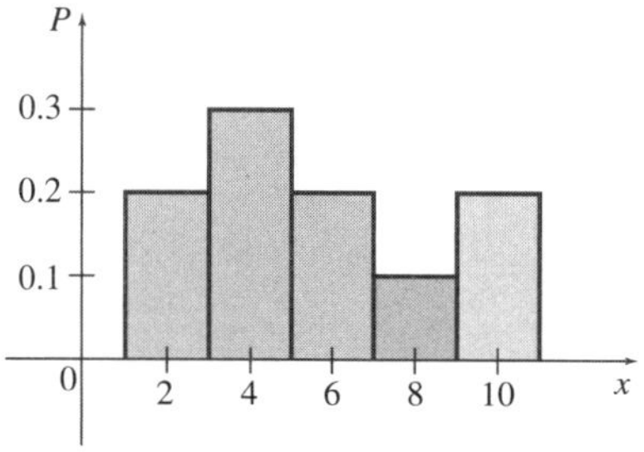

15.

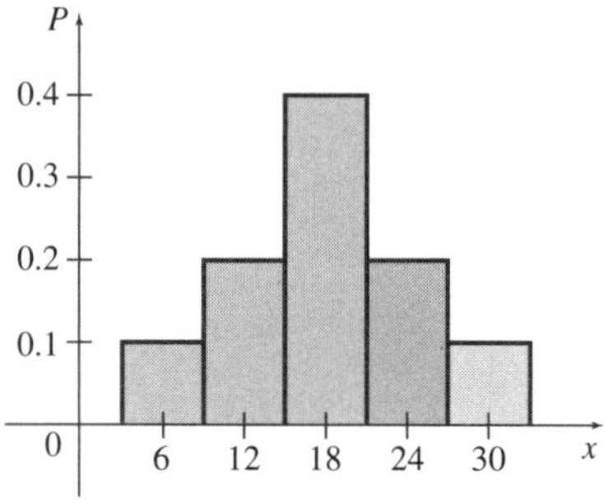

16.

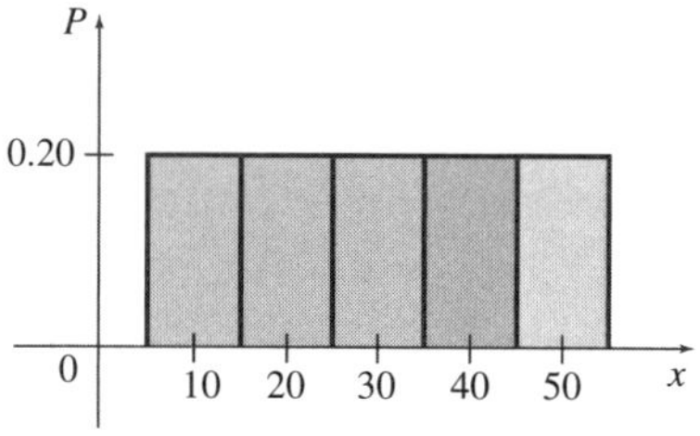

17. For the game in Example 4, find Mary's expected winnings. Is it a fair game?

18. Suppose one day Mary brings a 2-headed coin and uses it to toss for the coffee. Since Mary tosses, Donna calls.

a. Is this still a fair game?

b. What is Donna's expected gain if she calls heads?

c. What is Donna's expected gain if she calls tails?

Solve each exercise. Many of these exercises require the use of combinations.

19. Suppose 3 marbles are drawn from a bag containing 3 yellow and 4 white marbles.

a. Draw a histogram for the number of yellow marbles in the sample.

b. What is the expected number of yellow marbles in the sample?

20. Suppose 5 apples in a barrel of 25 apples are known to be rotten.

a. Draw a histogram for the number of rotten apples in a sample of 2 apples.

b. What is the expected number of rotten apples in a sample of 2 apples?

21. Suppose a die is rolled 4 times.

a. Find the probability distribution for the number of times 1 is rolled.

b. What is the expected number of times 1 is rolled?

22. A delegation of 3 is selected from a city council made up of 5 liberals and 6 conservatives.

a. What is the expected number of liberals in the delegation?

b. What is the expected number of conservatives in the delegation?

23. From a group of 3 women and 5 men, a delegation of 2 is selected. Find the expected number of women in the delegation.

24. In a club with 20 senior and 10 junior members, what is the expected number of junior members on a 4-member committee?

25. If 2 cards are drawn at one time from a deck of 52 cards, what is the expected number of diamonds?

26. Suppose someone offers to pay you $5 if you draw 2 diamonds in the game in Exercise 25. He says that you should pay 50 cents for the chance to play. Is this a fair game?

27. Your friend missed class the day probability distributions were discussed. How would you explain probability distribution to him?

28. Explain what expected value means in your own words.

29. Four slips of paper numbered 2, 3, 4, and 5 are in a hat. You draw a slip, note the result, and then draw a second slip and note the result (without replacing the first).

a. Find the probability distribution for the sum of the two slips.

b. Draw a histogram for the probability distribution in part a.

c. Find the odds that the sum is even.

d. Find the expected value of the sum.

Applications

BUSINESS AND ECONOMICS

30. *Complaints* A local used-car dealer gets complaints about his cars as shown in the table below.

Find the expected number of complaints per day.

Number of Complaints per Day	0	1	2	3	4	5	6
Probability	0.02	0.06	0.16	0.25	0.32	0.13	0.06

31. *Payout on Insurance Policies* An insurance company has written 100 policies for $100,000, 500 policies for $50,000, and 1000 policies for $10,000 for people of age 20. If experience shows that the probability that a person will die at age 20 is 0.0012, how much can the company expect to pay out during the year the policies were written?

32. *Rating Sales Accounts* Levi Strauss and Company* uses expected value to help its salespeople rate their accounts. For each account, a salesperson estimates potential additional volume and the probability of getting it. The product of these figures gives the expected value of the potential, which is added to the existing volume. The totals are then classified as A, B, or C, as follows: $40,000 or below, class C; from $40,000 up to and including $55,000, class B; above $55,000, class A. Complete the table on the next page for one salesperson.

*This example was supplied by James McDonald, Levi Strauss and Company, San Francisco.

Account Number	Existing Volume	Potential Additional Volume	Probability of Getting It	Expected Value of Potential	Existing Volume + Expected Value of Potential	Class
1	$15,000	$10,000	0.25	$2500	$17,500	C
2	$40,000	$0	—	—	$40,000	C
3	$20,000	$10,000	0.20			
4	$50,000	$10,000	0.10			
5	$5000	$50,000	0.50			
6	$0	$100,000	0.60			
7	$30,000	$20,000	0.80			

33. *Pecans* Refer to Exercise 41 in Section 8.4. Suppose that 60% of the pecan "blow outs" are good, and 80% of the "good nuts" are good.

a. If you purchase 50 pecans, what is the expected number of good nuts you will find if you purchase "blow outs"?

b. If you purchase 50 pecans, what is the expected number of bad nuts you will find if you have purchased "good nuts"?

34. *Device Failure* An insurance policy on an electrical device pays a benefit of $4000 if the device fails during the first year. The amount of the benefit decreases by $1000 each successive year until it reaches 0. If the device has not failed by the beginning of any given year, the probability of failure during that year is 0.4. What is the expected benefit under this policy?* (Choose one of the following.)

a. $2234 **b.** $2400 **c.** $2500 **d.** $2667 **e.** $2694

35. *Tour Bus* A tour operator has a bus that can accommodate 20 tourists. The operator knows that tourists may not show up, so he sells 21 tickets. The probability that an individual tourist will not show up is 0.02, independent of all other tourists. Each ticket costs $50, and is non-refundable if a tourist fails to show up. If a tourist shows up and a seat is not available, the tour operator has to pay $100 (ticket cost + $50 penalty) to the tourist. What is the expected revenue of the tour operator?† (Choose one of the following.)

a. $935 **b.** $950 **c.** $967 **d.** $976 **e.** $985

LIFE SCIENCES

36. *Animal Offspring* In a certain animal species, the probability that a healthy adult female will have no offspring in a given year is 0.29, while the probabilities of 1, 2, 3, or 4 offspring are, respectively, 0.23, 0.18, 0.16, and 0.14. Find the expected number of offspring.

37. *Ear Infections* Otitis media, or middle ear infection, is initially treated with an antibiotic. Researchers have compared two antibiotics, amoxicillin and cefaclor, for their cost effectiveness. Amoxicillin is inexpensive, safe, and effective. Cefaclor is also safe. However, it is considerably more expensive and it is generally more effective. Use the tree diagram below (where the costs are estimated as the total cost of medication, office visit, ear check, and hours of lost work) to answer the following.‡

a. Find the expected cost of using each antibiotic to treat a middle ear infection.

b. To minimize the total expected cost, which antibiotic should be chosen?

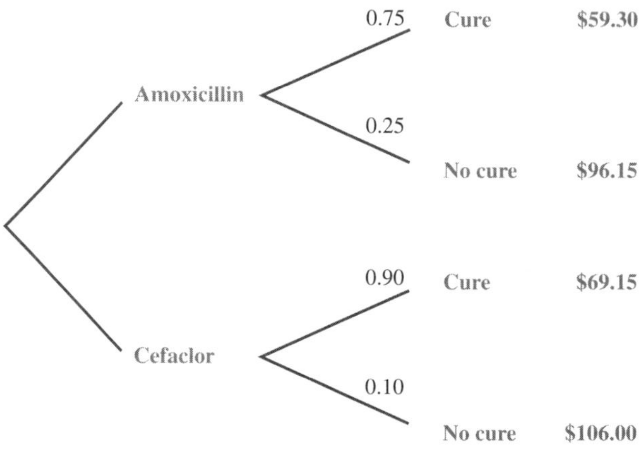

*Problem 48 from the 2005 Sample Exam P of the *Education and Examination Committee of the Society of Actuaries*. Reprinted by permission of the Society of Actuaries.

†Problem 96 from the 2005 Sample Exam P of the *Education and Examination Committee of the Society of Actuaries*. Reprinted by permission of the Society of Actuaries.

‡Weiss, Jeffrey and Shoshana Melman, based on "Cost Effectiveness in the Choice of Antibiotics for the Initial Treatment of Otitis Media in Children: A Decision Analysis Approach," *Journal of Pediatric Infectious Disease*, Vol. 7, No. 1, 1988, pp. 23–26.

38. *Hospitalization Insurance* An insurance policy pays an individual \$100 per day for up to 3 days of hospitalization and \$25 per day for each day of hospitalization thereafter. The number of days of hospitalization, X, is a discrete random variable with probability function

$$P(X = k) = \begin{cases} \dfrac{6 - k}{15} & \text{for } k = 1, 2, 3, 4, 5 \\ 0 & \text{otherwise.} \end{cases}$$

Calculate the expected payment for hospitalization under this policy.* (Choose one of the following.)

a. \$85 **b.** \$163 **c.** \$168 **d.** \$213 **e.** \$255

SOCIAL SCIENCES

39. *Education* Recall from Exercise 65 in Section 8.4 that a *Time* magazine "Numbers" section reported that 15.2% of low-birth-weight babies graduate from high school by age 19.[†] If 250 low-birth-weight babies are followed through high school, how many would you expect to graduate from high school?

40. *Cheating* Recall from Exercise 64 in Section 8.4 that a poll conducted by *U.S. News and World Report* reported that 84% of college students believe they need to cheat to get ahead in the world today.[‡] If 500 college students were surveyed, how many would you expect to say that they need to cheat to get ahead in the world today?

41. *Samuel Alito* When Supreme Court Justice Samuel Alito was on the U.S. Court of Appeals for the 3rd Circuit, he dissented in the successful appeal of a first-degree murder case. The prosecution used its peremptory challenges to eliminate all African Americans from the jury, as it had in three other first-degree murder trials in the same county that year. According to a majority of the judges, "An amateur with a pocket calculator can calculate the number of blacks that would have served had the State used its strikes in a racially proportionate manner. In the four capital cases there was a total of 82 potential jurors on the venires who were not removed for cause, of whom eight, or 9.76%, were black. If the prosecution had used its peremptory challenges in a manner proportional to the percentage of blacks in the overall venire, then only 3 of the 34 jurors peremptorily struck (8.82%) would have been black and 5 of the 48 actual jurors (10.42%) would have been black. Instead, none of the 48 jurors were black. Admittedly, there was no statistical analysis of these figures presented by either side in the post-conviction proceeding. But is it really necessary to have a sophisticated analysis by a statistician to conclude that there is little chance of randomly selecting four consecutive all white juries?"[§]

a. Using binomial probability, calculate the probability that no African Americans would be selected out of 48 jurors if the percentage African American is 9.76%.

b. Binomial probability is not entirely accurate in this case, because the jurors were selected without replacement, so the selections were not independent. Recalculate the probability in part a using combinations.

c. In his dissent, Judge Alito wrote, "Statistics can be very revealing—and also terribly misleading in the hands of 'an amateur with a pocket calculator.' . . . Although only about 10% of the population is left-handed, left-handers have won five of the last six presidential elections. Our 'amateur with a calculator' would conclude that 'there is little chance of randomly selecting' left-handers in five out of six presidential elections. But does it follow that the voters cast their ballots based on whether a candidate was right- or left-handed?" Given the figures quoted by Judge Alito, what is the probability that at least 5 out of the last 6 presidents elected would be left-handed?

d. The majority of the judges, in disagreeing with Judge Alito, said, "The dissent has overlooked the obvious fact that there is no provision in the Constitution that protects persons from discrimination based on whether they are right-handed or left handed." Furthermore, according to *Chance News,* only 2 of the last 6 men elected president were left-handed.[||] What is the probability that at least 2 out of the last 6 presidents elected would be left-handed?

PHYSICAL SCIENCES

42. *Seeding Storms* One of the few methods that can be used in an attempt to cut the severity of a hurricane is to *seed* the storm. In this process, silver iodide crystals are dropped into the storm. Unfortunately, silver iodide crystals sometimes cause the storm to *increase* its speed. Wind speeds

*Problem 36 from May 2003 Course 1 Examination of the *Education and Examination Committee of the Society of Actuaries*. Reprinted by permission of the Society of Actuaries.

[†]"Numbers," *Time*, July 17, 2000, p. 21.

[‡]Kleiner, Carolyn and Mary Lord, "The Cheating Game," *U.S. News and World Report*, Nov. 22, 1999, pp. 55–66.

[§]http://caselaw.lp.findlaw.com/scripts/getcase.pl?court=3rd&navby=case&no=989009v3&exact=1.

[||]*Chance News 9*, Nov. 1–27, 2005.

may also increase or decrease even with no seeding. Use the tree diagram below to answer the following.*

a. Find the expected amount of damage under each option, "seed" and "do not seed."

b. To minimize total expected damage, what option should be chosen?

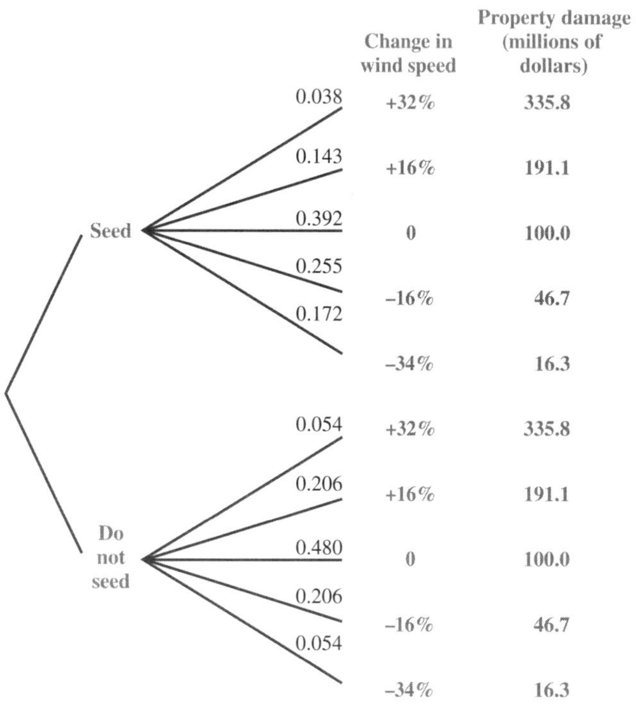

GENERAL INTEREST

43. *Cats* Kimberly Workman has four cats: Riley, Abby, Beastie, and Sylvester. Each cat has a 30% probability of climbing into the chair in which Kimberly is sitting, independent of how many cats are already in the chair with Kimberly.

a. Find the probability distribution for the number of cats in the chair with Kimberly.

b. Find the expected number of cats in the chair with Kimberly using the probability distribution in part a.

c. Find the expected number of cats in the chair with Kimberly using the formula for expected value of the binomial distribution.

44. *Postal Service* Mr. Statistics (a feature in *Fortune* magazine) investigated the claim of the U.S. Postal Service that 83% of first class mail in New York City arrives by the next day.† (The figure is 87% nationwide.) He mailed a letter to himself on 10 consecutive days; only 4 were delivered by the next day.

a. Find the probability distribution for the number of letters delivered by the next day if the overall probability of next-day delivery is 83%.

b. Using your answer to part a, find the probability that 4 or fewer out of 10 letters would be delivered by the next day.

c. Based on your answer to part b, do you think it is likely that the 83% figure is accurate? Explain.

d. Find the number of letters out of 10 that you would expect to be delivered by the next day if the 83% figure is accurate.

45. *Raffle* A raffle offers a first prize of $400 and 3 second prizes of $80 each. One ticket costs $2, and 500 tickets are sold. Find the expected payback for a person who buys 1 ticket. Is this a fair game?

46. *Raffle* A raffle offers a first prize of $1000, 2 second prizes of $300 each, and 20 third prizes of $10 each. If 10,000 tickets are sold at 50¢ each, find the expected payback for a person buying 1 ticket. Is this a fair game?

Find the expected payback for the games of chance described in Exercises 47–52.

47. *Lottery* A state lottery requires you to choose 4 cards from an ordinary deck: 1 heart, 1 club, 1 diamond, and 1 spade in that order from the 13 cards in each suit. If all four choices are selected by the lottery, you win $5000. It costs $1 to play.

48. *Lottery* If exactly 3 of the 4 choices in Exercise 47 are selected, the player wins $200. (Ignore the possibility that all 4 choices are selected. It still costs $1 to play.)

49. *Roulette* In one form of roulette, you bet $1 on "even." If 1 of the 18 even numbers comes up, you get your dollar

*The probabilities and amounts of property damage in the tree diagram for Exercise 42 are from Howard, R. A., J. E. Matheson, and D. W. North, "The Decision to Seed Hurricanes," *SCIENCE*, Vol. 176, No. 16, June 1972, pp. 1191–1202. Copyright © 1972 by The American Association for the Advancement of Science. Reprinted with permission from AAAS.

†Seligman, Daniel, "Ask Mr. Statistics," *Fortune*, July 24, 1995, pp. 170–171.

back, plus another one. If 1 of the 20 noneven (18 odd, 0, and 00) numbers comes up, you lose your dollar.

50. *Roulette* In another form of roulette, there are only 19 noneven numbers (no 00).

51. *Numbers Numbers* is a game in which you bet \$1 on any three-digit number from 000 to 999. If your number comes up, you get \$500.

52. *Keno* In one form of the game *Keno*, the house has a pot containing 80 balls, each marked with a different number from 1 to 80. You buy a ticket for \$1 and mark one of the 80 numbers on it. The house then selects 20 numbers at random. If your number is among the 20, you get \$3.20 (for a net winning of \$2.20).

53. *Contests* A magazine distributor offers a first prize of \$100,000, two second prizes of \$40,000 each, and two third prizes of \$10,000 each. A total of 2,000,000 entries are received in the contest. Find the expected payback if you submit one entry to the contest. If it would cost you 50¢ in time, paper, and stamps to enter, would it be worth it?

54. *Contests* A contest at a fast-food restaurant offered the following cash prizes and probabilities of winning on one visit. Suppose you spend \$1 to buy a bus pass that lets you go to 25 different restaurants in the chain and pick up entry forms. Find your expected value.

Prize	Probability
\$100,000	1/176,402,500
\$25,000	1/39,200,556
\$5000	1/17,640,250
\$1000	1/1,568,022
\$100	1/282,244
\$5	1/7056
\$1	1/588

55. *The Hog Game* In the hog game, each player states the number of dice that he or she would like to roll. The player then rolls that many dice. If a 1 comes up on any die, the player's score is 0. Otherwise, the player's score is the sum of the numbers rolled.*

a. Find the expected value of the player's score when the player rolls one die.

b. Find the expected value of the player's score when the player rolls two dice.

c. Verify that the expected nonzero score of a single die is 4, so that if a player rolls n dice that do not result in a score of 0, the expected score is $4n$.

d. Verify that if a player rolls n dice, there are 5^n possible ways to get a nonzero score, and 6^n possible ways to roll the dice. Explain why the expected value, E, of the player's score when the player rolls n dice is then

$$E = \frac{5^n(4n)}{6^n}.$$

56. *Football* After a team scores a touchdown, it can either attempt to kick an extra point or attempt a two-point conversion. During the 1999–2000 NFL season, two-point conversions were successful 37% of the time and the extra-point kicks were successful 94% of the time.†

a. Calculate the expected value of each strategy.

b. Which strategy, over the long run, will maximize the number of points scored?

c. Using this information, should a team always only use one strategy? Explain.

57. *Baseball* The 2005 National League batting champion was Derrek Lee, with an average of 0.335.‡ This can be interpreted as a probability of 0.335 of getting a hit whenever he bats. Assume that each time at bat is an independent event. Suppose he goes to bat four times in a game.

a. Find the probability distribution for the number of hits.

b. What is the expected number of hits that Derrek Lee gets in a game?

*Bohan, James and John Shultz, "Revisiting and Extending the Hog Game," *Mathematics Teacher*, Vol. 89, No. 9, Dec. 1996, pp. 728–733.

†Leonhardt, David, "In Football, 6 + 2 Often Equals 6," *The New York Times*, Sunday, Jan. 16, 2000, pp. 4–2.

‡http://www.infoplease.com/ipsa/A0932329.html.

Chapter 8 Review

Chapter Summary

In this chapter we continued our study of probability by introducing some elementary principles of counting. Our primary tool is the multiplication principle:

> If n choices must be made, with m_1 ways to make choice 1, and for each of these ways, m_2 ways to make choice 2, and so on, with m_n ways to make choice n, then there are $m_1 \cdot m_2 \cdot \ldots \cdot m_n$ ways to make the entire sequence of choices.

We learned two counting ideas to efficiently count the number of ways we can select a number of objects without replacement:

- permutations (when order matters), and
- combinations (when order doesn't matter).

We also considered distinguishable permutations, in which some of the objects are indistinguishable. All of these concepts were then used to calculate the numerator and denominator of various probabilities. We next explored binomial probability, in which the following conditions were satisfied:

- the same experiment is repeated a fixed number of times (n),
- there are only two possible outcomes (success and failure), and
- the trials are independent, so the probability of success remains constant (p).

We showed how to quickly calculate an entire set of combinations for binomial probability using Pascal's triangle. Finally, we introduced the following terms regarding probability distributions:

- random variable (a function assigning a real number to each outcome of an experiment),
- probability distribution (the possible values of a random variable, along with the corresponding probabilities),
- histogram (a bar graph displaying a probability distribution), and
- expected value (the average value of a random variable that we would expect in the long run).

In the next chapter, we will see how probability forms the basis of the field known as statistics.

COUNTING SUMMARY

Factorial Notation $n! = n(n-1)(n-2)\cdots(3)(2)(1)$

$0! = 1$

Permutations $P(n, r) = \dfrac{n!}{(n-r)!}$

Distinguishable Permutations If there are n_1 objects of type 1, n_2 of type 2, and so on for r different types, then the number of distinguishable permutations is

$$\frac{n!}{n_1!\, n_2! \cdots n_r!}.$$

Combinations $\dbinom{n}{r} = \dfrac{n!}{(n-r)!\, r!}$

PROBABILITY SUMMARY

Binomial Probability $P(x) = \binom{n}{x} p^x (1 - p)^{n-x}$

Expected Value $E(x) = x_1p_1 + x_2p_2 + x_3p_3 + \cdots + x_np_n$

For binomial probability, $E(x) = np$.

Key Terms

8.1 **multiplication principle**
factorial notation
permutations
distinguishable permutations

8.2 **combinations**

8.4 **Bernoulli trials**
binomial experiment
binomial probability
Pascal's triangle

8.5 **random variable**
probability distribution
probability function
histogram
expected value
fair game

Concept Check

Determine whether each of the following statements is true or false, and explain why.

1. Permutations provide a way of counting possibilities when order matters.

2. Combinations provide a way of counting possibilities when order doesn't matter.

3. The number of distinguishable permutations of n objects, when r are indistinguishable and the remaining $n - r$ are also indistinguishable, is the same as the number of combinations of r objects chosen from n.

4. Calculating the numerator or the denominator of a probability can involve either permutations or combinations.

5. The probability of at least 2 occurrences of an event is equal to the probability of 1 or fewer occurrences.

6. The probability of at least two people in a group having the same birthday is found by subtracting the probability of the complement of the event from 1.

7. The trials in binomial probability must be independent.

8. Binomial probability can be used when each trial has three possible outcomes.

9. A random variable can have negative values.

10. The expected value of a random variable must equal one of the values that the random variable can have.

11. The probabilities in a probability distribution must add up to 1.

12. A fair game can have an expected value that is greater than 0.

Chapter 8 Review Exercises

1. In how many ways can 6 shuttle vans line up at the airport?

2. How many variations in first-, second-, and third-place finishes are possible in a 100-yd dash with 6 runners?

3. In how many ways can a sample of 3 oranges be taken from a bag of a dozen oranges?

4. If 2 of the 12 oranges in Exercise 3 are rotten, in how many ways can the sample of 3 include

a. 1 rotten orange?

b. 2 rotten oranges?

c. no rotten oranges?

d. at most 2 rotten oranges?

5. In how many ways can 2 pictures, selected from a group of 5 different pictures, be arranged in a row on a wall?

6. In how many ways can the 5 pictures in Exercise 5 be arranged in a row if a certain one must be first?

7. In how many ways can the 5 pictures in Exercise 5 be arranged if 2 are landscapes and 3 are puppies and if

a. like types must be kept together?

b. landscapes and puppies are alternated?

8. In a Chinese restaurant the menu lists 8 items in column A and 6 items in column B.

a. To order a dinner, the diner is told to select 3 items from column A and 2 from column B. How many dinners are possible?

b. How many dinners are possible if the diner can select up to 3 from column A and up to 2 from column B? Assume at least one item must be included from either A or B.

9. A representative is to be selected from each of 3 departments in a small college. There are 7 people in the first department, 5 in the second department, and 4 in the third department.

a. How many different groups of 3 representatives are possible?

b. How many groups are possible if any number (at least 1) up to 3 representatives can form a group? (Each department is still restricted to at most one representative.)

10. Explain under what circumstances a permutation should be used in a probability problem, and under what circumstances a combination should be used.

11. Discuss under what circumstances the binomial probability formula should be used in a probability problem.

A basket contains 4 *black,* 2 *blue, and* 7 *green balls. A sample of* 3 *balls is drawn. Find the probabilities that the sample contains the following.*

12. All black balls

13. All blue balls

14. 2 black balls and 1 green ball

15. Exactly 2 black balls

16. Exactly 1 blue ball

17. 2 green balls and 1 blue ball

Suppose a family plans 6 *children, and the probability that a particular child is a girl is* 1/2. *Find the probabilities that the* 6*-child family has the following children.*

18. Exactly 3 girls

19. All girls

20. At least 4 girls

21. No more than 2 boys

Suppose 2 cards are drawn without replacement from an ordinary deck of 52. Find the probabilities of the following results.

22. Both cards are red.

23. Both cards are spades.

24. At least 1 card is a spade.

25. One is a face card and the other is not.

26. At least one is a face card.

27. At most one is a queen.

In Exercises 28 and 29, **(a)** *give a probability distribution,* **(b)** *sketch its histogram, and* **(c)** *find the expected value.*

28. A coin is tossed 3 times and the number of heads is recorded.

29. A pair of dice is rolled and the sum of the results for each roll is recorded.

In Exercises 30 and 31, give the probability that corresponds to the shaded region of each histogram.

30.

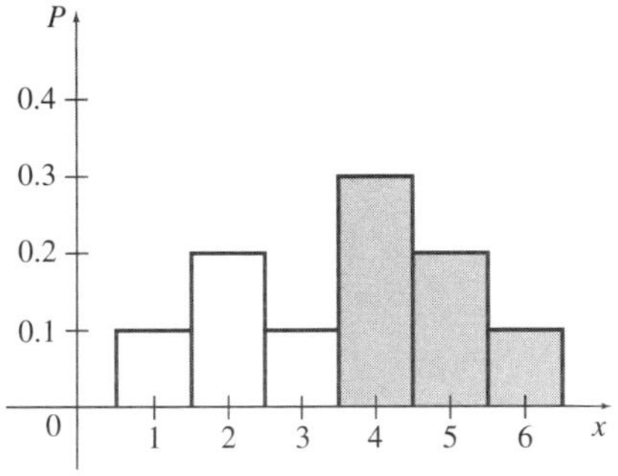

31.

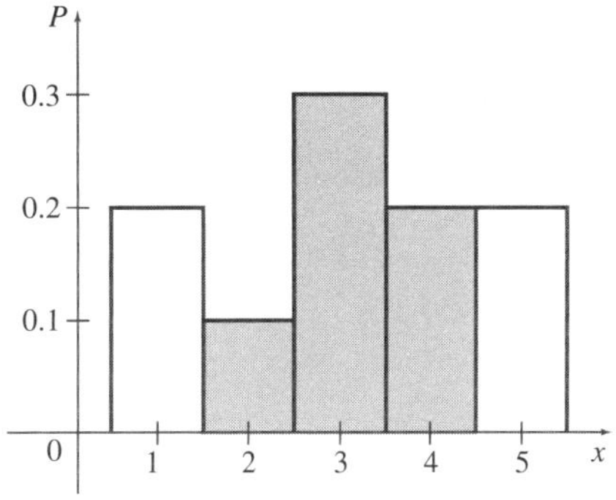

32. You pay \$6 to play in a game where you will roll a die, with payoffs as follows: \$8 for a 6, \$7 for a 5, and \$4 for any other results. What are your expected winnings? Is the game fair?

33. Find the expected number of girls in a family of 5 children.

34. Three cards are drawn from a standard deck of 52 cards.

a. What is the expected number of aces? **b.** What is the expected number of clubs?

35. Suppose someone offers to pay you \$100 if you draw 3 cards from a standard deck of 52 cards and all the cards are clubs. What should you pay for the chance to win if it is a fair game?

36. Six students will decide which of them are on a committee by flipping a coin. Each student flips the coin, and is on the committee if he or she gets a head. What is the probability that someone is on the committee, but not all 6 students?

37. In this exercise we study the connection between sets (from Chapter 7) and combinations (from Chapter 8).

a. Given a set with n elements, what is the number of subsets of size 0? of size 1? of size 2? of size n?

b. Using your answer from part a, give an expression for the total number of subsets of a set with n elements.

c. Using your answer from part b and a result from Chapter 7, explain why the following equation must be true:

$$\binom{n}{0} + \binom{n}{1} + \binom{n}{2} + \cdots + \binom{n}{n} = 2^n.$$

d. Verify the equation in part c for $n = 4$ and $n = 5$.

e. Explain what the equation in part c tells you about Pascal's triangle.

*In the following exercise, find the digit (0 through 9) that belongs in each box. This exercise is from the 1990 University Entrance Center Examination, given in Japan to all applicants for public universities.**

38. The numbers 1 through 9 are written individually on nine cards. Choose three cards from the nine, letting x, y, and z denote the numbers of the cards arranged in increasing order.

a. There are □□ such x, y, and z combinations.

b. The probability of having x, y, and z all even is $\frac{\square}{\square\square}$.

c. The probability of having x, y, and z be consecutive numbers is $\frac{\square}{\square\square}$.

d. The probability of having $x = 4$ is $\frac{\square}{\square\square}$.

e. Possible values of x range from □ to □. If k is an integer such that $\square \le k \le \square$, the probability that $x = k$ is $\frac{(\square - k)(\square - k)}{\square\square\square}$. The expected value of x is $\frac{\square}{\square}$.

Applications

BUSINESS AND ECONOMICS

Identity Theft *According to a survey by Javelin Strategy and Research, 1 out of 6 adults in Arizona were victims of identity theft.*[†] *Suppose that 12 adults are randomly selected from Arizona. Find the probabilities of each of the following results.*

39. None of the adults were victims of identity theft.

40. All of the adults were victims of identity theft.

41. Exactly 10 of the adults were victims of identity theft.

42. Exactly 2 of the adults were victims of identity theft.

43. At least 2 of the adults were victims of identity theft.

44. At most 3 of the adults were victims of identity theft.

45. Find the expected number of victims of identity theft in a sample of 12 adults in Arizona.

46. *Land Development* A developer can buy a piece of property that will produce a profit of \$26,000 with probability 0.7, or a loss of \$9000 with probability 0.3. What is the expected profit?

47. *Insurance Claims* An insurance company determines that N, the number of claims received in a week, is a random variable with $P(N = n) = 1/2^{n+1}$, where $n \ge 0$. The company also determines that the number of claims received in a given week is independent of the number of claims received in any other week. Determine the probability that exactly seven claims will be received during a given two-week period.[‡] (Choose one of the following.)

a. 1/256 **b.** 1/128 **c.** 7/512 **d.** 1/64 **e.** 1/32

48. *Injury Claims* The number of injury claims per month is modeled by a random variable N with

$$P(N = n) = \frac{1}{(n + 1)(n + 2)}, \quad \text{where } n \ge 0.$$

*"Japanese University Entrance Examination Problems in Mathematics," by Ling-Erl Eileen T. Wu, ed., Mathematical Association of America, 1993, p. 5. Copyright © 1993 from Wu's *Japanese University Entrance Examination Problems in Mathematics*, published by The Mathematical Association of America.

[†] *The New York Times*, May 30, 2006, p. A1.

[‡] Problem 16 from the 2005 Sample Exam P of the *Education and Examination Committee of the Society of Actuaries*. Reprinted by permission of the Society of Actuaries.

Determine the probability of at least one claim during a particular month, given that there have been at most four claims during that month.* (Choose one of the following.)

a. $1/3$ **b.** $2/5$ **c.** $1/2$ **d.** $3/5$ **e.** $5/6$

49. *Product Success* A company is considering the introduction of a new product that is believed to have probability 0.5 of being successful and probability 0.5 of being unsuccessful. Successful products pass quality control 80% of the time. Unsuccessful products pass quality control 25% of the time. If the product is successful, the net profit to the company will be $40 million; if unsuccessful, the net loss will be $15 million. Determine the expected net profit if the product passes quality control.† (Choose one of the following.)

a. $23 million **b.** $24 million **c.** $25 million
d. $26 million **e.** $27 million

50. *Sampling Fruit* A merchant buys boxes of fruit from a grower and sells them. Each box of fruit is either Good or Bad. A Good box contains 80% excellent fruit and will earn $200 profit on the retail market. A Bad box contains 30% excellent fruit and will produce a loss of $1000. The a priori probability of receiving a Good box of fruit is 0.9. Before the merchant decides to put the box on the market, he can sample one piece of fruit to test whether it is excellent. Based on that sample, he has the option of rejecting the box without paying for it. Determine the expected value of the right to sample.‡ (Choose one of the following. *Hint:* The a priori probability is the probability before sampling a piece of fruit. If the merchant samples the fruit, what are the probabilities of accepting a Good box, accepting a Bad box, and not accepting the box? What are these probabilities if he does not sample the fruit?)

a. 0 **b.** $16 **c.** $34 **d.** $72 **e.** $80

51. *Overbooking Flights* The March 1982 issue of *Mathematics Teacher* included "Overbooking Airline Flights," an article by Joe Dan Austin. In this article, Austin developed a model for the expected income for an airline flight. With appropriate assumptions, the probability that exactly x of n people with reservations show up at the airport to buy a ticket is given by the binomial probability formula. Assume the following: 6 reservations have been accepted for 3 seats, $p = 0.6$ is the probability that a person with a reservation will show up, a ticket costs $400, and the airline must pay $400 to anyone with a reservation who does not get a ticket. Complete the following table.

Number Who Show Up (x)	0	1	2	3	4	5	6
Airline's Income							
$P(x)$							

a. Use the table to find $E(I)$, the expected airline income from the 3 seats.

b. Find $E(I)$ for $n = 3$, $n = 4$, and $n = 5$. Compare these answers with $E(I)$ for $n = 6$. For these values of n, how many reservations should the airline book for the 3 seats in order to maximize the expected revenue?

LIFE SCIENCES

52. *Pharmacology* In placebo-controlled trials of Prozac®, a drug that is prescribed to fight depression, 23% of the patients who were taking the drug experienced nausea, whereas 10% of the patients who were taking the placebo experienced nausea.§

a. If 50 patients who are taking Prozac® are selected, what is the probability that 10 or more will experience nausea?

b. Of the 50 patients in part a, what is the expected number of patients who will experience nausea?

c. If a second group of 50 patients receives a placebo, what is the probability that 10 or fewer will experience nausea?

d. If a patient from a study of 1000 people, who are equally divided into two groups (those taking a placebo and those taking Prozac®), is experiencing nausea, what is the probability that he/she is taking Prozac®?

e. Since 0.23 is more than twice as large as 0.10, do you think that people who take Prozac® are more likely to experience nausea than those who take a placebo? Explain.

*Problem 24 from the 2005 Sample Exam P of the *Education and Examination Committee of the Society of Actuaries*. Reprinted by permission of the Society of Actuaries.

†*Course 130 Examination*, Operations Research, Nov. 1989. Reprinted by permission of the Society of Actuaries.

‡Ibid.

§Advertisement in *The New England Journal of Medicine*, Vol. 338, No. 9, Feb. 26, 1998, for Prozac®, developed and marketed by Eli Lilly and Company.

SOCIAL SCIENCES

53. *Education* In Exercise 38 of Section 8.3, we saw that a school in Bangkok requires that students take an entrance examination. After the examination, 5 students are randomly drawn from each group of 40 for automatic acceptance into the school regardless of their performance on the examination. The drawing consists of placing 35 red and 5 green pieces of paper into a box. If the lottery is changed so that each student picks a piece of paper from the box and then returns the piece of paper to the box, find the probability that exactly 5 of the 40 students will choose a green piece of paper.*

GENERAL INTEREST

In Exercises 54–57, **(a)** *give a probability distribution,* **(b)** *sketch its histogram, and* **(c)** *find the expected value.*

54. *Candy* According to officials of Mars, the makers of M&M Plain Chocolate Candies, 20% of the candies in each bag are orange.† Four candies are selected from a bag and the number of orange candies is recorded.

55. *Women Athletes* In 1992, the Big 10 collegiate sports conference moved to have women compose at least 40% of its athletes within 5 years.‡ Suppose they exactly achieve the 40% figure, and that 5 athletes are picked at random from Big 10 universities. The number of women is recorded.

56. *Race* In the mathematics honors society at a college, 2 of the 8 members are African American. Three members are selected at random to be interviewed by the student newspaper, and the number of African Americans is noted.

57. *Homework* In a small class of 10 students, 3 did not do their homework. The professor selects half of the class to present solutions to homework problems on the board, and records how many of those selected did not do their homework.

58. *Lottery* A lottery has a first prize of \$5000, two second prizes of \$1000 each, and two \$100 third prizes. A total of 10,000 tickets is sold, at \$1 each. Find the expected payback of a person buying 1 ticket.

59. *Contests* At one time, game boards for a United Airlines contest could be obtained by sending a self-addressed, stamped envelope to a certain address. The prize was a ticket for any city to which United flies. Assume that the value of the ticket was \$2000 (we might as well go first-class), and that the probability that a particular game board would win was 1/8000. If the stamps to enter the contest cost 41¢ and envelopes cost 4¢ each, find the expected payback for a person ordering 1 game board. (Notice that 2 stamps and envelopes were required to enter.)

60. *Lottery* On June 23, 2003, an interesting thing happened in the Pennsylvania Lottery's Big 4, in which a four-digit number from 0000 to 9999 is chosen twice a day.§ On this day, the number 3199 was chosen both times.

a. What is the probability of the same number being chosen twice in one day?

b. What is the probability of the number 3199 being chosen twice in one day?

61. *Lottery* In the Pennsylvania Lottery's Daily Number game, a three-digit number between 000 and 999 is chosen each day.§ The favorite number among players is 000, which on July 28, 2003, was the winning number for the tenth time since 1977. Find the number of times that 000 would be expected to win in 26 years of play. (Assume that the game is played 365 days a year, ignoring leap years and the fact that before 1990, the game was not played on Christmas or New Year's Day. Also ignore the fact that since February 2003, the game has been played twice a day.)

62. *Lottery* New York has a lottery game called Quick Draw, in which the player can pick anywhere from 1 up to 10 numbers from 1 to 80. The computer then picks 20 numbers, and how much you win is based on how many of your numbers match the computer's. For simplicity, we will only consider the two cases in which you pick 4 or 5 numbers. The payoffs for each dollar that you bet are given in the table below.

	How Many Numbers Match the Computer's Numbers					
	0	**1**	**2**	**3**	**4**	**5**
You Pick 4	0	0	1	5	55	
You Pick 5	0	0	0	2	20	300

*"Media Clips," *Mathematics Teacher*, Vol. 92, No. 8, 1999. Copyright 1999. Used with permission from the National Council of Teachers of Mathematics. All rights reserved.
†http://us.mms.com/us/about/products/milkchocolate.
‡*Chicago Tribune*, Apr. 28, 1993, p. 19.
§http://www.palottery.com.

a. According to the Quick Draw playing card, the "Overall Chances of Winning" when you pick 4 are "1:3.86," while the chances when you pick 5 are "1:10.34." Verify these figures.

b. Find the expected value when you pick 4 and when you pick 5, betting $1 each time.

c. Based on your results from parts a and b, are you better off picking 4 numbers or picking 5? Explain your reasoning.

63. *Murphy's Law* Robert Matthews wrote an article about Murphy's Law, which says that if something can go wrong, it will.* He considers Murphy's Law of Odd Socks, which says that if an odd sock can be created it will be, in a drawer of 10 loose pairs of socks.

a. Find the probability of getting a matching pair when the following numbers of socks are selected at random from the drawer.

i. 5 socks **ii.** 6 socks

b. Matthews says that it is necessary to rummage through 30% of the socks to get a matching pair. Using your answers from part a, explain precisely what he means by that.

c. Matthews claims that if you lose 6 socks at random from the drawer, then it is 100 times more likely that you will be left with the worst possible outcome—6 odd socks—than with a drawer free of odd socks. Verify this calculation by finding the probability that you will be left with 6 odd socks and the probability that you will have a drawer free of odd socks.

64. *Baseball* The number of runs scored in 16,456 half-innings of the 1986 National League Baseball season was analyzed by Hal Stern. Use the table in the next column to answer the following questions.[†]

a. What is the probability that a given team scored 5 or more runs in any given half-inning during the 1986 season?

b. What is the probability that a given team scored fewer than 2 runs in any given half-inning of the 1986 season?

c. What is the expected number of runs that a team scored during any given half-inning of the 1986 season? Interpret this number.

Runs	Frequency	Probability
0	12,087	0.7345
1	2451	0.1489
2	1075	0.0653
3	504	0.0306
4	225	0.0137
5	66	0.0040
6	29	0.0018
7	12	0.0007
8	5	0.0003
9	2	0.0001

65. *St. Petersburg Paradox* Suppose you play a gambling game in which you flip a coin until you get a head. If you get a head on the first toss, you win $2. You win $4 if the first head occurs on the second toss, $8 if it occurs on the the third toss, and so forth, with a prize of $\$2^n$ if the first head occurs on the nth toss. Show that the expected value of this game is infinite. Explain why this is a paradox.[‡]

66. *Pit* The card game of Pit was introduced by Parker Brothers in 1904 and is still popular. In the version owned by one of the authors of this book, there are 10 suits of 9 identical cards, plus the Bull and the Bear card, for a total of 92 cards. (Newer versions of the game have only 8 suits of cards.) For this problem, assume that all 92 cards are used, and you are dealt 9 cards.

a. What is the probability that you have one card from each of 9 different suits, but neither the Bull nor the Bear?

b. What is the probability that you have a pairs of cards from one suit and one card from each of 7 other suits, but neither the Bull nor the Bear?

c. What is the probability that you have two pair of cards from two different suits and one card from each of 5 other suits, but neither the Bull nor the Bear?

*Matthews, Robert, "Why Does Toast Always Land Butter-Side Down?" *Sunday Telegraph*, March 17, 1996, p. 4.

[†]J. Laurie Snell's report of Hal Stern's analysis in *Chance News 7.05*, Apr. 27–May 26, 1998.

[‡]Many articles have been written in an attempt to explain this paradox, first posed by the Swiss mathematician Daniel Bernoulli when he lived in St. Petersburg. For example, see Székely, Gábor and Donald St. P. Richards, "The St. Petersburg Paradox and the Crash of High-Tech Stocks in 2000," *The American Statistician*, Vol. 58, No. 3, Aug. 2004, pp. 225–231.

Optimal Inventory for a Service Truck

For many different items it is difficult or impossible to take the item to a central repair facility when service is required. Washing machines, large television sets, office copiers, and computers are only a few examples of such items. Service for items of this type is commonly performed by sending a repair person to the item, with the person driving to the location in a truck containing various parts that might be required in repairing the item. Ideally, the truck should contain all the parts that might be required. However, most parts would be needed only infrequently, so that inventory costs for the parts would be high.

An optimum policy for deciding on which parts to stock on a truck would require that the probability of not being able to repair an item without a trip back to the warehouse for needed parts be as low as possible, consistent with minimum inventory costs. An analysis similar to the one below was developed at the Xerox Corporation.*

To set up a mathematical model for deciding on the optimum truck-stocking policy, let us assume that a broken machine might require one of 5 different parts (we could assume any number of different parts—we use 5 to simplify the notation). Suppose also that the probability that a particular machine requires part 1 is p_1; that it requires part 2 is p_2; and so on. Assume also that failures of different part types are independent, and that at most one part of each type is used on a given job.

Suppose that, on the average, a repair person makes N service calls per time period. If the repair person is unable to make a repair because at least one of the parts is unavailable, there is a penalty cost, L, corresponding to wasted time for the repair person, an extra trip to the parts depot, customer unhappiness, and so on. For each of the parts carried on the truck, an average inventory cost is incurred. Let H_i be the average inventory cost for part i, where $1 \leq i \leq 5$.

Let M_1 represent a policy of carrying only part 1 on the repair truck, M_{24} represent a policy of carrying only parts 2 and 4, with M_{12345} and M_0 representing policies of carrying all parts and no parts, respectively.

For policy M_{35}, carrying parts 3 and 5 only, the expected cost per time period per repair person, written $C(M_{35})$, is

$$C(M_{35}) = (H_3 + H_5) + NL[1 - (1 - p_1)(1 - p_2)(1 - p_4)].$$

(The expression in brackets represents the probability of needing at least one of the parts not carried, 1, 2, or 4 here.) As further examples,

$$C(M_{125}) = (H_1 + H_2 + H_5) + NL[1 - (1 - p_3)(1 - p_4)],$$

while

$$\begin{aligned} C(M_{12345}) &= (H_1 + H_2 + H_3 + H_4 + H_5) + NL[1 - 1] \\ &= H_1 + H_2 + H_3 + H_4 + H_5, \end{aligned}$$

and

$$C(M_0) = NL[1 - (1 - p_1)(1 - p_2)(1 - p_3)(1 - p_4)(1 - p_5)].$$

To find the best policy, evaluate $C(M_0)$, $C(M_1)$, ..., $C(M_{12345})$, and choose the smallest result. (A general solution method is in the *Management Science* paper.)

EXAMPLE

Suppose that for a particular item, only 3 possible parts might need to be replaced. By studying past records of failures of the item, and finding necessary inventory costs, suppose that the following values have been found.

p_1	p_2	p_3
0.09	0.24	0.17

H_1	H_2	H_3
\$15	\$40	\$9

*Smith, Stephen, John Chambers, and Eli Shlifer, "Optimal Inventories Based on Job Completion Rate for Repairs Requiring Multiple Items," *Management Science*, Vol. 26, No. 8, Aug. 1980. © 1980 by The Institute of Management Sciences.

Suppose $N = 3$ and L is \$54. Then, as an example,

$$\begin{aligned} C(M_1) &= H_1 + NL[1 - (1 - p_2)(1 - p_3)] \\ &= 15 + 3(54)[1 - (1 - 0.24)(1 - 0.17)] \\ &= 15 + 3(54)[1 - (0.76)(0.83)] \\ &\approx 15 + 59.81 = 74.81. \end{aligned}$$

Thus, if policy M_1 is followed (carrying only part 1 on the truck), the expected cost per repair person per time period is \$74.81. Also,

$$\begin{aligned} C(M_{23}) &= H_2 + H_3 + NL[1 - (1 - p_1)] \\ &= 40 + 9 + 3(54)(0.09) = 63.58, \end{aligned}$$

so that M_{23} is a better policy than M_1. By finding the expected values for all other possible policies (see the exercises), the optimum policy may be chosen.

EXERCISES

1. Refer to the example and find the following.

   a. $C(M_0)$ **b.** $C(M_2)$ **c.** $C(M_3)$ **d.** $C(M_{12})$

   e. $C(M_{13})$ **f.** $C(M_{123})$

2. Which policy leads to the lowest expected cost?

3. In the example, $p_1 + p_2 + p_3 = 0.09 + 0.24 + 0.17 = 0.50$. Why is it not necessary that the probabilities add up to 1?

4. Suppose an item to be repaired might need one of n different parts. How many different policies would then need to be evaluated?

DIRECTIONS FOR GROUP PROJECT

Suppose you and three others are employed as service repair persons and that you have some disagreement with your supervisor as to the quantity and type of parts to have on hand for your service calls. Use the answers to Exercises 1–4 to prepare a report with a recommendation to your boss on optimal inventory. Make sure that you describe each concept well since your boss is not mathematically minded.

Answers to Selected Exercises

Answers to selected writing exercises are provided.

CHAPTER 6 Logic

Exercises 6.1 (page 273)

1. Statement, not compound **3.** Not a statement **5.** Statement, compound **7.** Not a statement **9.** Statement, compound **11.** Statement, not compound **13.** Statement, compound **15.** My favorite flavor is not chocolate. **17.** $y \le 12$ **19.** $q < 5$ **23.** I'm not getting better. **25.** I'm not getting better or my parrot is dead. **27.** It is not the case that both I'm getting better and my parrot is not dead. **29.** False **31.** True **33.** Both components are false. **35.** True **37.** True **39.** False **41.** True **43.** True **45.** True **47.** Disjunction **49.** True **51.** False **53.** True **55.** True **57.** False **59.** True **61.** True **63.** True **65.** a, c, d **67.** We may not charge a fee of $35 in each billing period if the New Balance on your statement exceeds your credit line. **69.** c, d **71.** $p \wedge q$ where p is the statement, "Tax rates are lower for a head of household than for a person filing as single," and q is the statement, "The standard deduction is higher." **73.** b, c, d **75.** c, d, e **77.** Most legal problems are not matters of civil law. **81.** $\sim n \vee \sim b$ **83.** $\sim n \wedge b$ **85.** $(n \vee b) \wedge [\sim(n \wedge b)]$ **87.** 81, 82, and 84

Exercises 6.2 (page 283)

1. 4 **3.** 16 **5.** 128 **7.** 6

9.

p	q	$\sim p$	$\sim p \wedge q$
T	T	F	F
T	F	F	F
F	T	T	T
F	F	T	F

11.

p	q	$p \wedge q$	$\sim(p \wedge q)$
T	T	T	F
T	F	F	T
F	T	F	T
F	F	F	T

13.

p	q	$\sim p$	$\sim q$	$q \vee \sim p$	$(q \vee \sim p) \vee \sim q$
T	T	F	F	T	T
T	F	F	T	F	T
F	T	T	F	T	T
F	F	T	T	T	T

In Exercises 15–23, we are using the alternate method to save space.

15.

p	q	$\sim q$	$\wedge$	$(\sim p \vee q)$
T	T	F	F	F T T
T	F	T	F	F F F
F	T	F	F	T T T
F	F	T	T	T T F
		①	④	② ③ ②

17.

p	q	$(p \vee \sim q)$	$\wedge$	$(p \wedge q)$
T	T	T T F	T	T T T
T	F	T T T	F	T F F
F	T	F F F	F	F F T
F	F	F T T	F	F F F
		① ② ①	⑤	③ ④ ③

19.

p	q	r	$(\sim p$	$\wedge$	$q)$	$\wedge$	r
T	T	T	F	F	T	F	T
T	T	F	F	F	T	F	F
T	F	T	F	F	F	F	T
T	F	F	F	F	F	F	F
F	T	T	T	T	T	T	T
F	T	F	T	T	T	F	F
F	F	T	T	F	F	F	T
F	F	F	T	F	F	F	F
			①	②	①	④	③

21.

p	q	r	$(\sim p$	$\wedge$	$\sim q)$	$\vee$	$(\sim r$	$\vee$	$\sim p)$
T	T	T	F	F	F	F	F	F	F
T	T	F	F	F	F	T	T	T	F
T	F	T	F	F	T	F	F	F	F
T	F	F	F	F	T	T	T	T	F
F	T	T	T	F	F	T	F	T	T
F	T	F	T	F	F	T	T	T	T
F	F	T	T	T	T	T	F	T	T
F	F	F	T	T	T	T	T	T	T
			①	②	①	⑤	③	④	③

23.

p	q	r	s	$\sim$	$(\sim p$	$\wedge$	$\sim q)$	$\vee$	$(\sim r$	$\vee$	$\sim s)$
T	T	T	T	T	F	F	F	T	F	F	F
T	T	T	F	T	F	F	F	T	F	T	T
T	T	F	T	T	F	F	F	T	T	T	F
T	T	F	F	T	F	F	F	T	T	T	T
T	F	T	T	T	F	F	T	T	F	F	F
T	F	T	F	T	F	F	T	T	F	T	T
T	F	F	T	T	F	F	T	T	T	T	F
T	F	F	F	T	F	F	T	T	T	T	T
F	T	T	T	T	T	F	F	T	F	F	F
F	T	T	F	T	T	F	F	T	F	T	T
F	T	F	T	T	T	F	F	T	T	T	F
F	T	F	F	T	T	F	F	T	T	T	T
F	F	T	T	F	T	T	T	F	F	F	F
F	F	T	F	F	T	T	T	T	F	T	T
F	F	F	T	F	T	T	T	T	T	T	F
F	F	F	F	F	T	T	T	T	T	T	T
				③	①	②	①	⑥	④	⑤	④

25. It's not summertime, or the living is not easy.

27. The door was locked and the thief didn't break a window.

29. I'm not ready to go, or Emily Portwood is.

31. $12 \leq 4$ and $8 \neq 9$

33. Neither Larry nor Moe is out sick today.

35.

p	q	$p \underline{\vee} q$
T	T	F
T	F	T
F	T	T
F	F	F

37. True **39. a.** False **b.** True **c.** False **d.** True

43. $s \vee (r \wedge \sim q)$

s	r	q	s	$\vee$	$(r$	$\wedge$	$\sim q)$
T	T	T	T	T	T	F	F
T	T	F	T	T	T	T	T
T	F	T	T	T	F	F	F
T	F	F	T	T	F	F	T
F	T	T	F	F	T	F	F
F	T	F	F	T	T	T	T
F	F	T	F	F	F	F	F
F	F	F	F	F	F	F	T
			①	④	②	③	②

The guarantee would be false if you are not completely satisfied, and they either don't refund your money or ask you questions.

45. p: The Pennsylvania Fish and Boat Commission is sensitive to the needs of the physically challenged. q: The Pennsylvania Fish and Boat Commission works to make our facilities accessible. *Negation*: Either the Pennsylvania Fish and Boat Commission is not sensitive to the needs of the physically challenged or it does not work to make our facilities accessible. **47.** p: The court won't do it for you. q: Hiring an attorney is usually not cost effective. *Negation*: Either the court will do it for you or it is cost effective to hire an attorney.

49. $(c \wedge e) \vee \sim r$

c	e	r	$(c \wedge e)$	$\vee$	$\sim r$
T	T	T	T T T	T	F
T	T	F	T T T	T	T
T	F	T	T F F	F	F
T	F	F	T F F	T	T
F	T	T	F F T	F	F
F	T	F	F F T	T	T
F	F	T	F F F	F	F
F	F	F	F F F	T	T
			① ② ①	④	③

The promise would be false if the senator runs for reelection and either doesn't cut taxes or eliminate the deficit.

51. The lady is behind Door 2 and the tiger is behind Door 1.

Exercises 6.3 (page 295)

1. True **3.** True **5.** True **9.** True **11.** True **13.** If she dances tonight, then I'm leaving early and he sings loudly. **15.** If he doesn't sing loudly, then she dances tonight or I'm not leaving early. **17.** $d \vee (f \rightarrow g)$ **19.** $\sim f \rightarrow g$ **21.** True **23.** False **25.** False **27.** True **29.** True

35.

p	q	p	$\rightarrow$	$\sim q$
T	T	T	F	F
T	F	T	T	T
F	T	F	T	F
F	F	F	T	T
		①	②	①

37.

p	q	$(\sim q \rightarrow \sim p)$	$\rightarrow$	$\sim q$
T	T	F T F	F	F
T	F	T F F	T	T
F	T	F T T	F	F
F	F	T T T	T	T
		① ② ①	④	③

39.

p	q	$(p \wedge \sim q)$	$\wedge$	$(p \rightarrow q)$
T	T	T F F	F	T T T
T	F	T T T	F	T F F
F	T	F F F	F	F T T
F	F	F F T	F	F T F
		① ② ①	⑤	③ ④ ③

It is a contradiction.

41.

p	q	r	r	$\rightarrow$	$(p \wedge \sim q)$
T	T	T	T	F	T F F
T	T	F	F	T	T F F
T	F	T	T	T	T T T
T	F	F	F	T	T T T
F	T	T	T	F	F F F
F	T	F	F	T	F F F
F	F	T	T	F	F F T
F	F	F	F	T	F F T
			①	④	② ③ ②

43.

p	q	r	s	(~r	→	s)	∨	(p	→	~q)
T	T	T	T	F	T	T	T	T	F	F
T	T	T	F	F	T	F	T	T	F	F
T	T	F	T	T	T	T	T	T	F	F
T	T	F	F	T	F	F	F	T	F	F
T	F	T	T	F	T	T	T	T	T	T
T	F	T	F	F	T	F	T	T	T	T
T	F	F	T	T	T	T	T	T	T	T
T	F	F	F	T	F	F	T	T	T	T
F	T	T	T	F	T	T	T	F	T	F
F	T	T	F	F	T	F	T	F	T	F
F	T	F	T	T	T	T	T	F	T	F
F	T	F	F	T	F	F	T	F	T	F
F	F	T	T	F	T	T	T	F	T	T
F	F	T	F	F	T	F	T	F	T	T
F	F	F	T	T	T	T	T	F	T	T
F	F	F	F	T	F	F	T	F	T	T
				①	②	①	⑤	③	④	③

45. one **47.** They cannot see me now or they'd never believe it. **49.** I am not you or I would watch out. **51.** I can make it there and I won't make it anywhere. **53.** He's my brother and he's heavy. **55.** Equivalent **57.** Equivalent **59.** Equivalent **61.** Not equivalent

63.

p	q	p	∨	q	~p	→	q
T	T	T	T	T	F	T	T
T	F	T	T	F	F	T	F
F	T	F	T	T	T	T	T
F	F	F	F	F	T	F	F
		①	②	①	③	④	③

The columns labeled 2 and 4 are identical.

65.

p	q	p	∧	q	q	∧	p
T	T	T	T	T	T	T	T
T	F	T	F	F	F	F	T
F	T	F	F	T	T	F	F
F	F	F	F	F	F	F	F
		①	②	①	③	④	③

The columns labeled 2 and 4 are identical.

67.

p	q	r	(p	∧	q)	∧	r	p	∧	(q	∧	r)
T	T	T	T	T	T	T	T	T	T	T	T	T
T	T	F	T	T	T	F	F	T	F	T	F	F
T	F	T	T	F	F	F	T	T	F	F	F	T
T	F	F	T	F	F	F	F	T	F	F	F	F
F	T	T	F	F	T	F	T	F	F	T	T	T
F	T	F	F	F	T	F	F	F	F	T	F	F
F	F	T	F	F	F	F	T	F	F	F	F	T
F	F	F	F	F	F	F	F	F	F	F	F	F
			①	②	①	④	③	⑤	⑧	⑥	⑦	⑥

The columns labeled 4 and 8 are identical.

69.

p	q	r	p	$\wedge$	$(q$	$\vee$	$r)$	$(p$	$\wedge$	$q)$	$\vee$	$(p$	$\wedge$	$r)$
T	T	T	T	T	T	T	T	T	T	T	T	T	T	T
T	T	F	T	T	T	T	F	T	T	T	T	T	F	F
T	F	T	T	T	F	T	T	T	F	F	T	T	T	T
T	F	F	T	F	F	F	F	T	F	F	F	T	F	F
F	T	T	F	F	T	T	T	F	F	T	F	F	F	T
F	T	F	F	F	T	T	F	F	F	T	F	F	F	F
F	F	T	F	F	F	T	T	F	F	F	F	F	F	T
F	F	F	F	F	F	F	F	F	F	F	F	F	F	F
			①	④	②	③	②	⑤	⑥	⑤	⑨	⑦	⑧	⑦

The columns labeled 4 and 9 are identical.

71.

p	q	$(p$	$\vee$	$q)$	$\wedge$	p
T	T	T	T	T	T	T
T	F	T	T	F	T	T
F	T	F	T	T	F	F
F	F	F	F	F	F	F
		①	②	①	④	③

The p column and the column labeled 4 are identical.

73. $p \wedge (r \vee q)$ **75.** $q \vee [p \wedge (q \vee {\sim}p)] \equiv q$ **77.** $({\sim}p \vee q) \vee ({\sim}p \vee {\sim}q) \equiv \text{T}$
79. $\{[({\sim}p \wedge {\sim}q) \vee (p \vee q)] \wedge (p \vee q)\} \wedge p \equiv p$
81. 

$({\sim}q \wedge {\sim}p) \vee ({\sim}p \vee q) \equiv {\sim}p \vee q$

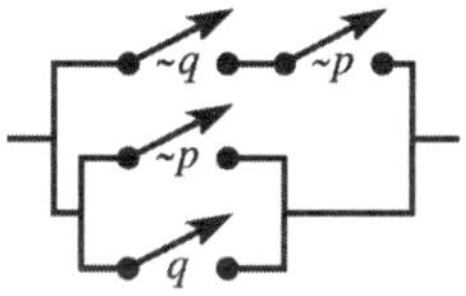

85. $[({\sim}p \wedge {\sim}r) \vee {\sim}q] \wedge ({\sim}p \wedge r) \equiv ({\sim}p \wedge r) \wedge {\sim}q$ **87.** ${\sim}p \rightarrow ({\sim}p \vee {\sim}q) \equiv \text{T}$

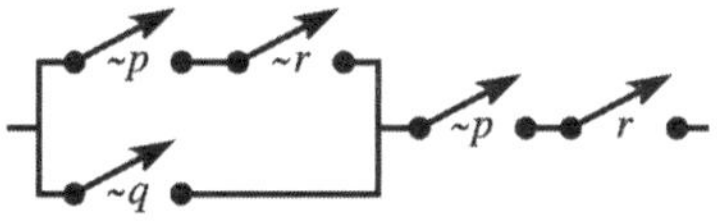

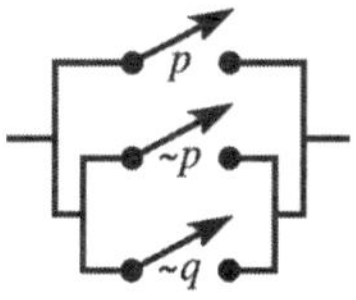

89. $[(p \wedge q) \vee (p \wedge q)] \vee (p \wedge r) \equiv p \wedge (q \vee r)$

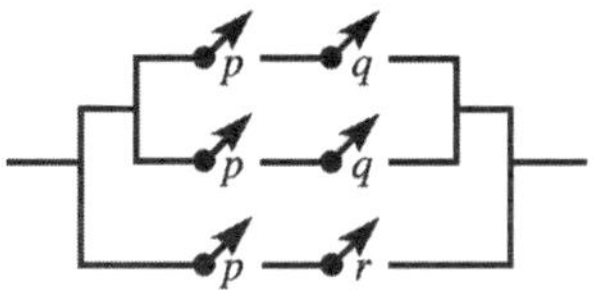

91. \$262.80 **93.** You are not wheezing persistently or you should see your doctor. *Negation*: You are wheezing persistently and you should not see your doctor.

Exercises 6.4 (page 303)

1. a. *Converse:* If I don't see it, then the exit is ahead. **b.** *Inverse:* If the exit is not ahead, then I see it. **c.** *Contrapositive:* If I see it, then the exit is not ahead. **3. a.** *Converse:* If I baked a cake, then I knew you were coming. **b.** *Inverse:* If I didn't know you were coming, I wouldn't have baked a cake. **c.** *Contrapositive:* If I didn't bake a cake, then I didn't know you were coming.

5. a. *Converse:* If a man doesn't wear plaid, then he's dead. **b.** *Inverse:* If a man is not dead, then he wears plaid. **c.** *Contrapositive:* If a man wears plaid, then he's not dead. **7. a.** *Converse:* $\sim q \rightarrow p$. **b.** *Inverse:* $\sim p \rightarrow q$. **c.** *Contrapositive:* $q \rightarrow \sim p$. **9. a.** *Converse:* $(q \vee r) \rightarrow p$. **b.** *Inverse:* $\sim p \rightarrow \sim(q \vee r)$ or $\sim p \rightarrow (\sim q \wedge \sim r)$. **c.** *Contrapositive:* $(\sim q \wedge \sim r) \rightarrow \sim p$. **13.** If you sign, then you accept the conditions. **15.** If you can take this course pass/fail, then you have prior permission. **17.** If the temperature is below 10°, then you can skate on the pond. **19.** If someone eats 10 hot dogs, then he or she will get sick. **21.** If you travel to France, then you have a valid passport. **23.** If a number has a real square root, then it is nonnegative. **25.** If someone is a bride, then she is beautiful. **27.** If the sum of a number's digits is divisible by 3, then it is divisible by 3. **29.** d **33.** True **35.** False **37.** False

39.

p	q	$(\sim p$	$\wedge$	$q)$	$\leftrightarrow$	$(p$	$\rightarrow$	$q)$
T	T	F	F	T	F	T	T	T
T	F	F	F	F	T	T	F	F
F	T	T	T	T	T	F	T	T
F	F	T	F	F	F	F	T	F
		①	②	①	⑤	③	④	③

41. a. If the award qualifies for full or partial tax-free treatment, then your employer must tell you. **b.** If medical expenses are "qualified," then they are incurred *after* the HSA has been established. **c.** If you elect to report the market discount annually as interest income, you can avoid this interest deduction limitation. **43.** *Converse:* If you may avoid paying the annual fee billed on this statement, then you close your account within 30 days from the date this statement was mailed. *Inverse:* If you do not close your account within 30 days from the date this statement was mailed, you may not avoid paying the annual fee billed on this s tatement. *Contrapositive:* If you may not avoid paying the annual fee billed on this statement, then you do not close your account within 30 days from the date this statement was mailed. The converse and the inverse are equivalent, and the contrapositive and the original statement are equivalent. **45. a.** $p \rightarrow (q \wedge r)$ **b.** If the most persistent does not stand to gain an extra meal or it does not eat at the expense of another, then there are not triplets. **47.** If liberty and equality are not best attained when all persons share alike in the government to the utmost, then they are not, as is thought by some, chiefly to be found in democracy. **49.** $(d \rightarrow l) \wedge \sim(l \rightarrow d)$ **51.** If there is an R.P.F. alliance, there there is a Modéré incumbent. *Converse:* If there is a Modéré incumbent, then there is an R.P.F. alliance. *Inverse:* If there is not an R.P.F. alliance, then there is not a Modéré incumbent. *Contrapositive:* If there is not a Modéré incumbent, then there is not an R.P.F. alliance. The contrapositive is equivalent to the original. **53.** "Worked on the weekend": Must be turned over to see whether the employee got a day off. "Did not work on the weekend": Need not be turned over, since it does not describe an employee who worked on the weekend. "Did get a day off": Need not be turned over, since it cannot describe an employee who worked on the weekend without getting a day off. "Did not get a day off": Must be turned over to see whether the other side says "worked on the weekend."

Exercises 6.5 (page 316)

1. Valid; Reasoning by Transitivity **3.** Valid; Modus Ponens **5.** Invalid; Fallacy of the Converse **7.** Valid; Modus Tollens **9.** Invalid; Fallacy of the Inverse **11.** Valid; Disjunctive Syllogism **13.** Invalid; $p = \text{T}, q = \text{T}$ **15.** Invalid; $p = \text{F}, q = \text{F}$

17. Valid.

1.	$\sim p \rightarrow \sim q$	Premise
2.	q	Premise
3.	p	1, 2, Modus Tollens

19. Valid.

1.	$p \rightarrow q$	Premise
2.	$\sim q$	Premise
3.	$\sim p \rightarrow r$	Premise
4.	$\sim p$	1, 2, Modus Tollens
5.	r	3, 4, Modus Ponens

21. Valid.

1.	$p \rightarrow q$	Premise
2.	$q \rightarrow r$	Premise
3.	$\sim r$	Premise
4.	$p \rightarrow r$	1, 2, Transitivity
5.	$\sim p$	3, 4, Modus Tollens

23. Valid.

1.	$p \rightarrow q$	Premise
2.	$q \rightarrow \sim r$	Premise
3.	p	Premise
4.	$r \vee s$	Premise
5.	q	1, 3, Modus Ponens
6.	$\sim r$	2, 5, Modus Ponens
7.	s	4, 6, Disjunctive Syllogism

25.

p	q	$(p \wedge q) \rightarrow p$
T	T	T T T T T
T	F	T F F T T
F	T	F F T T F
F	F	F F F T F
		① ② ① ③ ②

27.

p	q	$(p \wedge q) \rightarrow (p \wedge q)$
T	T	T T T T T T T
T	F	T F F T T F F
F	T	F F T T F F F
F	F	F F F T F F F
		① ② ① ⑤ ③ ④ ③

29. Valid.
1. c Premise
2. $f \rightarrow \sim c$ Premise
3. $\sim f \rightarrow s$ Premise
4. $\sim f$ 1, 2, Modus Tollens
5. s 3, 4, Modus Ponens

31. Invalid; $s =$ "you have strep throat" = T, $f =$ "you have a fever" = T, $c =$ "you have a serious cough" = F

33. Valid.
1. $y \vee \sim m$ Premise
2. $\sim m \rightarrow \sim n$ Premise
3. n Premise
4. m 2, 3, Modus Tollens
5. y 1, 4, Disjunctive Syllogism

35. Valid.
1. $(p \wedge q) \rightarrow \sim r$ Premise
2. r Premise
3. $\sim(p \wedge q)$ 1, 2, Modus Tollens
4. $\sim p \vee \sim q$ 3, De Morgan's Law

37. *Conclusion:* If I tell you the time, then my life will be miserable. **39. a.** $s \rightarrow l$ or $\sim l \rightarrow \sim s$ **b.** $\sim s \rightarrow \sim j$ **c.** $y \rightarrow \sim l$ **d.** $y \rightarrow \sim j$, *Conclusion:* If he is your son, then he is not fit to serve on a jury. In Lewis Carroll's words, "None of your sons are fit to serve on a jury." **41. a.** $a \rightarrow s$ or $\sim s \rightarrow \sim a$ **b.** $g \rightarrow i$ **c.** $i \rightarrow \sim s$ **d.** $g \rightarrow \sim a$, *Conclusion:* If it is a guinea pig, then it does not appreciate Beethoven. In Lewis Carroll's words, "Guinea pigs never really appreciate Beethoven."
43. a. $p \rightarrow b$ or $\sim b \rightarrow \sim p$ **b.** $\sim t \rightarrow \sim l$ **c.** $o \rightarrow \sim s$ **d.** $b \rightarrow l$ **e.** $w \rightarrow p$ **f.** $\sim s \rightarrow \sim t$ **g.** $o \rightarrow \sim w$, *Conclusion:* If he is an opium eater, then he doesn't wear white kid gloves. In Lewis Carroll's words, "Opium eaters never wear white kid gloves."

Exercises 6.6 (page 326)

1. a. $\exists x\,[b(x) \wedge s(x)]$ **b.** $\forall x\,[b(x) \rightarrow \sim s(x)]$ **c.** No books are bestsellers. **3. a.** $\forall x\,[c(x) \rightarrow \sim s(x)]$ **b.** $\exists x\,[c(x) \wedge s(x)]$ **c.** There is a CEO who sleeps well at night. **5. a.** $\forall x\,[l(x) \rightarrow b(x)]$ **b.** $\exists x\,[l(x) \wedge \sim b(x)]$ **c.** There is a leaf that's not brown.

7. a. $\forall x[g(x) \rightarrow f(x)]$ **b.**

$$\frac{g(t)}{f(t)}$$

Valid

9. a. $\forall x[p(x) \rightarrow c(x)]$ **b.**

$$\frac{c(o)}{p(o)}$$

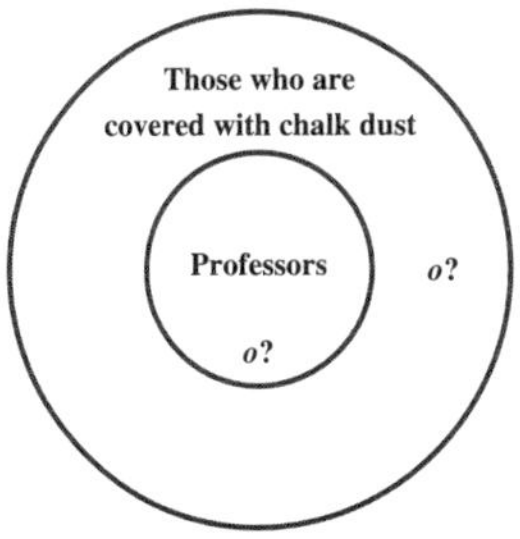

Invalid

11. $\forall x\,[c(x) \rightarrow p(x)]$ **b.**

$$\frac{\sim p(n)}{\sim c(n)}$$

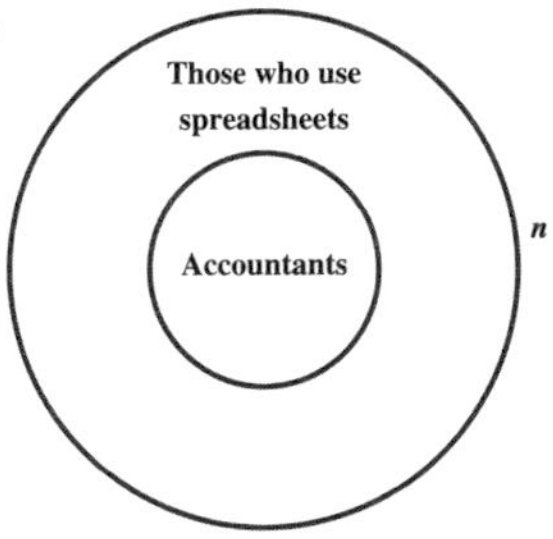

Valid

13. a. $\exists x\,[t(x) \wedge s(x)]$
$\forall x\,[t(x) \rightarrow b(x)]$
$\exists x\,[s(x) \wedge b(x)]$

b.

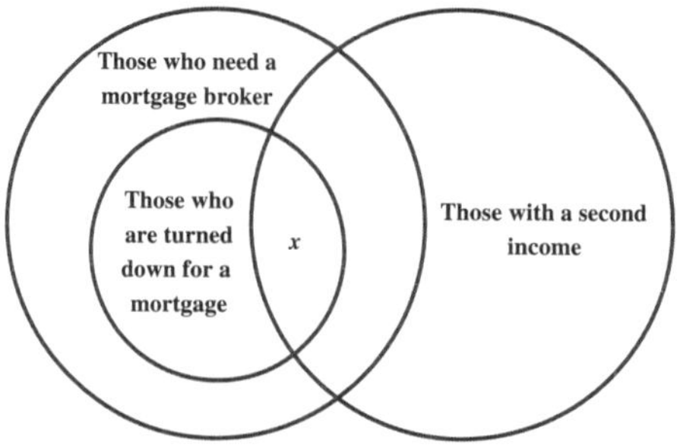

Valid

15. a. $\exists x\,[w(x) \wedge l(x)]$
$w(m)$
$l(m)$

b.

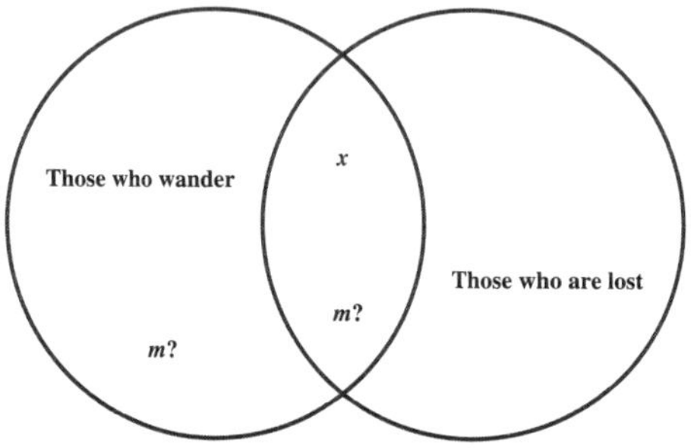

Invalid

17. a. $\exists x\,[p(x) \wedge u(x)]$
$\exists x\,[p(x) \wedge r(x)]$
$\exists x\,[u(x) \wedge r(x)]$

b.

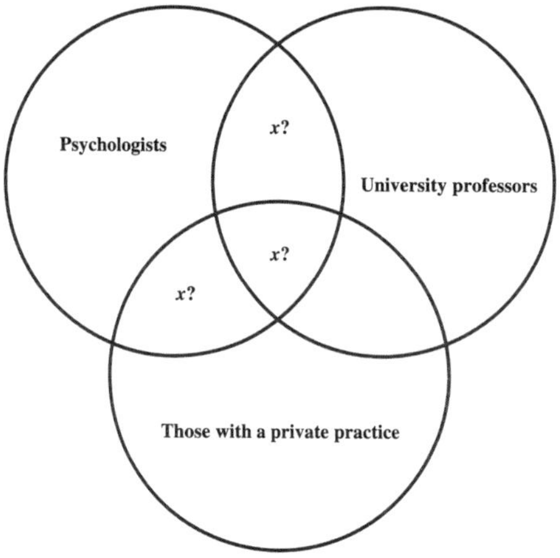

Invalid

19. a. $\forall x\,[a(x) \vee i(x)]$
$\exists x\,[\sim a(x)]$
$\exists x\,[i(x)]$

b.

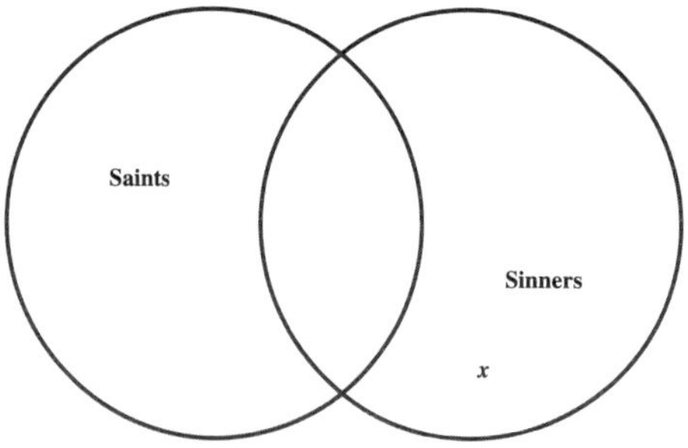

Valid

21. Yes

23. All major league baseball players earn at least $300,000 a year.
Ryan Howard is a major league baseball player.
Ryan Howard earns at least $300,000 a year.

25. Valid **27.** Invalid **29.** Invalid **31.** Invalid
37. a. $\forall x\,[l(x) \rightarrow g(x)]$ **b.** The power to collect taxes shall be vested in a Congress of the United States.
c.

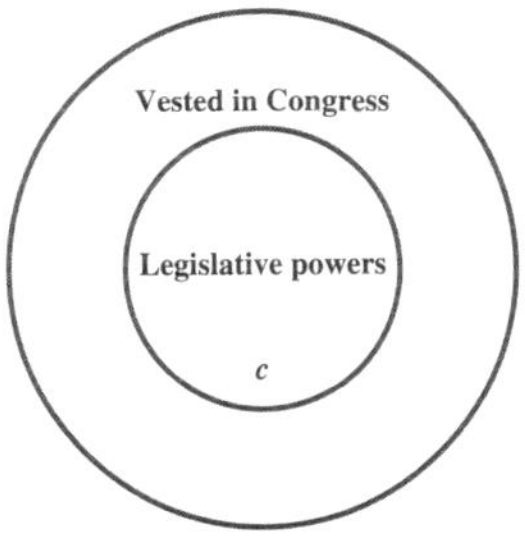

39. a. $\forall x\,\{[b(x) \vee e(x)] \rightarrow \sim p(x)\}$ **b.** The law forbidding members of the Communist Party to serve as an officer or as an employee of a labor union shall not be passed.
c.

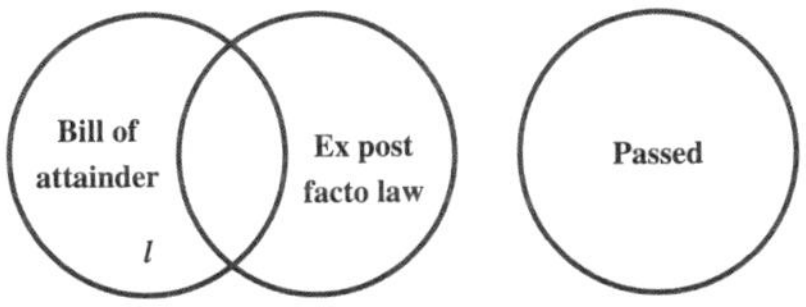

41. a, c, d **43.** Invalid **45.** Invalid **47.** Valid

Concept Check (page 331)

1. True **2.** False **3.** False **4.** False **5.** False **6.** True **7.** False **8.** True **9.** False **10.** False **11.** True **12.** True

Chapter 6 Review Exercises (page 331)

1. She pays me and I have enough cash. **3.** $l \wedge w$ **5.** $l \rightarrow \sim w$ **7.** He doesn't lose the election and he wins the hearts of the voters. **9.** True **11.** True

15.

p	q	p	$\wedge$	$(\sim p$	$\vee$	$q)$
T	T	T	T	F	T	T
T	F	T	F	F	F	F
F	T	F	F	T	T	T
F	F	F	F	T	T	F
		①	④	②	③	②

The statement is not a tautology.

17. If someone is a mathematician, then that person is loveable. **19.** If a system has a unique solution, then it has at least as many equations as unknowns. **21. a.** If we need to change the way we do business, then the proposed regulations have been approved. **b.** If the proposed regulations have not been approved, then we do not need to change the way we do business. **c.** If we do not need to change the way we do business, then the proposed regulations have not been approved.
23. $(p \wedge p) \wedge (\sim p \vee q) \equiv p \wedge q$ **25.**

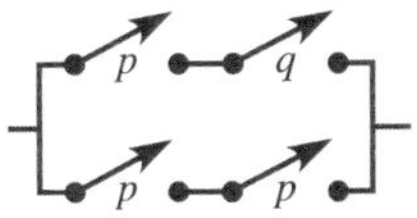

$(p \wedge q) \vee (p \wedge p) \equiv p$

27.

p	q	$p \veebar q$	$(p \vee q) \wedge \sim(p \wedge q)$
T	T	T F T	T T T F F T T T
T	F	T T F	T T F T T T F F
F	T	F T T	F T T T T F F T
F	F	F F F	F F F F T F F F
		① ② ①	③ ④ ③ ⑧ ⑦ ⑤ ⑥ ⑤

The columns labeled 2 and 8 are identical.

29. a. Yes **b.** No **31.** Valid; Modus Ponens **33.** Valid; Disjunctive Syllogism **35.** Invalid; Fallacy of the Converse

37. Valid.

1.	$h \to t$	Premise
2.	$r \to \sim t$	Premise
3.	r	Premise
4.	$\sim t$	2, 3, Modus Ponens
5.	$\sim h$	1, 4, Modus Tollens

39. Invalid; $p = \text{F}$, $q = \text{F}$

41. a. $\forall x\,[d(x) \to h(x)]$ **b.** $\exists x\,[d(x) \wedge \sim h(x)]$ **c.** There is a dog that doesn't go to heaven.

43. a. $\forall x\,[f(x) \to w(x)]$

$\underline{f(j)}$

$w(j)$

b.

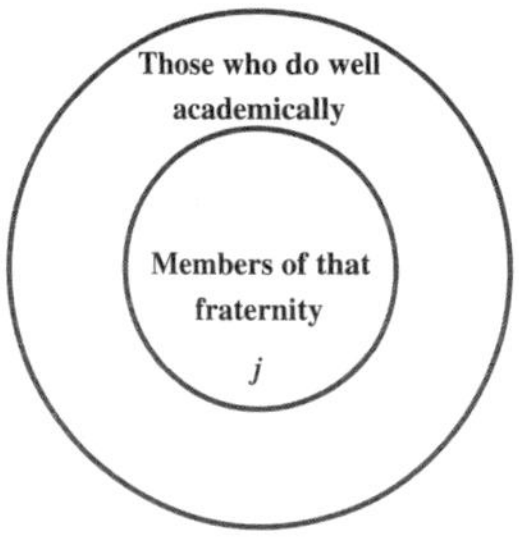

Valid

45.

p	q	r	$p \to$	$(q \to r)$	$(p \to q) \to r$
T	T	T	T T	T T T	T T T T T
T	T	F	T F	T F F	T T T F F
T	F	T	T T	F T T	T F F T T
T	F	F	T T	F T F	T F F T F
F	T	T	F T	T T T	F T T T T
F	T	F	F T	T F F	F T T F F
F	F	T	F T	F T T	F T F T T
F	F	F	F T	F T F	F T F F F
			① ④	② ③ ②	⑤ ⑥ ⑤ ⑧ ⑦

No

47. a.

p	q	$(p \wedge \sim p)$	$\to q$
T	T	T F F	T T
T	F	T F F	T F
F	T	F F T	T T
F	F	F F T	T F
		① ② ①	④ ③

49. a, b, c

51. If you do not have a net 28% rate gain or unrecaptured Section 1250 gain, then the Schedule D Tax Worksheet in the Schedule D instructions is not used. **53. a.** Regulations do not have both costs and benefits, or rules that are passed to solve a problem cannot make it worse. **b.** Shooters do not overwhelmingly have problems with alcoholism, or they do not have long criminal histories, particularly arrests for violent acts. **c.** They are not disproportionately involved in automobile crashes, or they are not much more likely to have their driver's license suspended or revoked. **55. a.**

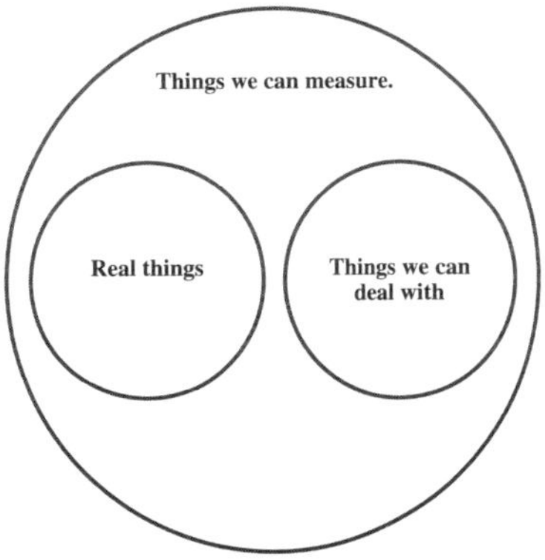

57. c **59. a.** $h \to o$ **b.** $a \to m$ **c.** $p \to \sim o \equiv o \to \sim p$ **d.** $m \to h$ **e.** If the bird is in this aviary, it does not live on mince pies. In Lewis Carroll's words, "No bird in this aviary lives on mince pies." **61. a.** $u \to c$ **b.** $p \to r$ **c.** $s \to h$ **d.** $\sim u \to \sim r \equiv r \to u$ **e.** $h \to p$ **f.** If the writer is Shakespeare, the writer is clever. In Lewis Carroll's words, "Shakespeare was clever."

CHAPTER 7 Sets and Probability

Exercises 7.1 (page 349)

1. False **3.** True **5.** True **7.** True **9.** False **11.** $\subseteq$ **13.** $\not\subseteq$ **15.** $\subseteq$ **17.** $\subseteq$ **19.** $\subset; \subset; \not\subset; \not\subset; \subset; \not\subset; \subset; \not\subset$ **21.** 16 **23.** 4 **25.** $\cap$ **27.** $\cap$ **29.** $\cap$ **31.** $\cup$ or $\cap$ **33.** $\{2, 4, 6\}$ **35.** $\{1, 3, 5, 7, 9\}$ **37.** $\{1, 7, 9\}$ **39.** $\{2, 3, 4, 6\}$ **41.** $\{4, 5, 6, 7, 8\}$ **43.** All students in this school not taking this course **45.** All students in this school taking accounting and zoology **47.** C and D, B and E, C and E, D and E **49.** B' is the set of all stocks on the list with a closing price below \$26 or above \$30; B' = {AT&T, CocaCola, Office Max Inc., Texas Instruments}. **51.** $(A \cap B)'$ is the set of all stocks on the list that do not have both a high price greater than \$34 and a closing price between \$26 and \$30; $(A \cap B)'$ = {AT&T, CocaCola, Disney, Office Max Inc., Texas Instruments}. **53. a.** True **b.** True **c.** False **d.** False **e.** True **f.** True **g.** False **55.** {Microsoft Corp., Proctor & Gamble} **57.** {ExxonMobil Corp., Citigroup, Inc., American International Group} **59.** $\{s, d, c\}$ **61.** $\{g\}$ **63.** $\{s, d, c\}$ **65.** 11 **67.** {The Disney Channel, Showtime, HBO, Encore} **69.** {Encore, Starz} **71.** {Showtime, HBO, Encore, Starz} **75. a.** The set of states who are not among those whose name contains the letter "e" or who are more than 4 million in population, and who also have an area of more than 40,000 square miles. **b.** {Alaska}

Exercises 7.2 (page 360)

1.

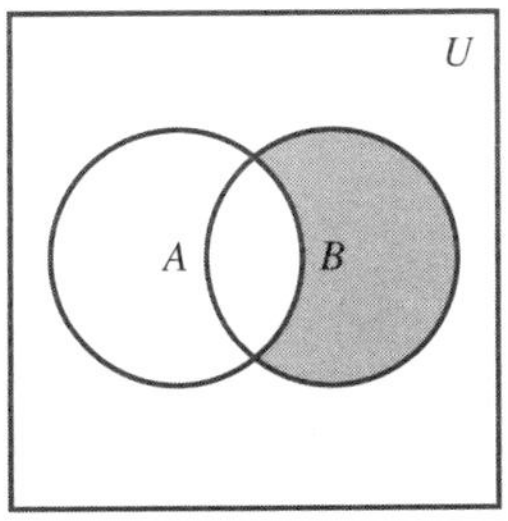

$B \cap A'$

3.

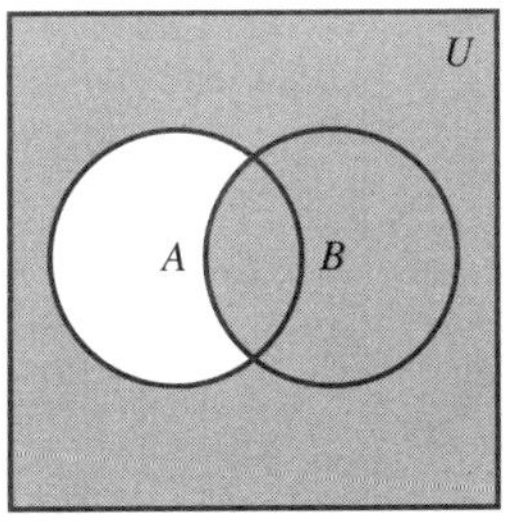

$A' \cup B$

5.

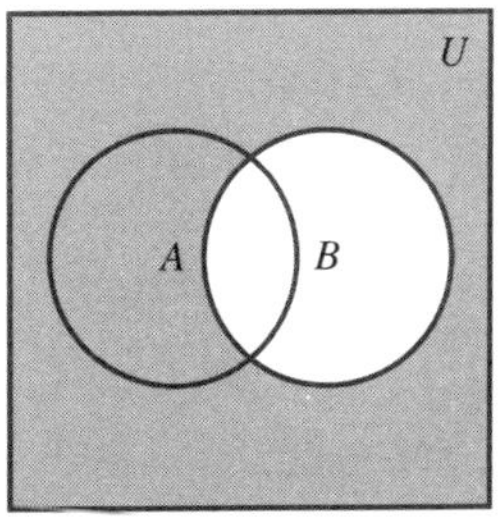

$B' \cup (A' \cap B')$

7.

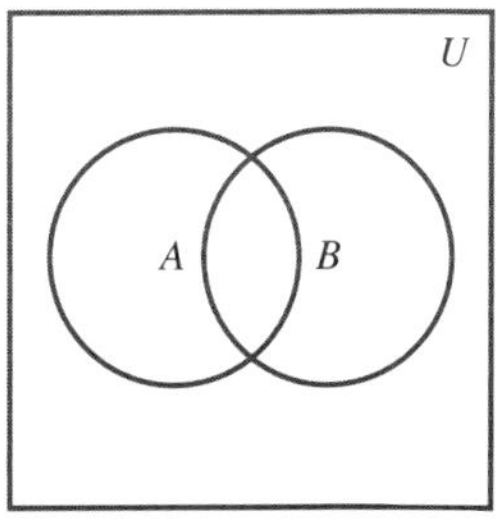

$U' = \emptyset$

9. 8

11.

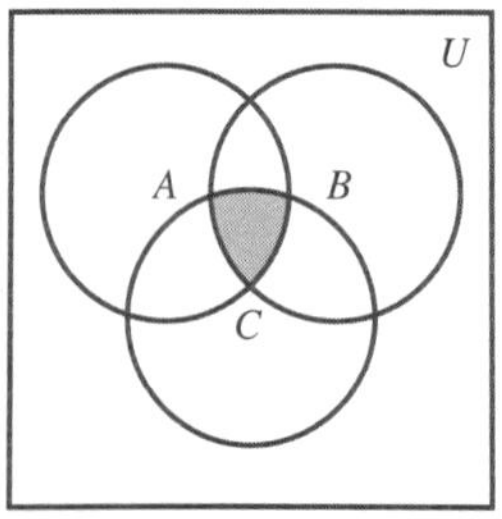

$(A \cap B) \cap C$

13.

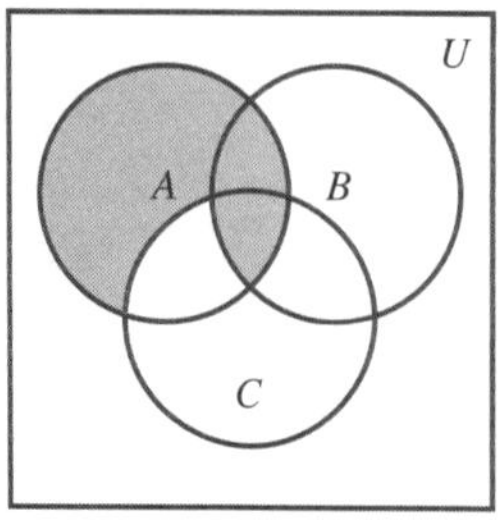

$A \cap (B \cup C')$

15.

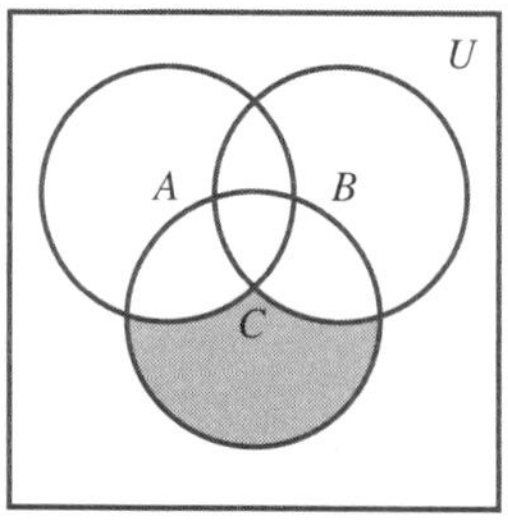

$(A' \cap B') \cap C$

17.

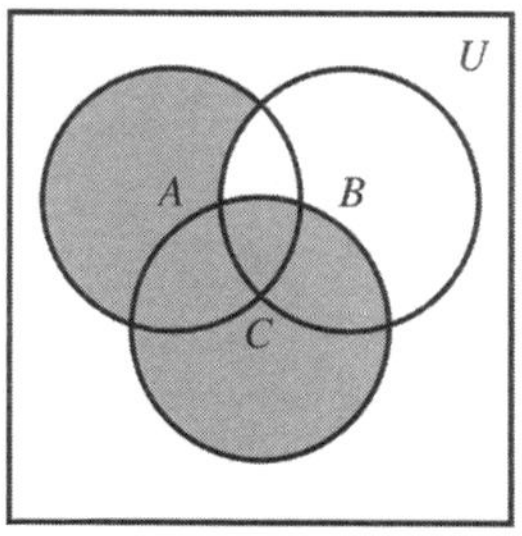

$(A \cap B') \cup C$

19.

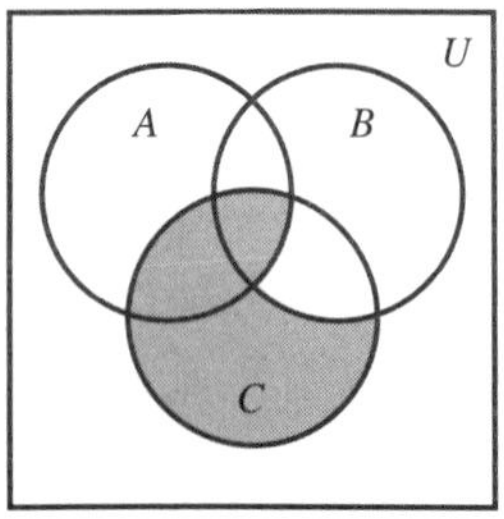

$(A \cup B') \cap C$

21. 13 **23.** 18 **25.**

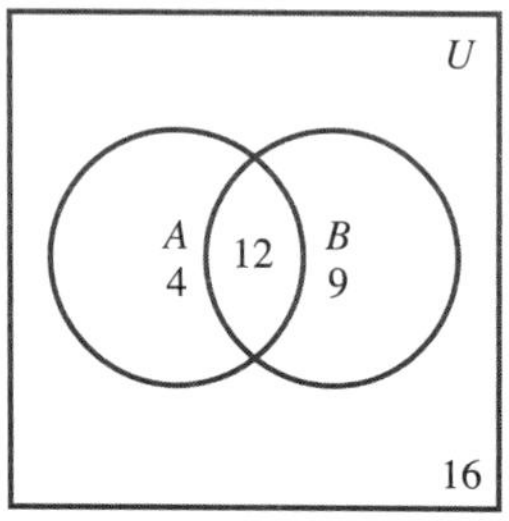

27.

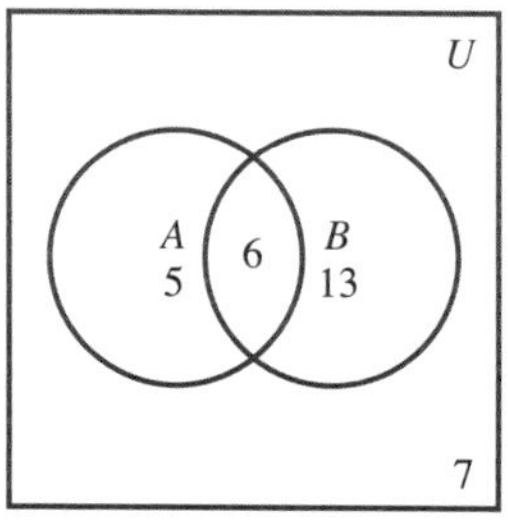

29.

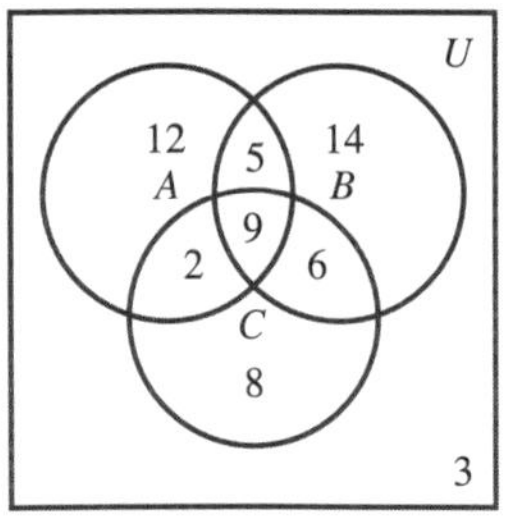

31.

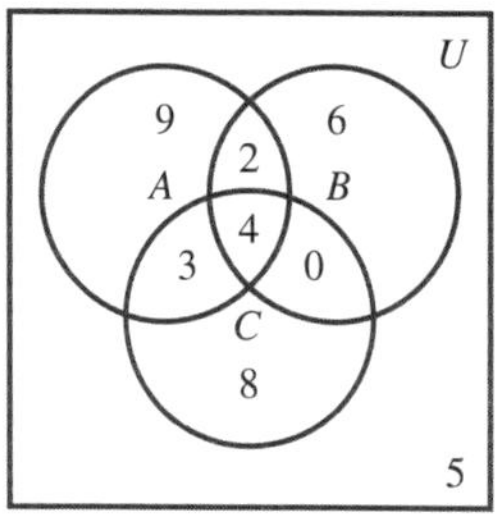

39. a. 12 **b.** 18 **c.** 37 **d.** 97 **41. a.** 2 **b.** 60 **c.** 8 **d.** 100 **e.** 27 **f.** All those who invest in stocks or bonds and are age 18–29 **43. a.** 17 **b.** 2 **c.** 14 **45. a.** 54 **b.** 17 **c.** 10 **d.** 7 **e.** 15 **f.** 3 **g.** 12 **h.** 1 **47. a.** 34 **b.** 16 **c.** 431 **d.** 481 **49.** 110.6 million **51.** 85.4 million **53.** 71.0 million **55.** 21,645,000 **57.** 194,083,000 **59. a.** 342 **b.** 192 **c.** 72 **d.** 86 **61. a.** 56 **b.** 32 **c.** 26 **d.** 30 **e.** 22 **f.** 21

Exercises 7.3 (page 371)

3. {January, February, March, …, December} **5.** {0, 1, 2, 3, …, 79, 80} **7.** {go ahead, cancel}
9. $\{(h,1),(h,2),(h,3),(h,4),(h,5),(h,6),(t,1),(t,2),(t,3),(t,4),(t,5),(t,6)\}$ **13.** {AB, AC, AD, AE, BC, BD, BE, CD, CE, DE}, 10, yes **a.** {AC, BC, CD, CE} **b.** {AB, AC, AD, AE, BC, BD, BE, CD, CE} **c.** {AC}
15. $\{(1,2),(1,3),(1,4),(1,5),(2,3),(2,4),(2,5),(3,4),(3,5),(4,5)\}$, 10, yes **a.** $\{(2,4)\}$ **b.** $\{(1,2),(1,4),(2,3),(2,5),(3,4),(4,5)\}$ **c.** $\emptyset$ **17.** {*hh*, *thh*, *hth*, *tthh*, *thth*, *htth*, *ttth*, *ttht*, *thtt*, *httt*, *tttt*}, 11, no **a.** {*tthh*, *thth*, *htth*, *ttth*, *ttht*, *thtt*, *httt*, *tttt*} **b.** {*hh*, *thh*, *hth*, *tthh*, *thth*, *htth*} **c.** {*tttt*} **19.** 1/6 **21.** 2/3 **23.** 1/3 **25.** 1/13 **27.** 1/26 **29.** 1/52 **31.** 2/13 **33.** 7/13 **35.** 3/20 **37.** 1/4 **39.** 3/5 **41.** The outcomes are not equally likely. **43. a.** Worker is male. **b.** Worker is female and has worked less than 5 years. **c.** Worker is female or does not contribute to a voluntary retirement plan. **d.** Worker has worked 5 years or more. **e.** Worker has worked less than 5 years or has contributed to a voluntary retirement plan. **f.** Worker has worked 5 years or more and does not contribute to a voluntary retirement plan. **45. a.** 8/15 **b.** 8/15

47. a. Person smokes or has a family history of heart disease, or both. **b.** Person does not smoke and has a family history of heart disease. **c.** Person does not have a family history of heart disease or is not overweight, or both. **49. a.** 0.1631 **b.** 0.2450 **c.** 0.1287 **d.** 0.1360 **51. a.** 0.3151 **b.** 0.3219 **c.** 0.3670 **d.** Calvary **e.** I Corps

Exercises 7.4 (page 382)

3. No **5.** Yes **7.** No **9. a.** 5/36 **b.** 1/9 **c.** 1/12 **d.** 0 **11. a.** 5/18 **b.** 5/12 **c.** 1/3 **13.** 2/9 **15. a.** 2/13 **b.** 7/13 **c.** 3/26 **d.** 3/4 **e.** 11/26 **17. a.** 5/13 **b.** 7/13 **c.** 3/13 **19. a.** 1/10 **b.** 2/5 **c.** 7/20 **21. a.** 0.51 **b.** 0.25 **c.** 0.10 **d.** 0.84 **25.** 1 to 1 **27.** 5 to 1 **29.** 7 to 2 **31.** No. The probability the reader wins is 5/12, while the probability his opponent wins is 7/12. **33.** 1/3; 1/5; 1/9001; 1/10; 9/19 **35.** Empirical **37.** Not empirical **39.** Empirical **41.** Empirical **43.** Possible **45.** Not possible; the sum of the probabilities is less than 1. **47.** Not possible; a probability cannot be negative. (*Note:* For 49 and 51, theoretical answers are given; actual answers will vary.) **49. a.** 0.2778 **b.** 0.4167 **51. a.** 0.0463 **b.** 0.2963 **53. a.** 25/57 **b.** 32/57 **c.** 4/19 **55.** 0.84 **57. a.** 0.07 **b.** 0.93 **c.** 0.49 **59.** 13 to 37 **61. a.** 0.961 **b.** 0.487 **c.** 0.513 **d.** 0.509 **e.** 0.004 **f.** 0.548 **63. a.** 1/4 **b.** 1/2 **c.** 1/4 **65.** a **67.** c **69. a.** 0.4 **b.** 0.1 **c.** 0.6 **d.** 0.9 **71.** 0 **73.** 2/5 **75. a.** 0.866 **b.** 0.478 **77.** 4/11

Exercises 7.5 (page 401)

1. 0 **3.** 1 **5.** 1/6 **7.** 4/17 **9.** 11/51 **11.** 8/663 **13.** 25/102 **17.** Dependent **19.** Independent **21. a.** 1/4 **b.** 0 **27. a.** It is better to switch, for a probability of winning of 1/6, as opposed to a probability of 1/7 of winning if you don't switch. **29.** No; yes **31. a.** 0.2 **b.** 0.4 **33.** The probability of a customer cashing a check, given that the customer made a deposit, is 3/4. **35.** The probability of a customer not cashing a check, given that the customer did not make a deposit, is 1/4. **37.** The probability of a customer not both cashing a check and making a deposit is 1/2. **39.** 0.999985; fairly realistic **41.** 0.03 **43.** 0.09 **45.** 1/4 **47.** 1/4 **49.** 1/7 **51.** They are independent. **53.** 0.039 **55.** 0.491 **57.** 0.072 **59.** Yes **61. a.** 0.323 **b.** 0.582 **c.** No **63.** e **65. a.** 0.0005 **b.** 0.9995 **c.** $(1999/2000)^a$ **d.** $1 - (1999/2000)^a$ **e.** $(1999/2000)^{Nc}$ **f.** $1 - (1999/2000)^{Nc}$ **67.** 7/229 **69.** 191/229 **71. a.** 0.7456 **b.** 0.1872 **73. a.** 0.58 **75. a.** 0.2713 **b.** 0.3792 **c.** 0.6246 **d.** 0.3418 **e.** 0.6897 **f.** 0.5137 **g.** Not independent **77. a.** 0.052 **b.** 0.476 **79. a.** 7/10 **b.** 2/15 **81.** 10^{-12} **83.** No **85. c.** They are the same. **d.** The 2-point first strategy has a smaller probability of losing.

Exercises 7.6 (page 413)

1. 1/3 **3.** 3/19 **5.** 21/38 **7.** 8/17 **9.** 85% **11.** 0.0727 **13.** 0.1765 **15.** 0.3636 **17.** 2/7 **19.** d **21.** c **23.** 0.0478 **25. a.** About 0.16 **b.** About 0.99 **c.** About 59 **27.** d **29.** b **31. a.** 0.53 **b.** 0.1623 **33.** 0.1457 **35.** 0.1118 **37.** 0.2441 **39.** 9.9×10^{-5}

Concept Check (page 420)

1. True **2.** True **3.** False **4.** False **5.** False **6.** True **7.** False **8.** False **9.** False **10.** True **11.** False **12.** True

Chapter 7 Review Exercises (page 421)

1. False **3.** False **5.** True **7.** False **9.** False **11.** 32 **13.** {a, b, g} **15.** {c, d} **17.** {a, b, e, f, g, h} **19.** *U* **21.** All female employees in the accounting department All employees who are in the accounting department or who have MBA degrees **25.** All male employees who are not in the sales department

27.

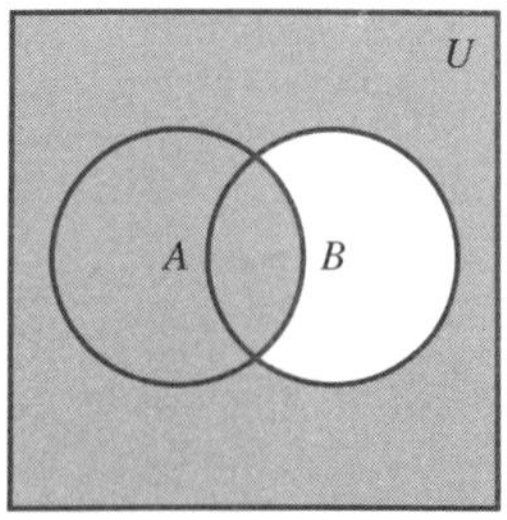

$A \cup B'$

29.

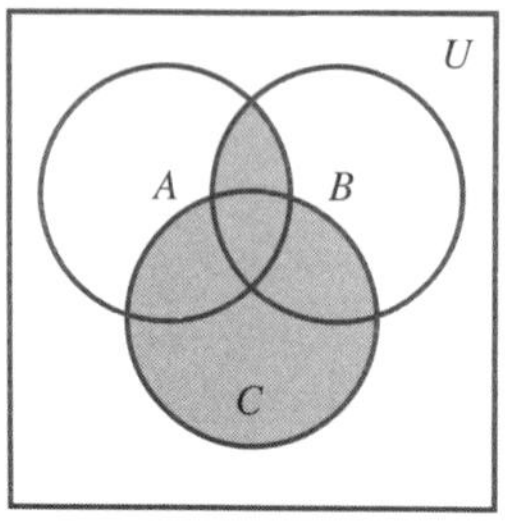

$(A \cap B) \cup C$

31. $\{1, 2, 3, 4, 5, 6\}$ **33.** $\{0, 0.5, 1, 1.5, 2, \ldots, 299.5, 300\}$ **35.** $\{(3, r), (3, g), (5, r), (5, g), (7, r), (7, g), (9, r), (9, g), (11, r), (11, g)\}$ **37.** $\{(3, g), (5, g), (7, g), (9, g), (11, g)\}$ **39.** 1/4 **41.** 3/13 **43.** 1/2 **45.** 1 **51.** The probability is 2/3 if you switch and 1/3 if you don't switch. The contestant should switch doors. **53.** 1 to 25 **55.** 5/36 **57.** 1/6 **59.** 1/6 **61.** 2/11 **63. a.** 0.66 **b.** 0.29 **c.** 0.71 **d.** 0.34 **65.** 4/9 **67.** 0.8736 **69.** 0.3077 **71.** 0.87 **73. a.** $(E \cup F)'$ or $E' \cap F'$ **b.** $E \cup F$ **75. a.** 0.0297 **b.** 0.0909 **c.** 0.2626 **d.** No **77.** b **79.** b **81. a.**

	N_2	T_2
N_1	N_1N_2	N_1T_2
T_1	T_1N_2	T_1T_2

b. 1/4 **c.** 1/2 **d.** 1/4 **83.** b **85. a.** 0.4614 **b.** 0.5497 **c.** 0.4708 **87. b.** 7/10 **89.** Independence **91.** 0.7145; 0.5688; 0.4101; 0.3212; 0.2708 **93. a.** 51 **b.** 31 **c.** 18 **d.** 15 **e.** 33 **f.** 23 **95. a.** 0.0025 **b.** 0.9975 **c.** 0.9753 **d.** 0.0247; no **e.** Independence

CHAPTER 8 Counting Principles; Further Probability Topics

Exercises 8.1 (page 439)

1. 720 **3.** 1.308×10^{12} **5.** 156 **7.** 1.024×10^{25} **9.** 1 **11.** n **13.** 36 **15.** 20 **19.** 3 **21.** 540,540 **23. a.** 8.718×10^{10} **b.** 4,354,560 **c.** 120,120 **d.** 84 **e.** 504 **25.** Multiply by 10 **27. a.** 2 **b.** 6 **c.** 18 **29.** 1120 **31.** No; use at least 4 initials **33.** 720 **35.** 120 **37.** 3.352×10^{10} **39. a.** 120 **b.** 48 **41.** 280 **43. a.** 78,125 **b.** 900,000 **c.** 90,000 **d.** 10,000 **e.** 544,320 **45.** 800 **47.** 1,000,000,000; yes **49.** 1,000,000,000 **51.** 4.096×10^{15} **53. a.** 362,880 **b.** 181,440 **c.** No

Exercises 8.2 (page 449)

3. 792 **5.** 1.134×10^{11} **7.** 1 **9.** n **11. a.** 10 **b.** 7 **13. a.** 9 **b.** 6 **c.** 3; yes, from both **17.** Combinations; **a.** 126 **b.** 462 **c.** 4620 **19.** Permutations; 479,001,600 **21.** Combinations; **a.** 120 **b.** 1820 **c.** 36 **23.** Combinations; **a.** 10 **b.** 0 **c.** 1 **d.** 10 **e.** 30 **f.** 15 **g.** 0 **27.** 28 **29. a.** 720 **b.** 360 **31. a.** 40 **b.** 20 **c.** 7 **33.** 4,115,439,900 **35. a.** 84 **b.** 10 **c.** 40 **d.** 74 **37.** No; 14,190 **39. a.** 48 **b.** 658,008 **c.** 652,080 **d.** 844,272 **e.** 79,092 **41.** 50 **43. a.** 15,504 **b.** 816 **45. a.** 1,120,529,256 **b.** 806,781,064,320 **47. a.** 1,048,576 **b.** 524,800 **49. a.** 255 **b.** 16 **c.** 24 **d.** 225 **e.** 128 **51. a.** 6.402×10^{15} **b.** 3.135×10^{10} **53. a** and **b.** 26

Exercises 8.3 (page 463)

1. 7/33 **3.** 14/55 **5.** 1326 **7.** 33/221 **9.** 52/221 **11.** 130/221 **13.** 8.417×10^{-8} **15.** 18,975/28,561 **19.** $1 - P(365, 42)/365^{42}$ **21.** 1 **23.** 0.6501 **25.** 3/5 and 2/5 **27.** 2.165×10^{-4} **29.** 0.0000289 **31.** 36/55 **33.** 21/55 **35.** 7/22 **37. a.** 225/646 **b.** 15/323 **c.** 225/2584 **d.** 0 **e.** 1/2 **f.** 175/2584 **g.** 503/646 **39.** 1.539×10^{-6} **41.** 2.401×10^{-4} **43.** 0.0211 **45.** 0.4226 **47.** 0.0438 **49.** 0.0133 **51.** 8.924×10^{-10} **53.** 4.507×10^{-34}; no **55. a.** 1/146,107,962 **b.** 0.3915 **57.** No, 1.48% **59. a.** 3.291×10^{-6} **b.** 7.962×10^{-12} **c.** 5.927×10^{-11} **d.** 5.524×10^{26}

Exercises 8.4 (page 472)

1. 5/16 **3.** 1/32 **5.** 3/16 **7.** 13/16 **9.** 4.594×10^{-10} **11.** 0.2692 **13.** 0.8748 **15.** 1/64 **17.** 21/32 **23.** The potential callers are not likely to have birthdates that are evenly distributed throughout the twentieth century. **25.** 0.2059 **27.** 0.4510

29. 0.1737 **31.** 0.7925 **33.** 0.2458 **35.** 0.0170 **37.** 0.3585 **39. a.** 0.1488 **b.** 0.0213 **c.** 0.9787 **41. a.** 0.0222 **b.** 0.1797 **c.** 0.7766 **43.** 0.0874 **45.** 0.9126 **47.** 0.2183 **49.** 0.0025 **51. a.** 0.0478 **b.** 0.9767 **c.** 0.8971 **53. a.** 1 chance in 1024 **b.** About 1 chance in 1.1×10^{12} **c.** About 1 chance in 2.587×10^{6} **55.** 0.9523 **57.** e **59.** 0.0305 **61.** 0.8676 **63. a.** 0.1990 **b.** 0.5684 **c.** 0.1332 **d.** 0.7936 **65. a.** 0.9995 **67. a.** 0.125, 0.25, 0.3125, 0.3125 **b.** 0.2893, 0.3222, 0.2353, 0.1531

Exercises 8.5 (page 485)

1.

Number of Heads	0	1	2	3	4
Probability	1/16	1/4	3/8	1/4	1/16

3.

Number of Aces	0	1	2	3
Probability	0.7826	0.2042	0.0130	0.0002

5.

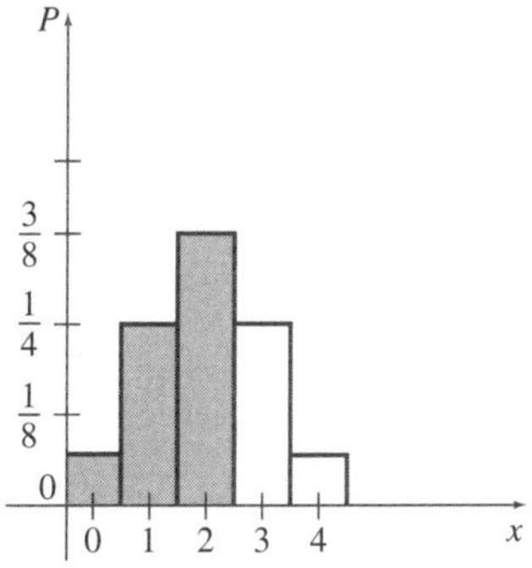

7.

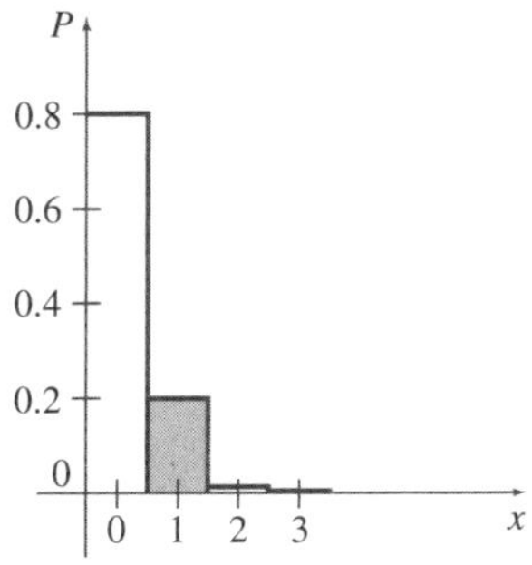

9. 3.6 **11.** 14.49 **13.** 2.7 **15.** 18 **17.** 0; yes

19. a.

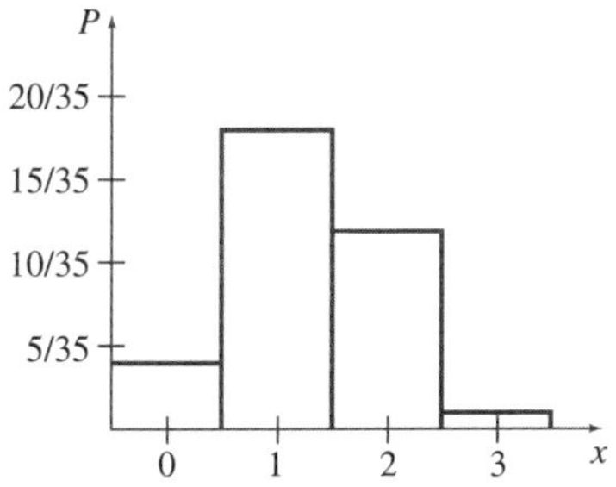

b. 9/7

21. a.

x	0	1	2	3	4
$P(x)$	625/1296	125/324	25/216	5/324	1/1296

b. 2/3

23. 3/4 **25.** 1/2

29. a.

Sum	5	6	7	8	9
Probability	1/6	1/6	1/3	1/6	1/6

b.

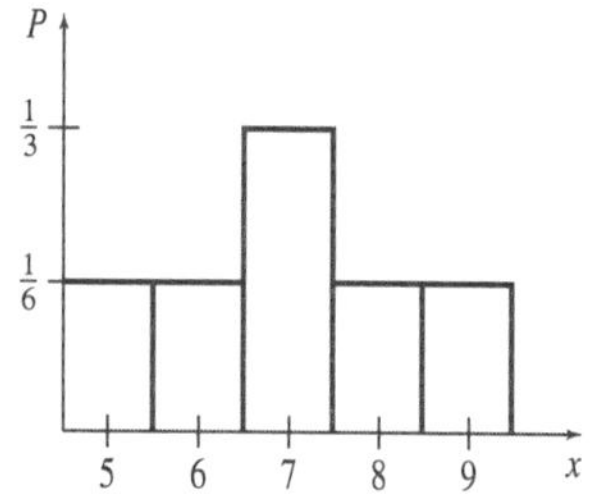

c. 1 to 2 **d.** 7

31. \$54,000 **33. a.** 30 **b.** 10

35. e **37. a.** \$68.51; \$72.84 **b.** Amoxicillin **39.** 38 **41. a.** 0.007230 **b.** 5.094×10^{-4} **c.** 5.5×10^{-5} **d.** 0.1143

43. a.

Number of Cats	0	1	2	3	4
Probability	0.2401	0.4116	0.2646	0.0756	0.0081

b. 1.2 **c.** 1.2

45. −\$0.72; no **47.** −\$0.82 **49.** −\$0.053 **51.** −\$0.50 **53.** \$0.10; no **55. a.** 10/3 **b.** 50/9

57. a.

x	0	1	2	3	4
$P(x)$	0.1956	0.3941	0.2978	0.1000	0.0126

b. 1.34

Concept Check (page 493)

1. True **2.** True **3.** True **4.** True **5.** False **6.** True **7.** True **8.** False **9.** True **10.** False **11.** True **12.** False

Chapter 8 Review Exercises (page 493)

1. 720 **3.** 220 **5.** 20 **7. a.** 24 **b.** 12 **9. a.** 140 **b.** 239 **13.** 0 **15.** 27/143 **17.** 21/143 **19.** 1/64 **21.** 11/32 **23.** 1/17 **25.** 80/221 **27.** 220/221

29. a.

Number	2	3	4	5	6	7	8	9	10	11	12
Probability	1/36	1/18	1/12	1/9	5/36	1/6	5/36	1/9	1/12	1/18	1/36

b.

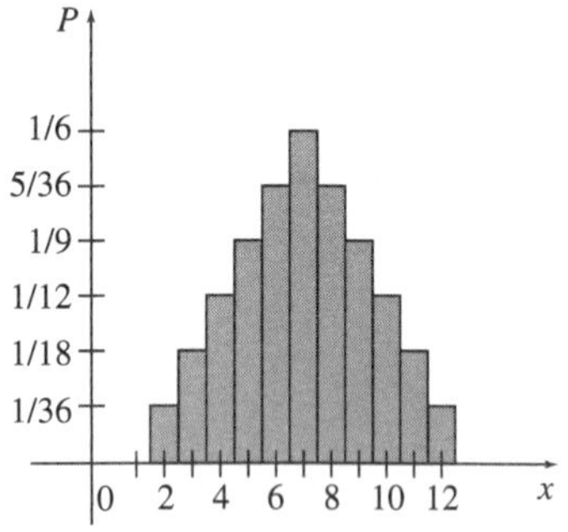

c. 7 **31.** 0.6 **33.** 2.5 **35.** \$1.29 **37. a.** $\binom{n}{0}$, or 1; $\binom{n}{1}$, or n; $\binom{n}{2}$; $\binom{n}{n}$, or 1 **b.** $\binom{n}{0} + \binom{n}{1} + \binom{n}{2} + \cdots + \binom{n}{n}$ **e.** The sum of the elements in row n of Pascal's triangle is 2^n. **39.** 0.1122 **41.** 7.580×10^{-7} **43.** 0.6187 **45.** 2 **47.** d **49.** e

51.

x	0	1	2	3	4	5	6
Income	0	400	800	1200	800	400	0
P(x)	0.0041	0.0369	0.1382	0.2765	0.3110	0.1866	0.0467

a. \$780.60 **b.** \$720; \$856.32; \$868.22; 5

53. 0.1875

55. a.

Number of Women	0	1	2	3	4	5
Probability	0.0778	0.2592	0.3456	0.2304	0.0768	0.0102

b.

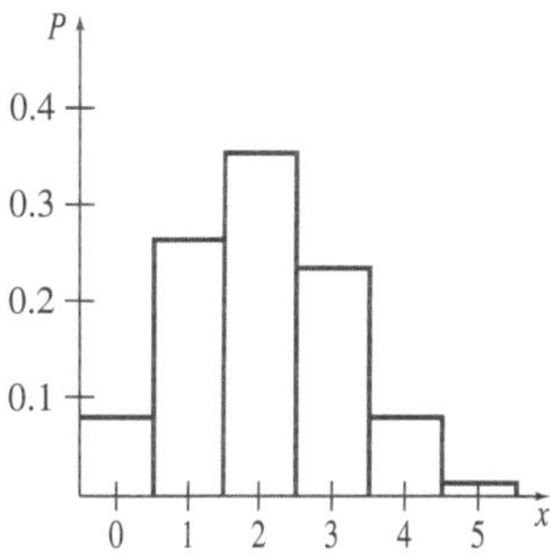

c. 2

57. a.

Number Who Did Not Do Homework	0	1	2	3
Probability	1/12	5/12	5/12	1/12

b.

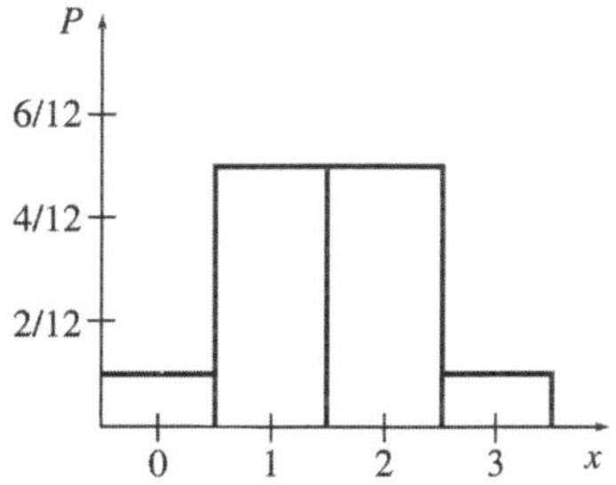

c. 3/2

59. −\$0.65 **61.** 9.5 **63. a.** 0.4799; 0.6533 **c.** 0.3467; 0.003096

6.1, 6.3 Truth Tables

The following truth table defines the logical operators in this chapter.

p	q	$\sim p$	$p \wedge q$	$p \vee q$	$p \rightarrow q$
T	T	F	T	T	T
T	F	F	F	T	F
F	T	T	F	T	T
F	F	T	F	F	T

7.3 Basic Probability Principle

Let S be a sample space of equally likely outcomes, and let event E be a subset of S. Then the probability that event E occurs is

$$P(E) = \frac{n(E)}{n(S)}.$$

7.4 Union Rule

For any two events E and F from a sample space S,

$$P(E \cup F) = P(E) + P(F) - P(E \cap F).$$

7.4 Odds

If $P(E') \neq 0$, the odds in favor of an event E are defined as the ratio of $P(E)$ to $P(E')$, or

$$\frac{P(E)}{P(E')}.$$

7.4 Properties of Probability

Let S be a sample space consisting of n distinct outcomes, $s_1, s_2, \ldots, s_n$. An acceptable probability assignment consists of assigning to each outcome s_i a number p_i (the probability of s_i) according to these rules.

1. The probability of each outcome is a number between 0 and 1.

$$0 \leq p_1 \leq 1, \quad 0 \leq p_2 \leq 1, \ldots, \quad 0 \leq p_n \leq 1$$

2. The sum of the probabilities of all possible outcomes is 1.

$$p_1 + p_2 + p_3 + \cdots + p_n = 1$$

7.5 Product Rule

If E and F are events, then $P(E \cap F)$ may be found by either of these formulas.

$$P(E \cap F) = P(F) \cdot P(E|F) \qquad \text{or} \qquad P(E \cap F) = P(E) \cdot P(F|E)$$

7.6 Bayes' Theorem

$$P(F_i|E) = \frac{P(F_i) \cdot P(E|F_i)}{P(F_1) \cdot P(E|F_1) + P(F_2) \cdot P(E|F_2) + \cdots + P(F_n) \cdot P(E|F_n)}$$

8.1 Multiplication Principle

Suppose n choices must be made, with

m_1 ways to make choice 1,

and for each of these ways,

m_2 ways to make choice 2,

and so on, with

m_n ways to make choice n.